European Coal Geology

Geological Society Special Publications
Series Editor A. J. FLEET

GEOLOGICAL SOCIETY SPECIAL PUBLICATION NO. 82

European Coal Geology

EDITED BY

M. K. G. Whateley
Department of Geology,
University of Leicester

and

D. A. Spears
Department of Earth Sciences,
University of Sheffield

1995
Published by
The Geological Society
London

THE GEOLOGICAL SOCIETY

The Society was founded in 1807 as The Geological Society of London and is the oldest geological society in the world. It received its Royal Charter in 1825 for the purpose of 'investigating the mineral structure of the Earth'. The Society is Britain's national society for geology with a membership of 7500 (1993). It has countrywide coverage and approximately 1000 members reside overseas. The Society is responsible for all aspects of the geological sciences including professional matters. The Society has its own publishing house which produces the Society's international journals, books and maps, and which acts as the European distributor for publications of the American Association of Petroleum Geologists and the Geological Society of America.

Fellowship is open to those holding a recognized honours degree in geology or cognate subject and who have at least two years' relevant postgraduate experience, or who have not less than six years' relevant experience in geology or a cognate subject. A Fellow who has not less than five years' relevant postgraduate experience in the practice of geology may apply for validation and, subject to approval, may be able to use the designatory letters C. Geol (Chartered Geologist).

Further information about the Society is available from the Membership Manager, The Geological Society, Burlington House, Piccadilly, London W1V 0JU, UK.

Published by The Geological Society from:
The Geological Society Publishing House
Unit 7
Brassmill Enterprise Centre
Brassmill Lane
Bath BA1 3JN
UK
(*Orders:* Tel. 0225 445046
       Fax 0225 442836)

Registered Charity No. 210161

First published 1995

The publisher makes no representation, express or implied, with regard to the accuracy of the information contained in this book and cannot accept any legal responsibility for any errors or omissions that may be made.

© The Geological Society 1995. All rights reserved. No reproduction, copy or transmission of this publication may be made without prior written permission. No paragraph of this publication may be reproduced, copied or transmitted save with the provisions of the Copyright Licensing Agency, 90 Tottenham Court Road, London W1P 9HE. Users registered with the Copyright Clearance Center, 27 Congress Street, Salem, MA 01970, USA: the item-fee code for this publication is 0305-0394 $07.00.

British Library Cataloguing in Publication Data
A catalogue record for this book is available from the British Library
ISBN 1-897799-19-5

Typeset by Bath Typesetting Ltd
Bath, England

Printed in Great Britain by
Alden Press, Oxford

Distributors
USA
  AAPG Bookstore
  PO Box 979
  Tulsa
  Oklahoma 74101-0979
  USA
(*Orders:* Tel. (918) 584-2555
       Fax (918) 584-0469)

Australia
  Australian Mineral Foundation
  63 Conyngham Street
  Glenside
  South Australia 5065
  Australia
(*Orders:* Tel. (08) 379-0444
       Fax (08) 379-4634)

India
  Affiliated East-West Press PVT Ltd
  G-1/16 Ansari Road
  New Delhi 110 002
  India
(*Orders:* Tel. (11) 327-9113
       Fax (11) 326-0538)

Japan
  Kanda Book Trading Co.
  Tanikawa Building
  3-2 Kanda Surugadai
  Chiyoda-Ku
  Tokyo 101
  Japan
(*Orders:* Tel. (03) 3255-3497
       Fax (03) 3255-3495)

Contents

Preface vii

Exploration and evaluation techniques
FLINT, S., AITKEN, J. & HAMPSON, G. Application of sequence stratigraphy to coal-bearing coastal plain successions: implications for the UK Coal Measures 1
FULTON, I. M., GUION, P. D. & JONES, N. S. Application of sedimentology to the development and extraction of deep-mined coal 17
GUION, P. D., FULTON, I. M. & JONES, N. S. Sedimentary facies of the coal-bearing Westphalian A and B north of the Wales–Brabant High 45
HAMPSON, G. Discrimination of regionally extensive coals in the Upper Carboniferous of the Pennine Basin, UK using high resolution sequence stratigraphic concepts 79
HARRIS, I. H. Newly developed techniques to determine proportions of undersized (friable) coal during prospective site investigations 99
JONES, N. S., GUION, P. D. & FULTON, I. M. Sedimentology and its applications within the UK opencast coal mining industry 115
SPEARS, D. A. & LYONS, P. C. An update on British tonsteins 137
QUEROL, X. & CHENERY, S. Determination of trace element affinities in coal by laser ablation microprobe–inductively coupled plasma mass spectrometry 147

Geophysical exploration
GOULTY, N. R. Review of borehole seismic methods developed for opencast coal exploration 159

Resources, environment and energy policies
KOUKOUZAS, C. & KOUKOUZAS, N. Coals of Greece: distribution, quality and reserves 171
MEREFIELD, K., STONE, I., JARMAN P., REES, G., ROBERTS, J., JONES, J. & DEAN, A. Environmental dust analysis in opencast mining areas 181
PEŠEK, J. & PEŠKOVÁ, J. Coal production and coal reserves of the Czech Republic and former Czechoslovakia 189
WRIGHT, P. European Community energy policy: import dependency and the ineffectual consensus 195

Case Histories
ALLEN, M. J. Exploration and exploitation of the East Pennine Coalfield 207
DREESEN, R., BOSSIROY, D., DUSAR, M., FLORES, R. M. & VERKAEREN, P. Overview of the influence of syn-sedimentary tectonics and palaeo-fluvial systems on coal seam and sand body characteristics in the Westphalian C strata, Campine Basin, Belgium 215
GAYER, R., HATHAWAY, T. & DAVIS, J. Structural geological factors in open pit coal mine design, with special reference to thrusting: case study from the Ffyndaff sites in the South Wales Coalfield 233
BAILY, H. E., GLOVER, B. W., HOLLOWAY, S. & YOUNG, S. R. Controls on coalbed methane prospectivity in Great Britain 251
OPLUŠTIL, S. & VÍZDAL, P. Pre-sedimentary palaeo-relief and compaction: controls of peat deposition and clastic sedimentation of the Radnice Member, Kladno Basin, Bohemia 267
READ, W. A. Sequence stratigraphy and lithofacies geometry in an early Namurian coal-bearing succession in central Scotland 285
TOMSCHEY, O. Unusual enrichment of U, Mo and V in an Upper Cretaceous coal seam of Hungary 299
WHATELEY, M. K. G. & TUNCALI, E. Origin and distribution of sulphur in the Neogene Beypazari Lignite Basin, Central Anatolia, Turkey 307

Index 325

Preface

Although there has been extensive closure of underground mines throughout the European Union, coal remains a major energy source. The growing energy demands of the developing countries mean that the world's coal consumption continues to increase. The European Union has long established coalfields and considerable research experience in coal exploitation and utilization. This knowledge base will be important in the future development of coal. This volume of 17 papers on many aspects of European coal geology illustrates the depth and breadth of this research. The chapters cover a wide spectrum of interests in European coal geology from sedimentological, geochemical and exploration models, to exploration drilling and economic evaluation of coal deposits, on a local and country-wide scale, as well as the environmental aspects of coal burning and disposal of CO_2.

These papers were put together to fill a gap that exists in the literature on regional coal deposits. A forum for pure research papers already exists, catering for papers which are usually biased towards detailed geochemistry, petrography or sedimentology of coal and coal-bearing strata. This volume encouraged papers in which the description of coal deposits was specifically orientated towards the application of the research in the exploration for, exploitation of and environmental considerations required in the study of coal. Such an extensive spectrum of papers will appeal to a wide audience, ranging from researchers, lecturers and students to professionals in industry. The well-balanced content of the book should provide a particularly attractive read for those who seek an update on some of the coal deposits of Europe.

The low number of papers relating to geophysical exploration is unrepresentative of the amount of active research in this field. Several papers were expected from the British Coal Technical Services Research Executive at Bretby, but they were not able to publish their state-of-the-science findings because of the upheaval in the British coal industry at the time of going to press.

The editors are particularly grateful to Ms Gail Williamson for helping with the many essential administrative tasks associated with putting together a volume of this size. Naturally, they thank the authors and the referees of the manuscripts for providing the material for this book. They are grateful to those companies and organizations who provided finance either directly or indirectly. In particular the editors would like to acknowledge Atlas Exhibitions International Ltd, Blackwell Scientific Publications Ltd, British Coal plc, British Geological Survey, British Drilling and Freezing Co Ltd, Golder Associates, Hall & Watts Systems Ltd, KRJA Ltd, Leicester University Bookshop, Peter J Norton Associates, Pergamon Press, Roberston Geologging Ltd, John Wiley & Son Ltd and World Mining Equipment. British Coal Opencast kindly supplied the photograph used to illustrate the front cover

Exploration and evaluation techniques

Application of sequence stratigraphy to coal-bearing coastal plain successions: implications for the UK Coal Measures

STEPHEN FLINT, JOHN AITKEN & GARY HAMPSON

Sequence Stratigraphy Research Group, Department of Earth Sciences, University of Liverpool, PO Box 147, Liverpool L69 3BX, UK

Abstract: Advances in the understanding of coal depositional environments have led to the notion that thick, low ash deposits in close proximity to siliciclastic sediment inputs are most commonly restricted to the products of raised mires. These mires are initiated, sustained and preserved in conditions of slowly rising base level (relative sea-level). Hence it is possible to consider the stratigraphic significance of economic coal seams within the concept of unconformity bounded depositional sequences (*sensu* Vail *et al.*). Time-equivalent clastic deposits tend to be minor (commonly with a heterolith fill) channel deposits, with brackish to freshwater lake and crevasse splay sediments, excluded from the peat mire either via the topographically raised nature of the mire or by the landward dislocation of fluvial facies. In this paradigm, spatially related major fluvial channel deposits, major seam splits and even washout deposits may not be time-equivalent to the coal seam. Regional correlations suggest that thick coal seams are time-correlative to significant flooding surfaces at the coeval coastline. Thus coals may be correlated with initial flooding surfaces over incised valley systems and sometimes with parasequence and parasequence set boundaries. This factor may be important in coal exploration. Application of these ideas to the classic Westphalian 'Coal Measures' of England indicates that the 'marine bands' may be regarded as flooding surfaces or condensed sections, with several being good candidates for maximum flooding surfaces. Between these marine bands the stratigraphy is punctuated by thick, stacked, regionally extensive fluvial units, incised into previously deposited sediments. These are interpreted as incised valley complexes. Systems tract assignment, based on changing accommodation space, indicates that the main coal zones fall dominantly within transgressive systems tracts.

Depositional modelling of coal-bearing strata was largely initiated by work in the Carboniferous coal fields of the Central Appalachian Basin (e.g. Horne *et al.* 1978; Howell & Ferm 1980). These models assumed that the nature of coal deposits (thickness, lateral extent, ash content and quality) was controlled by the autocyclicity of clastic environments which surrounded the peat-forming environments. Horne *et al.* (1978) suggested that coals developed in back-barrier environments are narrow, elongate units parallel to depositional strike. Coal seams from lower delta plains were reported as occurring in laterally widespread but thin units, whereas coals from upper delta plain and fluvial environments occurred in laterally restricted, locally thick beds aligned parallel to depositional dip. It was concluded that the thickest and most laterally extensive coals occurred in the transition between lower and upper delta plain environments. This model has not been proved to be very predictive, even within the Appalachian Basin (e.g. Ferm & Staub 1984), and it has been shown that there is a considerable amount of variability in coal-bearing successions, independent of depositional environment (e.g. Fielding 1984, 1985, 1986, 1987), such that these models are not accurate predictors of coal development.

The development of sequence stratigraphy, which emphasizes the allocyclic controls of tectonism, climate and eustasy, has improved our understanding of the nature of depositional sedimentary systems. Sequence stratigraphic principles provide guidelines from which to ascertain sand body stacking patterns and thus the areal distribution of genetic facies units within sedimentary successions (Fig. 1). It is therefore potentially a powerful predictive tool to the hydrocarbon, coal and other extractive industries. Apart from the conceptual framework of Posamentier & Vail (1988) and the field based studies of Shanley & McCabe (1991a, b, 1993) and Shanley *et al.* (1992) few attempts (notably Read and Forsyth 1991) have been made to apply sequence stratigraphic concepts to fluvial/delta plain successions such as those of the UK Silesian.

The concept of accommodation (Jervey 1988) is critical to sequence stratigraphic models.

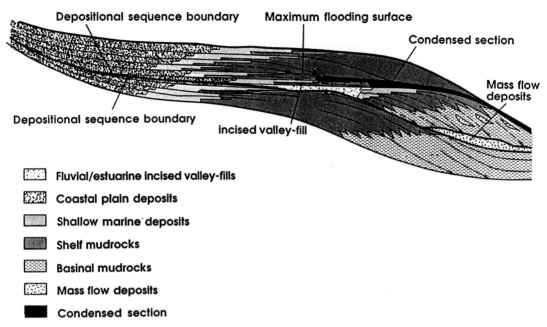

Fig. 1. Basic accommodation envelope or 'slug model' of an unconformity bounded sequence, showing systems tracts and predicted progradational, retrogradational and aggradational stacking patterns of clastic sediments. After Van Wagoner *et al.* (1990).

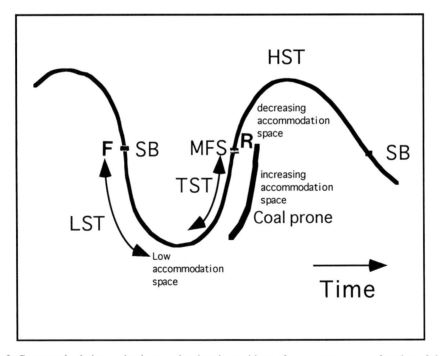

Fig. 2. Conceptual relative sea-level curve showing the positions of systems tracts as a function of the rate of creation and destruction of accommodation space necessary to preserve sediment (modified after Posamentier and Vail 1988). Preservation of thick coal accumulations are favoured by high accommodation space (see text). LST, Lowstand systems tract; TST, transgressive systems tract; HST, highstand systems tract; SB, sequence boundary; and MFS, maximum flooding surface.

Facies architecture is dependent on variations in the rate of sediment supply and rate of change in accommodation (Posamentier & Vail 1988; Posamentier et al. 1988; Van Wagoner et al. 1990; Fig. 2). In a mire, accommodation space can be defined as the maximum height to which a peat could build (McCabe 1993). Properties of coal, such as thickness, ash content and maceral type, are controlled by the type of vegetation, humification rates, sediment supply rates and rates of base level change. In turn, these are controlled by the allocyclic processes of eustasy, climate and tectonics which determine the sequence stratigraphy.

This paper summarizes a scheme for the sequence stratigraphic interpretation of delta plain deposits in terms of key surface and stacking pattern analysis. For this we draw on published data and our studies of exceptionally exposed coal-bearing successions in the Cretaceous of the US Western Interior Basin and the Westphalian of eastern Kentucky. These models are then discussed with general reference to parts of the Coal Measures of the UK.

Stratigraphic models for coal depositional environments

The environments of deposition of coal have been discussed in detail by McCabe (1984, 1987, 1991) and Haszeldine (1989), among others, and so are only briefly summarized here in terms of delta plain and fluvial settings. Most modern mires are low-lying (McCabe 1984; Haszeldine 1989) and drape and infill the underlying topography. The water-table in a low-lying mire approximates the regional groundwater-table. By contrast, raised mires have convex-upward surfaces and can be raised several metres above the surrounding topograpy (e.g. Anderson 1964; Styan & Bustin 1983). They require abundant year-round precipitation to maintain a local water-table above the regional ground-water-table. In addition, minor amounts of peat can accumulate in lakes and floating mires (McCabe 1984, 1987) (Fig. 3).

Many facies models for coal-bearing strata show peat accumulation in close association with active clastic depositional environments such as on floodplains (e.g. Gersib & McCabe 1981), in interdistributary areas and on levees in deltaic settings (e.g. Baganz et al. 1975). However, good quality, potentially coal-forming peat is unlikely to develop in such areas because of the regular introduction of overbank siliciclastic material. Raised mires, which by their raised nature self-exclude clastics, can develop good quality, potentially coal-forming peat in close proximity to active clastic depositional systems.

Humification (decay of organic material within the peat profile) limits the development of coals to about 3 m thickness with a static water-table (Clymo 1987). It has therefore been argued that the preservation of thick coals is dependent on a rising water-table, which itself is best achieved by base level rise (McCabe & Parrish 1992). Base level in a mire is a surface or water level above which humification exceeds organic supply. Mires developed under conditions of rising base level may maintain themselves through many thousands of years. The destruction of mires, both raised and low-lying, is commonly due to abrupt falls in the regional water-table, which lead to the water-table of the mire declining. Without a constant source of water the swamp dies. Mire destruction has also been attributed to climatic change, tectonics or increased rates of subsidence (Haszeldine 1989).

It is suggested that the initiation and termination of mires may be related to accommodation space and its rate of change, as for all other sediments in fluvial to marine settings. Mires require high rates of creation of accommodation space and the exclusion of clastic material to grow. These conditions may be predicted theoretically within the transgressive systems tract (TST) and early highstand systems tract (HST) of an Exxon-style sequence (Vail et al. 1977; Posamentier et al. 1988). Peats will not accumulate during periods of falling base level, although freshly exposed strata may provide a platform on which peat may grow (McCabe 1993). Peats deposited before a fall in base level will probably be destroyed through oxidation and decay, or be incised as fluvial systems attempt to return to grade during falling base level. Typically, during lowstands, incised valleys are cut and partially filled. Laterally equivalent interfluve surfaces are subjected to subaerial exposure, weathering and the development of palaeosols and/or thin coals (e.g. Krystinik & Blakeney-DeJarnett 1991; Aitken & Flint in press a, in press b). The water-table is lowered during lowstand, thus palaeosols will evolve, with time, towards well-drained variants, although they may subsequently become gleyed during the succeeding base level rise (Aitken & Flint, in press b). During base level rise the valleys will back-fill with fluvial (up-dip) or estuarine (down-dip) sediments (Allen & Posamentier 1993; Shanley & McCabe 1993) to the point where they are full and sediment is free to spill over the interfluves (initial flooding surface or transgressive surface of Van Wagoner et al.

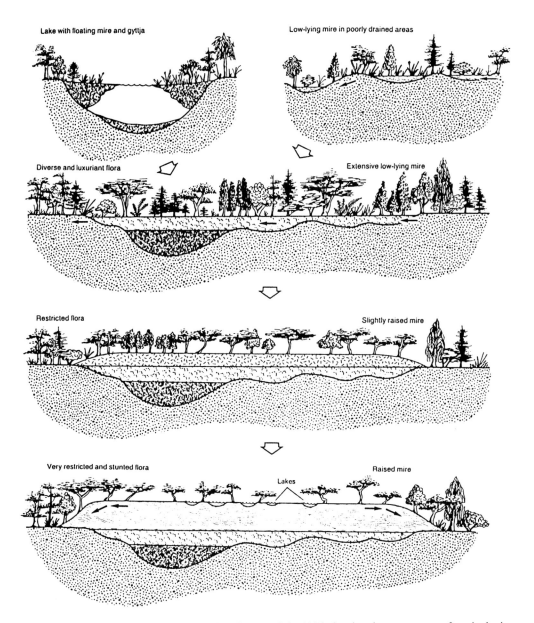

Fig. 3. Modern environments of peat formation (from McCabe 1987) showing the components of a raised mire.

1988). However, the rate of rise of base level is increasing throughout the TST and clastic sediment supply is effectively retarded as rivers attempt to regrade to the rising base level (Posamentier *et al.* 1988). Fluvial gradients decrease and the suspension load to bedload ratio increases. The increase in accommodation space moves landwards. The net effect is that widespread areas with a high water-table and little or no introduction of clastic material are produced. These areas are ideal nucleation sites for mires (Fig. 2).

Sequence stratigraphy of coal-bearing strata: ancient examples

In the world-wide rock record we see corroborative stratigraphic and geometrical relationships which lend support to the above theoretical models. Several workers have shown that thick,

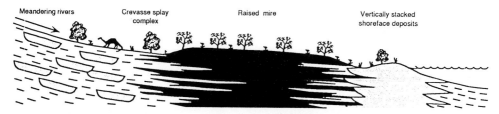

Fig. 4. Schematic representation of the position of stacked low ash/low sulphur coals, inferred to have been formed in long-term raised mires in the Santonian John Henry Member, Kaiparowits Plateau, Utah. These mires were maintained by a slowly rising base level in the early highstand systems tract. From McCabe and Shanley (1992).

economic coals in the Cretaceous of the Western Interior Basin, USA are restricted to transgressive or landward-stepping stratigraphic intervals (Sears et al. 1941; Ryer 1981, 1983; Cross 1988). In the Kaiparowits Plateau, Utah, Shanley & McCabe (1993) have documented the restriction of economic coals to the TST and early HST of the John Henry Member. Here, raised mires were developed within a few kilometres of the coeval wave-dominated shoreline (Fig. 4); the low sulphur, low ash composition of the coals testifies to the efficiency of the raised mires to self-exclude clastic material. Indeed, the raised mires appear to have modified parasequence stacking patterns within the highstand (McCabe & Shanley 1992). Sequence stratigraphy has also been applied to Carboniferous deltaic and coal-bearing rocks of the Appalachian Basin (Gastaldo et al. 1993; Aitken & Flint in press a, in press b) and the UK Namurian (Read 1991, Maynard 1992, Hampson, this volume). Although more difficult in the absence or paucity of marine rocks, all these workers have documented systematic changes in stacking patterns of clastic strata and trends in coal development which fit with sequence stratigraphic principles. All coals preserved during rising base level may not represent raised mires. Thin, widespread coals at initial flooding surfaces may have developed in low-lying mires, where a low ash content results from the landward translation of clastics during the sudden increase in accommodation space and flooding when incised valleys have finished backfilling.

Detailed work in Kentucky has clearly indicated that laterally extensive and thick coals are largely restricted to the TST in high frequency (fourth-order) unconformity bounded sequences (Aitken & Flint in press a, in press b). In Utah and Kentucky the coal seam stratigraphy seems to reflect accurately increments of base level rise, towards the maximum flooding surface, which represents the maximum rate in base level rise (the R inflection point; Fig. 2). In the Late Permian of the Sydney basin, Australia, Arditto (1987, 1991) has documented several basinwide coals which he termed 'transgressive cycle coals'. These coals occur within the TST and are capped by marine shale beds, interpreted by Arditto as maximum flooding surfaces to high frequency sequences.

Sequence boundary recognition

Incised valleys

Recognition of sequence bounding unconformities in predominantly alluvial strata can be difficult because basinward shifts in facies tracts and related incision can easily be confused with more local channel scour. Sequence boundaries can be identified by an abrupt facies tract dislocation, an abrupt increase in vertical sandstone amalgamation and an increase in mean grain size and/or change in petrographic composition (Shanley & McCabe 1991a, b, 1993; Flint 1993). Such criteria, however, can only be used to identify sequence boundaries if the changes are of regional (mappable) extent, such that there is incision in many places at the same stratigraphic level.

In the Breathitt Group of eastern Kentucky, at a number of stratigraphic levels, medium- to coarse-grained, multi-storey channel sand bodies with basal pebble lag conglomerates are incised into marine siltstone, fine-grained mouth bar sandstone, interdistributary siltstone, crevasse splay packages, coals and single-storey, laterally accreted, channel sandstone bodies (Fig. 5a). It is (a) the degree of facies dislocation; (b) the marked contrast in grain size; (c) the degree of incision; (d) the contrast in depth of individual channel-fills (maximum of c. 3–4 m) in relation to the total depth of incision (15–20m); and (e) the regional, mappable extent of the surfaces that indicates fluvial incision related to a significant fall in base level, rather than simple

Fig. 5. (a) Photograph showing an interpreted sequence boundary in the Westphalian of east Kentucky. A stacked, multi-storey/multilateral fluvial sandstone complex is incised into marine shale, removing any shoreline facies. Daniel Boone Parkway, west of Hazard, Kentucky. (b) Photograph showing an interfluve sequence boundary, characterized by a bleached palaeosol lying on marine siltstone and overlain by a coal. Four Corners on KY 80, near Hazard Kentucky. See Fig. 6 for stratigraphic setting.

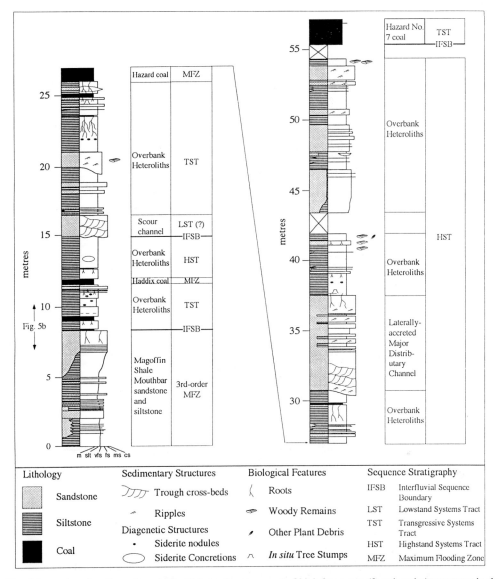

Fig. 6. Logged section through part of a highstand sequence set of high frequency (fourth-order) sequences in the Westphalian Breathitt Group of East Kentucky (from Aitken and Flint in press *b*). Note the upward increase in thickness of coal beds towards a maximum, interpreted as the up-dip equivalent of a maximum flooding surface (highest accommodation space and water-table). The upper part of the log shows a (rarely) fully preserved highstand systems tract, capped by an interfluve sequence boundary. The HST is characterized by sandy, single-storey channel-fills and few or no coals, due to decreasing accommodation space for peat preservation.

channel switching or other 'normal progradation' phenomena. Hence the multi-storey, multi-lateral fluvial deposits are interpreted as incised valley-fills.

Interfluves

Interfluvial sequence boundaries are more difficult to identify than sequence boundaries associated with incised valley-fills (Van Wagoner *et al.* 1990). Outside of the incised valleys, the sequence boundary is a surface of subaerial exposure with either minor or no deposition, developed during a time gap when accommodation space was at a minimum (O'Byrne & Flint 1993, 1994). Hence, during lowstand, these interfluvial sequence boundaries will be characterized by palaeosol development and negli-

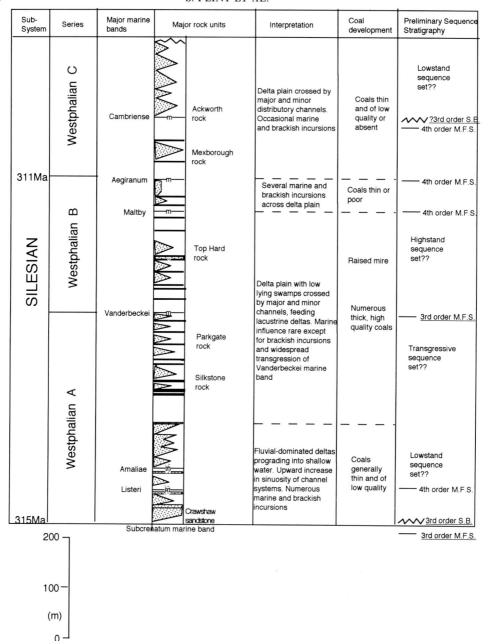

Fig. 7. General lithostratigraphy of the Westphalian A–C of the East Midlands Coalfields, adapted from Guion and Fulton (1993), with a *preliminary* sequence stratigraphic interpretation added.

gible aggradation. In the Breathitt Group of eastern Kentucky, it is occasionally possible to trace the sequence boundary at the base of an incised valley-fill onto the interfluve. In these situations, the interfluves are represented by thin coals, carbonaceous siltstone, seat earths, rooted surfaces (Figs 5b and 6) and/or slight bleaching and mottling (Aitken & Flint in press *a*). However, taken in isolation, these deposits are difficult to distinguish from similar sediments deposited in the TST and HST.

Transgressive and highstand systems tracts

Although the transgressive surface (Van Wagoner et al. 1988) is commonly difficult to identify, the lowstand systems tract (LST) passes vertically upward into thin, isolated, high sinuosity, single-storey channel deposits, within a framework of fine-grained overbank deposits in relatively up-dip areas, or into marine siltstone in relatively down-dip areas. This transition is interpreted to reflect coastal transgression and the equivalent up-dip increase in accommodation space, and characterizes the TST (Fig. 1). Floodplain sediments are poorly drained with brackish intervals. In many basins, tidally influenced fluvial deposits occur within the TST, related to high rates of base level rise, with tidal influence known to occur up to 100 km landward of the shoreline in modern rivers (Allen 1984) and at least 40 km landward in ancient deposits (Shanley et al. 1992).

Conventionally, the maximum flooding surface could be defined by a change in parasequence stacking pattern from retrogradational to aggradational and eventually progradational (Posamentier & Vail 1988; Posamentier et al. 1988; Van Wagoner et al. 1990). However, it is difficult to distinguish the HST from the underlying TST in relatively up-dip locations because both the TST and HST are represented by similar, single-storey, channel-fills, high proportions of crevasse splays, coals and thick successions of floodplain siltstone. It is, therefore, more appropriate to identify a zone rather than a surface of maximum flooding in these settings. As parasequences are generally not identifiable in up-dip delta plain successions, other criteria are required to delineate the TST from the HST. Such a criterion is provided by the concept of accommodation potential and its relationship with the development of coal.

In the Breathitt Group, at positions where sequences are bounded by interfluve sequence boundaries and thus fully preserved, there is a clear vertical organization in terms of the thickness and lateral extent of coal (Fig. 6). The deposits directly overlying the interfluvial sequence boundaries are characterized by coals which commonly approach 1–2 m in thickness, have lateral extents of tens of kilometres and drape over the tops of incised valley-fills. Higher in the sequences coals are less common, much thinner (commonly a few centimetres thick) and laterally restricted (typically <5 km laterally). Carbonaceous siltstones and poorly developed silty coals are common. The transition between well developed and more poorly developed coals is interpreted as representing the transition zone between the TST and HST. In the Campanian Blackhawk Formation of the Book Cliffs, Utah, marine shales developed due to small rises in relative sea-level can be traced landward (up-dip) into coeval thin coals. The coals thus represent a terrestrial temporal response to an episode of base level rise, i.e. a slowly accumulated sediment, starved of clastic input, with a high water-table (O'Byrne & Flint 1994).

Initial application of a sequence stratigraphic model to the UK Coal Measures

Marine bands (faunal concentrate horizons)

In the Coal Measures of central and northern England marine bands (marine mudstones with a diagnostic goniatite-dominated fauna, commonly correlatable across NW Europe) include the *Subcrenatum*, *Listeri* and *Vandebeckei* units (Ramsbottom 1977, 1979) (Fig. 7). Other marine bands such as the *Amaliae* contain a more restricted fauna or are only reported from the central part of the basin (Guion & Fielding 1988). All these marine bands are correlated with shales bearing marginal marined to non-marine faunas (*Lingula*, *Carbonicola* and *Estheria* bands) at the basin margins (Calver 1968). Additional *Lingula*, *Carbonicola* and *Estheria* bands with no open marine correlatives are also reported (Eden 1954; Calver 1968). We believe that the marine bands represent the most condensed sedimentation in the basin at the time and therefore equate with flooding surfaces (Van Wagoner et al. 1990). It is critical to appreciate that maximum flooding surfaces (MFS) are developed only once in an Exxon sequence, at the point of maximum rate of base level rise (the *R* inflection point, not at the eustatic sea-level highstand; Fig.2) The apparent differences in lateral extent and faunal content ('marineness') of the different bands can be explained in terms of palaeogeographical position coupled with the amount of base level rise. Thus there is an apparent hierarchy among the different bands; a marginal marine *Lingula* band with no open marine correlative may represent a parasequence set boundary with less extensive flooding than an open marine band representing a MFS. Also a marine band with a restricted fauna, such as the *Amaliae* band, may represent a MFS within a high-frequency sequence, associated with less extensive flooding than a fully developed marine band, such as the *Subcrenatum* band. This latter may represent a MFS within a low frequency composite sequence (*sensu* Mitchum & Van Wagoner 1991). This

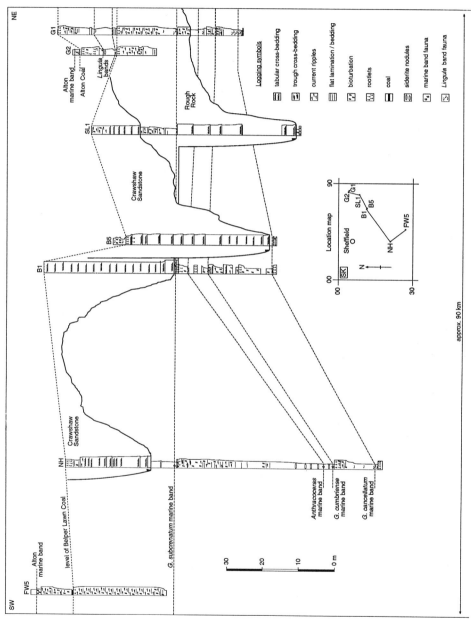

Fig. 8. Cross-section through the south east Central Pennine basin based on our logged sections of core material. The Crawshaw sandstone complex is clearly seen to be incised into the underlying deposits, removing almost the entire uppermost Namurian Rough Rock Group succession, down to the *Cancellatum* marine band. Incision is interpreted to have been localized along synsedimentary fault lines. Core sections are abbreviated as follows: FW5, Farleys Wood No. 5; NH, Nether Heage; B1, Bothamsall No. 1; B5, Bothamsall No 5; SL1, South Leverton No 1; G2, Gainsborough No 2; and G1, Gainsborough No 1.

theoretical discussion is verified to a certain extent by the relative abundance and vertical arrangement of different bands within the strata (e.g. Fig. 13.12 of Guion & Fielding 1988).

Incised valley-fills/lowstand deposits

The acceptance of at least some of the marine bands as eustatically controlled condensed sections by almost all workers suggests at least a consensus that at times rates of relative sea-level rise varied; it is logical to argue that sea-level also periodically fell, yet this thesis is hardly considered in existing published work on the UK Namurian/Westphalian stratigraphy (see review in Guion & Fielding 1988).

The basal Westphalian (Fig. 7) is well known for its abundance of coarse-grained, channelized fluvial deposits similar to those such as the Rough Rock in the underlying Namurian. Work by Maynard (1992) and Hampson (this volume) has convincingly shown a sequence boundary at the base of the Rough Rock Group and a further sequence boundary within the Rough Rock (Hampson, this volume). We suggest that the Crawshaw Sandstone may well represent the first lowstand complex of the Westphalian, based on its highly incised character, truncating several underlying marine bands, and the synchroneity of incision as far apart as the East Midlands and the western Pennines (Fig. 8). Detailed palaeocurrent analyses confirm a low sinuosity, relatively high gradient fill (Guion & Fielding 1988).

Throughout the Westphalian A the stratigraphy is dominated by regional, stacked fluvial channel deposits, individual units being up to 10 km wide, bifurcating downstream. These units, in most instances, erode out underlying delta front and mouth bar deposits, again confirming widespread incision. This basinward facies shift has previously been interpreted as due to progradation into shallow water. However, true shallow water lobate deltas such as the Atchafalaya system and ancient examples (e.g. Flint *et al.* 1989) usually preserve mouth bar rocks between the distributaries and, moreover, have widths of a few kilometres, whereas associated coeval distributary channels are less than 1 km wide. We believe that the abrupt superposition of low sinuosity, broad, multi-storey fluvial complexes on subaqueous delta front sediments over several hundred square kilometres is more logically explained by a regional fall in relative sea-level. The downstream bifurcation is documented in exceptionally exposed incised valley-fills such as the Cretaceous Castlegate Sandstone of Utah (Van Wagoner *et al.* 1991). The case for a possible high frequency set of stacked lowstand incised valley-fills (a lowstand sequence set) can be suggested for the basal Westphalian A through synthesis of available published data (summarized succinctly by Guion & Fielding 1988).

Following this reasoning, the well-documented upward transition from the coarse-grained, low sinuosity channel belts to smaller scale, more sinuous, finer grained fluvial systems during the Westphalian (Fig. 7) is completely consistent with a background rising base level and the concomitant reduction in fluvial gradient (Posamentier *et al.* 1988; Shanley & McCabe 1993; Aitken & Flint in press *a*). The existing model for the UK Westphalian suggests that this change is due to the passage from lower to upper delta plain (Guion & Fielding 1988). It is difficult to visualize finer grained, sinuous fluvial systems feeding contemporaneous coarser grained, lower sinuosity systems, as implied by this model. In modern distributive systems there is little evidence for a downstream increase in bedload at a constant or even reducing gradient.

Hierarchy of surfaces and sequences

Exxon-type, unconformity bounded sequences have been defined in terms of approximate duration. Third-order sequences have a recurrence frequency of 2–5 Ma, whereas higher frequency fourth-order sequences represent approximately 0.25–1 Ma (Vail *et al.* 1977). Applying the estimated dates of Hess and Lippolt (1986) for the base Westphalian (315 Ma) and top Westphalian B (311 Ma), the two candidate low frequency sequences we propose (Fig. 7) overlie important sequence boundaries of possibly third-order magnitude and duration. Following this argument, the less widespread marine bands within these sequences, such as the *Amalie* and *Maltby* marine bands, may represent higher frequency (fourth-order) maximum flooding surfaces, with the intervening major stacked fluvial complexes representing fourth-order incised valley-fills. We interpret the less fully marine intervening *Lingula* bands as likely higher order flooding surfaces. These may represent parasequence and parasequence set boundaries within the fourth-order sequences or, as suggested by Read (1993 pers. comm.), potential MFSs, to even higher order discrete sequences. Modelling by Maynard and Leeder (1992) indicated widespread, very high frequency relative sea-level changes through the Late Carboniferous, consistent with potential Milankovitch-scale cyclicity in the study strata. We are

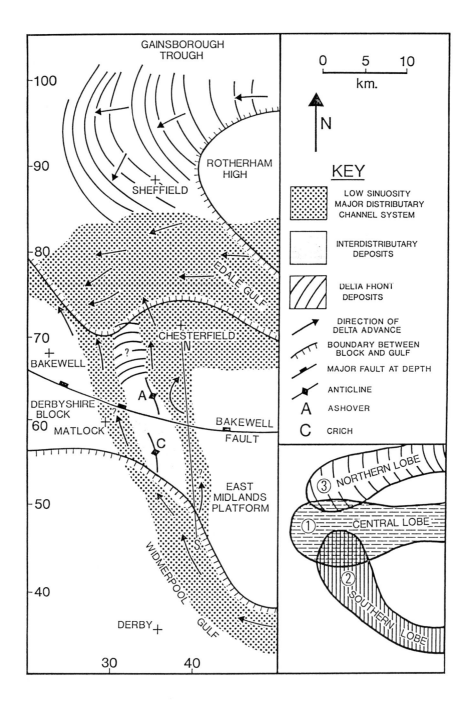

Fig. 9. General palaeogeography for the Westphalian of England (from Guion & Fielding, 1988) showing general palaeocurrent directions.

currently unable to delineate reliably such fifth-order complete sequences on a regional scale.

Mid-Westphalian A to late Westphalian B coal-bearing strata of England

Guion & Fielding (1988) argue for an 'upper delta plain' setting for this approximately 500 m thick succession, based on the association of major distributary channel belts (up to 5 km wide) and associated freshwater lake deposits and coals (Fig. 9). Palaeocurrents are more variable than in the basal Westphalian A and there is more evidence for mixed load channel-fills, which has been attributed to 'climatic factors in the source area' (Guion 1984; Guion & Fielding 1988).

These characteristics are remarkably similar to those of intervals in the Westphalian of Kentucky summarized earlier (Aitken & Flint in press *a*, in press *b*), characterized by large volumes of fine-grained, brackish to lacustrine deposits and heterolithic channel-fills, with major coal zones. These zones are interpreted as resulting from rising base level with a high and slowly rising water-table. Many studies, as summarized earlier, show that these conditions are ideal for the establishment and maintenance of mires over large areas. Accordingly, coals are more abundant, thicker and more laterally continuous in the mid-Westphalian A to late Westphalian B UK coal measures, and most of the productive seams lie within this interval (Guion 1987; Fulton & Williams 1988), with individual seams generally low in ash (<10%; Fielding 1987). We question the notion of Guion & Fielding (1988) that sediment supply *per se* was the main control on the stratigraphic architecture of this interval. Sediment supply is not an independent variable; it is changed either by tectonic uplift or base level change, both of which modify the graded river profile (Posamentier *et al.* 1988), or by climatic change, itself commonly linked to tectonic or eustatic changes (McCabe & Shanley 1993). Coal seams in the UK Westphalian split progressively towards the basin depocentre. This has been interpreted by Fielding (1987) as due to basinward increases in subsidence (presumably assuming a static sea-level). The geometric relationships are, however, consistent with the landward amalgamation of flooding surfaces at parasequence boundaries into coals and are comparable with the up-dip amalgamation of several coals documented in the Cretaceous of Utah (O'Byrne & Flint 1994). One such 'amalgamated' coal seam is the Warwickshire Thick Coal, up to 8.5 m thick, which formed as a persistent topographic high. Its palynological characteristics indicate deposition in a raised mire (Fulton 1987). A similar interpretation, based on lycopod successions, has been made for the Low Barnsley seam of Yorkshire (Bartram 1987).

In the Dutch offshore Cleaver Bank area, the Coal Measures comprise over 1200 m of coal-bearing delta plain strata. Quirk (1993) has argued for changes in accommodation space, driven by base level changes, to have governed facies architecture. Quirk (1993) also makes the important point that, although no biostratigraphically proved marine shales were found in any of the 32 wells in the study area, the systematic development of brackish or freshwater mudstones was controlled by base level and would be expected to correlate with marine bands nearer the basin centre. This is exactly the same situation as we envisage in the UK Coal Measures. Importantly, there is no evidence of seismic-scale faulting within the Westphalian of the Cleaver Bank, yet stratal architecture appears to be similar to that of the UK Coal Measures. This indicates to us that tectonic processes may have been previously overemphasized as controls on UK Westphalian stratigraphy.

Conclusions

1. Economic coals, spatially associated with clastic deposits, probably accumulated in raised mires.

2. The establishment, maintenance and preservation of regionally extensive mires (both low-lying and raised varieties) required a slowly rising water-table, and thus base level, on a regional scale. This is possible to some extent in all systems tracts but is a characteristic of the TST.

3. Economic coals occur dominantly in the TST of high frequency sequences. Thick seams may be up-dip time-correlative rocks to flooding surfaces.

4. Models for the sequence stratigraphy of coal-bearing strata, developed from the Cretaceous of the US Western Interior and the Pennsylvanian of the Appalachian basins, can be applied to the English Coal Measures, although more detailed studies of this type are required.

5. Coal exploration may profitably be associated with the TST of high frequency (fourth-order) Exxon-type sequences, as this is where the thickest and most laterally extensive coals occur. During a lower order, higher magnitude (third

or second-order) sequence this pattern may be accentuated within the transgressive or highstand sequence set.

The ideas in this paper have been influenced by discussions with P. McCabe, K. Shanley, J. Van Wagoner and members of the Liverpool Sequence Stratigraphy Research Group. We thank W. Read for a careful and perceptive review.

References

AITKEN, J. F. & FLINT, S. High frequency sequences and the nature of incised valley fills in fluvial systems of the Breathitt Group (Pennsylvanian), Appalachian basin, eastern Kentucky. *In:* BOYD, R., DALRYMPLE, R. W. & ZAITLAN, B. (eds) *Incised Valley Systems: Origin and Sedimentary Sequences.* Society of Economic Paleontologists and Mineralogists, Special Publication, **51** in press.

—— & —— The application of sequence stratigraphy to fluvial systems: an example from the Late Carboniferous of the Appalachians. *Sedimentology*, in press.

ALLEN, G. P. 1984. Tidal processes in estuaries: a key to interpreting fluvial-tidal facies transitions. *In:* PURSER, B. *et al.* (eds) *Recent Developments in Fluvial Sedimentology Proceedings 5th European Meeting.* International Association of Sedimentologists, Marseille, 23–24 [abstract].

—— & POSAMENTIER, H. W. 1993. Sequence stratigraphy and facies models of an incised valley fill: the Gironde estuary, France. *Journal of Sedimentary Petrology*, **63**, 378–391.

ANDERSON, J. A. R. 1964. The structure and development of the peat swamps of Sarawak and Brunei. *Journal of Tropical Geography*, **18**, 7–16.

ARDITTO, P. A. 1987. Eustasy, sequence stratigraphic analysis and peat formation: a model for widespread late Permian coal deposition in the Sydney basin, N.S.W. *In: Advanced Studies of the Sydney Basin. 21st Newcastle Symposium Proceedings*, 11–17.

—— 1991. A sequence stratigraphic analysis of the Late Permian succession in the Southern Coalfield, Sydney Basin, New South Wales. *Australian Journal of Earth Sciences*, **38**, 125–137.

BAGANZ, B. P., HORNE, J. C. & FERM, J. C. 1975. Carboniferous and recent Mississippi lower delta plains: a comparison. *Transactions of the Gulf Coast Association of Geological Societies*, **25**, 183–191.

BARTRAM, K. M. 1987. Lycopod successions in coals: an example from the Low Barnsley seam (Westphalian B), Yorkshire, England. *In:* SCOTT, A. C. (ed.) *Coal and Coal-bearing Strata: Recent Advances.* Geological Society, London, Special Publication, **32**, 187–199.

CALVER, M. A. 1968. Distribution of Westphalian marine faunas in northern England and adjoining areas. *Proceedings of the Yorkshire Geological Society*, **37**, 1–72.

CLYMO, R. S. 1987. Rainwater-fed peat as a precursor of coal. *In:* SCOTT, A. C. (ed.) *Coal and Coal-bearing Strata: Recent Advances.* Geological Society, London, Special Publication, **32**, 17–22.

CROSS, T. A. 1988. Controls on coal distribution in transgressive–regressive cycles, Upper Cretaceous, Western Interior, U.S.A. *In:* WILGUS, C.H., HASTINGS, B. S., KENDALL, C. G. ST. C., POSAMENTIER, H. W., ROSS, C. A. & VAN WAGONER, J. C. (eds) *Sea Level Changes—an Integrated Approach.* Society of Economic Paleontologists and Mineralogists, Special Publication, **42**, 371–380.

EDEN, R. A. 1954. The Coal Measures of the Anthraconaia lenisulcata zone. *Bulletin of the Geological Survey of Great Britain*, **5**, 81–106.

FERM, J. C. & STAUB, J. R. 1984. Depositional controls of minable coal bodies. *In:* RAHMANI, R. A. & FLORES, R. M. (eds) *Sedimentology of Coal and Coal-bearing Sequences.* International Association of Sedimentologists, Special Publication, 7, Blackwell Scientific Publications, Oxford, 275–290.

FIELDING, C. R. 1984. Upper delta plain lacustrine and fluviolacustrine facies from the Westphalian of the Durham coalfield, NE England. *Sedimentology*, **31**, 547–567.

—— 1985. Coal depositional models and the distinction between alluvial and delta plain environments. *Sedimentary Geology*, **42**, 41–48.

—— 1986. Fluvial channel and overbank deposits from the Westphalian of the Durham coalfield, NE England. *Sedimentology*, **33**, 119–140.

—— 1987. Coal depositional models for deltaic and alluvial plain sequences. *Geology*, **15**, 661-664.

FLINT, S. 1993. The application of sequence stratigraphy to ancient fluvial systems. *In:* FIELDING, C. R. (ed.) *Recent Developments in Fluvial Sedimentology. Proceedings of the 5th International Conference on Fluvial Sedimentology.* University of Queensland Press, 16 pp.

——, STEWART D. J. & VAN RIESSEN, E. D. 1989. Reservoir geology of the Sirikit oilfield, Thailand: lacustrine-deltaic sedimentation in a Tertiary intermontane basin. *In:* WHATELEY, M. & PICKERING, K. (eds) *Deltas: Sites and Traps for Fossil Fuels.* Geological Society, London, Special Publication, **41**, 223–237.

FULTON, I. M. 1987. Genesis of the Warwickshire Thick Coal: a group of long residence histosols. *In:* SCOTT, A. C. (ed.) *Coal and Coal-bearing Strata: Recent Advances.* Geological Society, London, Special Publication, **32**, 201–218.

—— & WILLIAMS, H. 1988. Palaeogeographical change and controls on Namurian and Westphalian A/B sedimentation at the southern margin of the Pennine basin, central England. *In:* BESLY, B. & KELLING, G. (eds) *Sedimentation in a Synorogenic Basin Complex—the Upper Carboniferous of NW Europe.* Blackie, Glasgow, 178–199.

GASTALDO, R. A., DEMKO, T. M. & LIU, Y. 1993. Application of sequence and genetic stratigraphic concepts to Carboniferous coal-bearing strata: an example from the Black Warrior basin, USA. *Geologische Rundschau*, **82**, 212–226.

GERSIB, G. A. & MCCABE, P. J. 1981. Continental coal-bearing sediments of the Port Hood Formation (Carboniferous), Cape Linzee, Nova Scotia, Canada. *In:* ETHRIDGE, F. G. & FLORES, R. M. (eds) *Recent and Ancient Nonmarine Depositional Environments: Models for Exploration.* Society of Economic Paleontologists and Mineralogists, Special Publication, **31**, 95–108.

GUION, P. D. 1984. Crevasse splay deposits and roof-rock quality in the Threequarters seam (Carboniferous) in the East Midlands coalfield, U.K. *In:* RAHMANI, R. A. & FLORES, R. M. (eds) *Sedimentology of Coal and Coal-bearing Sequences.* International Association of Sedimentologists, Special Publication, **7**, 291–308.

—— 1987. Palaeochannels in mine workings in the High Hazles coal (Westphalian B), Nottinghamshire Coalfield, England. *Journal of the Geological Society of London*, **144**, 471–488.

—— & FIELDING, C. R. 1988. Westphalian A and B sedimentation in the Pennine Basin, U.K. *In:* BESLY, B. & KELLING, G. (eds) *Sedimentation in a Synorogenic Basin Complex—the Upper Carboniferous of NW Europe.* Blackie, Glasgow, 153–177.

—— & FULTON, I. M. 1993. The importance of sedimentology in deep-mined coal extraction. *Geoscientist*, **3**, 25–33.

HAMPSON, G. 1995. Discrimination of regionally extensive coals in the Upper Carboniferous of the Pennine Basin, UK, using high-resolution sequence stratigraphic concepts. *In:* WHATELEY, M. K. G. & SPEARS, D. A. (eds) *European Coal Geology.* Geological Society, London, Special Publication, **82**, 79–97.

HASZELDINE, R. S. 1989. Coal reviewed: depositional controls, modern analogues and ancient climates. *In:* WHATELEY, M. K. G. & PICKERING, K. T. (eds) *Deltas: Sites and Traps for Fossil Fuels.* Geological Society, London, Special Publication, **41**, 289–308.

HESS, J. C. & LIPPOLT, H. J. 1986. $^{40}Ar/^{39}Ar$ ages of tonstein and tuff sanidines: new calibration points for the improvement of the upper Carboniferous time scale. *Chemical Geology*, **59**, 143–154.

HORNE, J. C., FERM, J. C., CARUCCIO, F. T. & BAGANZ, B. P. 1978. Depositional models in coal exploration and mine planning in Appalachian region. *Bulletin of the American Association of Petroleum Geologists*, **62**, 2379–2411.

HOWELL, D. J. & FERM, J. C. 1980. Exploration model for Pennsylvanian upper delta plain coals, southwest West Virginia. *Bulletin of the American Association of Petroleum Geologists*, **64**, 938–941.

JERVEY, M. T. 1988. Quantitative modelling of siliciclastic rock sequences and their seismic expression. *In:* WILGUS, C. H., HASTINGS, B. S., KENDALL, C. G. St. C., POSAMENTIER, H. W., ROSS, C. A. & VAN WAGONER, J. C. (eds) *Sea Level Changes—an Integrated Approach.* Society of Economic Paleontologists and Mineralogists, Special Publication, **42**, 47–70.

KRYSTINIK, L. F. & BLAKENEY-DEJARNETT, A. 1991. Sequence stratigraphy and sedimentologic character of valley fills, Lower Pennsylvanian Morrow Formation, Eastern Colorado and western Kansas. *In: Proceedings of the 1991 Nuna Conference on High Resolution Sequence Stratigraphy, Banff, August 1991*, 24–26.

MAYNARD, J. R. 1992. Sequence stratigraphy of the Upper Yeadonian of northern England. *Marine and Petroleum Geology*, **9**, 197–207.

—— & LEEDER, M. R. 1992. On the periodicity and magnitude of Late carboniferous glacio-eustatic sea-level changes. *Journal of the Geological Society of London*, **149**, 303–311.

MCCABE, P. J. 1984. Depositional models of coal and coal-bearing strata. *In:* RAHMANI, R. A. & FLORES, R. M. (eds) *Sedimentology of Coal and Coal-bearing Sequences.* International Association of Sedimentologists, Special Publication, **7**, 13–42.

—— 1987. Facies studies of coal and coal-bearing strata. *In:* SCOTT, A. C. (ed.) *Coal and Coal-bearing Strata: Recent Advances.* Geological Society, London, Special Publication, **32**, 51–66.

—— 1991. Geology of coal: environments of deposition. *In:* GLUSKOTER, H. J., RICE, D. D., TAYLOR, R. B. (eds) *Economic Geology, U.S. The Geology of North America*, V P-2. Geological Society of America. Boulder, Colorado.

—— 1993. Sequence stratigraphy of coal-bearing strata. *In:* BREYER, J. A., ARCHER, A. W. & MCCABE, P. J. (eds) *Sequence Stratigraphy of Coal-Bearing Strata: Field Trip Guidebook and Short Course Supplement.* Energy Minerals Division, American Association of Petroleum Geologists, Tulsa.

—— & PARRISH, J. T. 1992. Tetonic and climatic controls on the distribution and quality of Cretaceous coals. *In:* MCCABE, P. J. & PARRISH, J. T., (eds) *Controls on the Distribution and Quality of Cretaceous Coals.* Geological Society of America, Special Paper **267**, 1–15.

—— & SHANLEY, K. W. 1992. An organic control on shoreface stacking patterns: bogged down in the mire. *Geology*, **20**, 741–744.

MITCHUM, R. M. J. & VAN WAGONER, J. C. 1991. High-frequency sequences and their stacking patterns: sequence stratigraphic evidence of high-frequency eustatic cycles. *Sedimentary Geology*, **70**, 131–160.

O'BYRNE, C. J. & FLINT, S. 1993. Sequence stratigraphy of Cretaceous shallow marine sandstones, Book Cliffs, Utah: application to reservoir modelling. *First Break*, **11**, 445–459

—— & —— 1994. Sequence, parasequence and intraparasequence architecture of the Grassy Member (Campanian), Book Cliffs, Utah. *American Association of Petroleum Geologists. Methods in Exploration Series*, in press.

POSAMENTIER, H. W. & VAIL, P. R., 1988. Eustatic controls on clastic deposition II—sequence and systems tract models. *In:* WILGUS, C. H., HASTINGS, B. S., KENDALL, C. G. St. C., POSAMENTIER, H. W., ROSS, C. A. & VAN WAGONER, J. C. (eds) *Sea Level Changes—an Integrated Approach.* Society of Economic Paleontologists and Mineralogists, Special Publication, **42**, 125–154.

——, JERVEY, M. T. & VAIL, P. R. 1988. Eustatic controls on clastic depositional-conceptual framework. *In:* WILGUS, C. H., HASTINGS, B. S., KENDALL, C. G. St. C., POSAMENTIER, H. W., ROSS, C. A. & VAN WAGONER, J. C. (eds) *Sea Level Changes—an Integrated Approach.* Society of Economic Paleontologists and Mineralogists, Special Publication, **42**, 109–124.

QUIRK, D. G. 1993. Interpreting the Upper Carboniferous of the Dutch Cleaver Bank High. *In:* PARKER, J. R. (ed.) *Petroleum Geology of Northwest Europe: Proceedings of the 4th Conference.* Geological Society, London, 696–706.

RAMSBOTTOM, W. H. 1977. Major cycles of transgression and regression (mesothems) in the Namurian. *Proceedings of the Yorkshire Geological Society*, **41**, 261–291.

—— 1979. Rates of transgression and regression in the Carboniferous of NW Europe. *Journal of the Geological Society of London*, **136**, 147–153.

READ, W. A. 1991. The Millstone Grit (Namurian) of the southern Pennines viewed in the light of eustatically controlled sequence stratigraphy. *Geological Journal*, **26**, 157–165.

—— & FORSYTH, I. H. 1991. Allocycles in the upper part of the Limestone Coal Group (Pendleian E1) of the Glasgow–Stirling Region, viewed in the light of sequence stratigraphy. *Geological Journal*, **26**, 85–89.

RYER, T. A. 1981. Deltaic coals of Ferron sandstone member of Mancos shale-predictive model for Cretaceous coal-bearing strata of western interior. *Bulletin of the American Association of Petroleum Geologists*, **65**, 2323–2340.

—— 1983. Transgressive–regressive cycles and the occurrence of coal in some Upper Cretaceous strata of Utah. *Geology*, **11**, 207–210

SEARS, J. D., HUNT, C. B. & HENDRICKS, T. A. 1941. Transgressive and regressive Cretaceous deposits in southern San Juan Basin, New Mexico. *United States Geological Survey*, **193-F**.

SHANLEY, K. W. & MCCABE, P. J. 1991a. Perspectives on the sequence stratigraphy of continental strata. *Proceedings of the 1991 Nuna Conference on High Resolution Sequence Stratigraphy, Banff, August 1991.*

—— & —— 1991b. Predicting facies architecture through sequence stratigraphy—an example from the Kaiparowits Plateau, Utah. *Geology*, **19**, 742–745.

—— & —— 1993. Alluvial architecture in a sequence stratigraphic framework—case history from the upper Cretaceous of southern Utah, U.S.A. *In:* FLINT, S. S. & BRYANT, I. D. (eds) *The Geological Modelling of Hydrocarbon Reservoirs.* International Association of Sedimentologists, Special Publication, **15**, 21–56.

——, —— & HETTINGER, R. D. 1992. Tidal influence in Cretaceous fluvial strata from Utah, U.S.A.: a key to sequence stratigraphic interpretation. *Sedimentology*, **39**, 905–946

STYAN, W. B. & BUSTIN, R. M. 1983. Sedimentology of Fraser River Delta peat deposits: a modern analogue for some deltaic coals. *International Journal of Coal Geology*, **3**, 101–143.

VAIL, P. R., MITCHUM, R. M. Jr., TODD, R. G. WIDMEIR, J. M., THOMPSON, S. III., SANGREE, J. B., BUBB, J. N. & HATLELID, W. G. 1977. Seismic stratigraphy and global changes of sea-level *In:* CLAYTON, C. E. (ed.) *Seismic Stratigraphy and Applications to Hydrocarbon Exploration.* American Association of Petroleum Geologists, Memoir, **26**, 49–212.

VAN WAGONER, J. C., MITCHUM, R. M., CAMPION, K. M. and RAHMANIAN, V. D. 1990. *Siliciclastic Sequence Stratigraphy in Well Logs, Cores and Outcrops.* American Association of Petroleum Geologists methods in Exploration, Series, 7, 55p.

——, NUMMEDAL, D., JONES, C. R. TAYLOR, D. R., JENNETTE, D. C. & RILEY, G. W. 1991. Sequence stratigraphy applications to shelf sandstone reservoirs, outcrop to subsurface examples. *American Association of Petroleum Geologists. Sequence Stratigraphy Field Conference Guide.*

——, POSAMENTIER, H. W., MITCHUM, R. M. Jr., VAIL, P. R., SARG, J. F., LOUTIT, T. S. & HARDENBOL, J. 1988. An overview of the fundamentals of sequence stratigraphy and key definitions. *In:* WILGUS, C. H., HASTINGS, B. S., KENDALL, C. G. St. C., POSAMENTIER, H. W., ROSS, C. A. & VAN WAGONER, J. C. (eds) *Sea Level Changes—an Integrated Approach.* Society of Economic Paleontologists and Mineralogists, Special Publication, **42**, 39–45.

Application of sedimentology to the development and extraction of deep-mined coal

IAIN M. FULTON,[1,2] PAUL D. GUION[3] & NEIL S. JONES[3,4]

[1] *British Coal, Technical Services and Research Executive, Stanhope Bretby, Burton-upon-Trent DE15 0QD, UK*
[2] *Present address: Golder Associates (UK) Ltd, Landmere Lane, Edwalton, Nottingham NG12 4DG, UK*
[3] *Geology and Cartography Division, Oxford Brookes University, Oxford OX3 0BP, UK*
[4] *Present address: British Geological Survey, Kingsley Dunham Centre, Keyworth, Nottingham NG12 5GG, UK*

Abstract: Most deep-mined coal in the UK comes from the Westphalian A and B (Langsettian–Duckmantian) of the coalfields of the Pennine Basin, in which several sedimentary facies have been identified. Deep mining of coal may be divided into four main operations: shaft sinking and driving of access drifts; drivage of underground roadways; excavation of coal faces; and the provision of temporary underground storage. The prediction of the lithological, geometrical and geotechnical properties of the various sedimentary facies likely to be encountered during mining is important for the success of all these operations.

Roadway location may be influenced by the geotechnical attributes of the sedimentary facies, which also govern the type of machinery and methods used to cut and support the roads. The thickness, quality and continuity of the coal, and the properties of the roof and floor, are important in selecting the location of longwall faces and in their productivity. Facies composed of strong sedimentary rocks are likely to be the most suitable for the location of staple shafts and bunkers for coal storage. The geotechnical properties of each of the sedimentary facies is described, with particular reference to their excavation and support characteristics, and the discontinuities of sedimentary origin that are likely to be present. Distal lacustrine delta facies appear to provide the optimum roof conditions for coal faces and roadways.

A simple scheme for the prediction of the sedimentary facies which may influence mining is presented, and a variety of mapping techniques, including facies maps, isopach maps and palaeocurrent analysis, are outlined. An improved knowledge of the properties and distribution of the various sedimentary facies at a mine site should lead to improved productivity and a reduction in risks and costs.

Most of the deep-mined coal in the UK is currently obtained from collieries in the coalfields of the Pennine Basin (Fig. 1), in which most of the seams mined are of Westphalian A and B (Langsettian–Duckmantian) age (Fig. 2). Exploration for coal within the Pennine Basin was intense from the mid-1970s to the mid-1980s. As a consequence of this, new mining areas were opened at Selby in Yorkshire and Asfordby in Leicestershire, and many extensions were developed from collieries in existing coalfields. Economic conditions have since resulted in a shrinkage in the demand for coal and a corresponding reduction in mined tonnage, with current exploration and development being restricted to existing mines.

The object of this paper is to show that sedimentology can be applied to all stages of deep-mined coal exploration and development, by reference to the coalfields of the Pennine Basin. In particular, it concentrates on how the lithological, geometrical and geotechnical attributes of the various facies described by Guion *et al.* (this volume), affect the extraction of deep-mined coal. The accurate prediction of both the properties and distribution of facies is important for the success of many mining operations, and a scheme is presented for the prediction of facies distribution.

Although sedimentological analysis is used routinely by British Coal, its usage has not been widely reported. Studies describing the applications of sedimentology to deep coal mining in the UK include Clarke (1963), Elliott (1965,

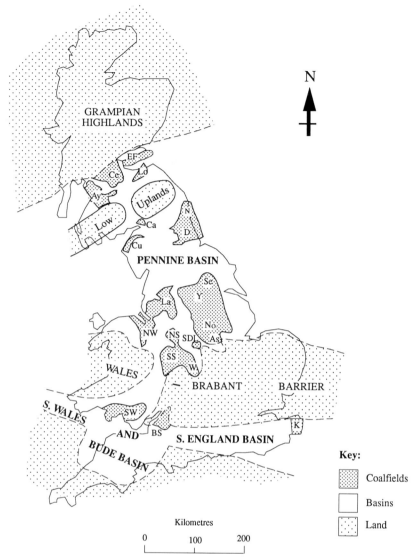

Fig. 1. Coalfields of Great Britain, showing a generalized late Westphalian A (Langsettian) palaeogeographical reconstruction. Modified from Guion (1992). Key to the coalfields: As, Asfordby; Ay, Ayrshire; BS, Bristol and Somerset; Ca, Canonbie; Ce, Central Scottish; Cu, Cumbria; D, Durham; EF, East Fife; K, Kent; La, Lancashire; Lo, Lothian; N, Northumberland; No, Nottinghamshire; NS, North Staffordshire; NW, North Wales; Se, Selby; SDL, South Derbyshire and Leicestershire; SS, South Staffordshire; SW, South Wales; W, Warwickshire; and Y, Yorkshire.

1969, 1974, 1979, 1984), Smith (1975), Guion (1984, 1987a), Rippon (1984), Williams (1986), Fulton (1987a), Rippon & Spears (1989) and Guion & Fulton (1993).

Documentation of the applications of sedimentology to deep coal mining world-wide is extensive: examples include Horne et al. (1978a, b), McCabe & Pascoe (1978), Voelker (1978), Mathewson & Gowan (1981), Moebs (1981), Stoppel & Bless (1981), Taylor (1981), Houseknecht & Iannaccione (1982), Moebs & Ellenburger (1982), Chase & Sames (1983), Nelson (1983), Weisenfluh & Ferm (1991), contributors to Peters (1991) and Dreesen (1993).

Coal Measures facies

Ten facies, grouped into five facies associations, have been identified in the Coal Measures of the Pennine Basin (Guion et al. this volume). Each

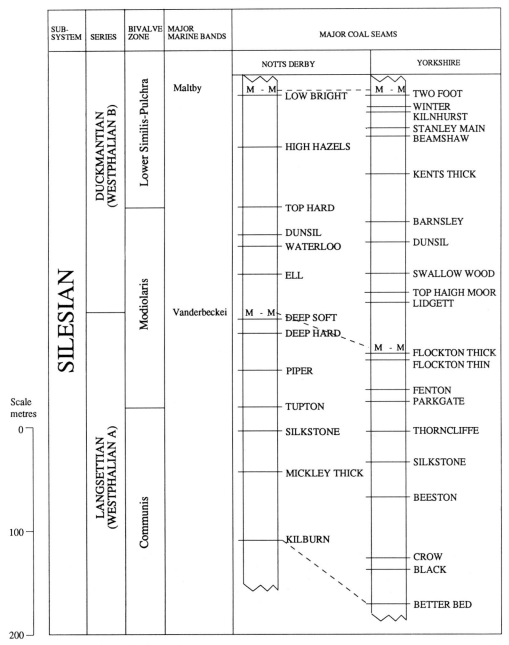

Fig. 2. Seam nomenclature of the main Westphalian A and B coal seams (Langsettian–Duckmantian) of the East Pennine Coalfields (modified after Ramsbottom *et al.* 1978).

facies can be recognized consistently from borehole cores and underground exposures (Table 1) and is characterized by a particular geometry and range of dimensions. The use of these and other diagnostic characteristics in the recognition of facies is discussed later. The relationships between facies are often predictable because of the processes controlling their origin, e.g. crevasse splay deposits occur adjacent to channels. An understanding of these relationships has allowed the construction of a facies model for the Pennine Basin Coal

Table 1. *Characteristics of the major facies of the Westphalian (mid-Langsettian to late Duckmantian) coal-bearing upper delta plain deposits of the Pennine Basin.*

Facies associations and facies	Lithology	Sedimentary structures	Geometry	Fossils
Pedogenic				
Mire (mainly coal)	Coal, impure coal	Lamination, banding	Extensive sheets greater than 0.1 m thick; may split or die out laterally	Plant remains
Palaeosol (mainly seat earth)	Grey to white, brown/cream or red claystone, siltstone or sandstone depending on substrate and drainage conditions. Thin impure coals	Irregular laminations disturbed by rootlets, or lamination totally destroyed. Common mottling. Abundant polished ('listric') surfaces, siderite nodules, sphaerosiderite or iron oxides, depending on drainage conditions	Generally extensive sheets	*In situ* rootlets, *Stigmaria*, *Calamites* roots in some palaeosols
Marine	Dark grey to black fissile carbonaceous claystone	Well-developed thin lamination, sometimes massive	Extensive sheets	Marine goniatites, bivalves, brachiopods, gastropods, crinoids, bryozoa, forams, *Lingula*, *Orbiculoidea*, plant fragments, trace fossils
Lake fill				
Lacustrine	Medium grey to black claystone or siltstone, carbonaceous mudstone, cannel, boghead, coal	Flat lamination, rare sandy laminae and scours	Sheet-like	Non-marine bivalves, fish, ostracods, *Spirorbis*, plant fragments, trace fossils
Lacustrine delta (a) distal	Siltstones, interlaminated siltstone/sandstone forming upward-coarsening sequences	Flat lamination, current- and wave-ripple cross-lamination, climbing ripples, backflow ripples, ripple form sets, soft sediment deformation	Sheet-like to lobate deposits occur distally and beneath proximal lacustrine delta deposits, generally shows upward coarsening above lacustrine deposits	Trace fossils, plant debris, rare non-marine bivalves
(b) Proximal	Sandstone, interlaminated sandstone/siltstone, forming upward-coarsening sequences	Current-ripple cross-lamination, cross-bedding at tops of sequences, climbing ripples, wave-ripple cross-lamination, occasional trough-like scour surfaces, flat lamination, ripple form sets, backflow ripples, soft sediment deformation	Lobate to sheet-like deposits up to about 10 km across and 8 m thick, with gradational or sharp bases, forming upward-coarsening sequences above distal lacustrine delta	Plant debris, trace fossils
Channel				
Major channel	Thick erosively based sandstone bodies, often with horizons of breccia of conglomerate. Subordinate siltstone, claystone	Erosion surfaces, trough and planar cross-bedding, inclined heterolithic surfaces, ripple and cross-lamination, soft sediment deformation	Elongate belts, typically 1–20 km wide, 10s of km long, greater than 8 m thick	Plant fragments and debris, often abundant, rare trace fossils
Minor channel	Sandstone, siltstone, claystone, typically heterolithic, breccia, conglomerate; very variable	Erosion surfaces, ripple cross-lamination, trough and planar cross-bedding inclined heterolithic surfaces, flat lamination, soft sediment deformation, bank collapse deposits	Elongate belts typically up to 1 km wide, several km long, up to 8 m thick	Plant fragments and debris, often abundant, rare trace fossils
Near-channel				
Overbank	Siltstone, claystone; sometimes interbedded sandstone	Massive or weakly laminated siltstone, passing distally into laminated claystone. Occasionally interbedded siltstone/sandstone with low dips away from adjacent channel	Elongate belts adjacent and parallel to channel margins	Abundant plant leaves and stems, *in situ* trees
Crevasse splay	Sharp based sandstones interbedded with siltstones or claystones	Current-ripple cross-lamination, flat lamination, cross-bedding, wave rippled tops	Lobate deposits adjacent to channels. Individual splays generally < 1 m thick, often vertically stacked. Thin distally away from channel	Plant debris. *Pelecypodichnus* escape shafts

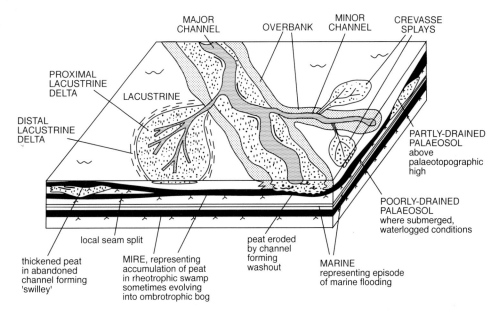

Fig. 3. Schematic facies model of the Westphalian (mid-Langsettian to late Duckmantian) coal-bearing upper delta plain environment, showing relationships of major facies.

Measures (Fig. 3) which is of assistance in the prediction of the location of the various facies, and ultimately in the prediction of the properties of the rocks encountered during mining.

Stages in deep mine exploration and development

The exploration and development of deep-mined coal may be divided into three stages (Fig. 4). At each stage an understanding of different facies may be required to aid in geological assessments associated with various exploration and development activities (Fig. 4). Initially, exploration is needed to identify new prospects, which are usually extensions to existing coalfields, but which may be remote from previous sites. In this initial investigative stage, the properties of the coal seams (mire facies) and the relationships between the mire and channel facies (major channels can significantly reduce coal thickness) are important. During exploration, the deposit boundaries are identified and an indication of thickness and volume, continuity and chemical characteristics of the seams are obtained. This enables the calculation of *in situ* resources (coal in place).

Should these criteria prove to be satisfactory, the second feasibility stage involves a preliminary evaluation of the geological environment in which mining will take place and an assessment of recoverable reserves, leading eventually to the selection of a mine site. At this stage all facies are examined and preliminary facies maps for seam roofs and floors produced.

The final productive stage involves the detailed planning and construction of a working colliery, including the conduit between the surface and target seams and the tunnels and faces needed to mine the coal. This requires an understanding of the hydrogeology, engineering geology and sedimentology of the coal-bearing strata. As mining proceeds and more data become available, all facies can be studied in increasing detail to refine and update calculations of recoverable reserves and to aid mine planning. This paper concentrates on the application of sedimentology to the development and extraction of coal in the productive stage of deep mining.

The principal techniques used in coal exploration are the drilling of boreholes, both from the surface and underground, and seismic surveys. A summary of exploration methods is given in Guion & Fulton (1993). In the initial development phase boreholes may be the only method of exploration, but as development proceeds, a small number of seismic lines may be shot. The intensity of exploration increases as development progresses, with up to five surface boreholes and up to several kilometres of seismic line per square kilometre at the productive stage.

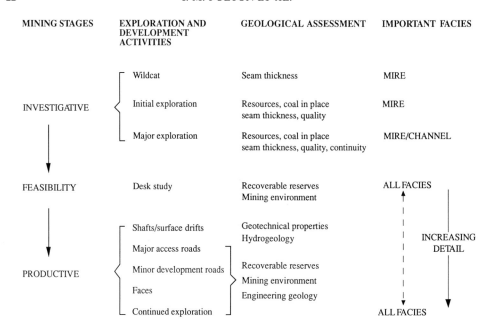

Fig. 4. Stages in the evolution of a mine showing, for each stage, the nature of exploration and development, the geological assessments required (where the nature of sedimentary facies is significant) and the most important facies.

Sedimentological attributes of the rocks may be deduced from a range of data types including cored boreholes, geophysical wireline logs and both conventional and three-dimensional seismic surveys. Each combination of data type allows a different quality of sedimentological interpretation. In addition, underground workings provide an ever-increasing amount of sedimentological data and roads associated with faces provide a continuous, but vertically limited, sedimentary record.

Deep mining operations

Deep mining of coal may be divided into four main operations. Shaft sinking and drift drivage enable coal-bearing rocks to be reached below the surface; drivage of underground roadways allows access to blocks of coal; excavation of faces is the means of extraction of coal; and staple shafts and bunkers provide temporary underground storage (Fig. 5). Each of these activities has different engineering requirements because the excavations have different purposes, shapes, dimensions, orientations and lengths of time they are required for use.

The geological factors which affect the choice of mine site and the construction of access roads and faces have been discussed by Guion & Fulton (1993). Underground roads, faces and storage facilities are described briefly in the following text.

Roadways may be up to several kilometres in length and narrow in cross-section, and are usually driven by roadheaders. They are supported by a variety of means including roof bolts, steel arches and block linings. Major arterial roads may be driven at a preferred horizon or, more often, may be driven 'cross-measures' to access major areas of coal at different stratigraphic horizons. They are required to last for long periods of time and must be engineered to a high standard to minimize repair work. Ideally, strata in which these are driven should be weak enough to allow rapid excavation, but strong enough to maximize support. Minor roads, usually driven in coal, give access to smaller blocks of coal and are used for shorter lengths of time. A conventionally driven cross-measures road, 3.66 m high and supported above and to the sides by D-shaped steel arches costs between £2000 and £2500 per metre at 1993 costs.

Coal in the UK is normally worked by faces using the longwall method (Figs 5–7). Faces are wider in cross-section than roads, typically 200–250 m, fairly long and their usage is relatively short-lived. They may be operated either as advancing or retreat faces (Fig. 6). In advancing faces the two parallel roadways or 'gates', which

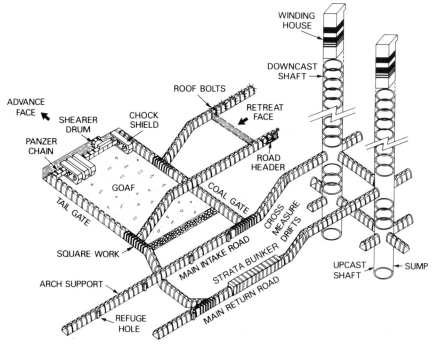

Fig. 5. Isometric representation of a typical deep coal mine, illustrating major excavations including shafts, access roadways, bunkers and faces. Modified from Guion & Fulton (1993).

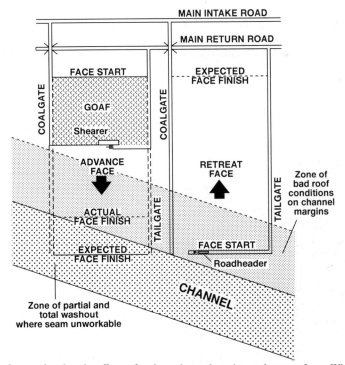

Fig. 6. Typical face layout showing the effects of a channel on advancing and retreat faces. With an advancing face, the face is stopped prematurely; with a retreat face, the roadways are driven to the margins of the channel and the limits of workable coal can be defined before mining commences. Modified from Guion & Fulton (1993).

give access to the face, are driven with the face as it advances. In retreat faces, the two gates are driven first and then the face is set up, which retreats back towards the main access roadways. Retreat faces have the advantage of proving geological problems such as those caused by channel facies before the face is set up, whereas with advancing faces, the face may be stopped prematurely (Fig. 6). On longwall faces, coal is usually cut by one or more shearers, which cut between the two roadways (Fig. 7). Faces are supported by hydraulic chocks which move forwards as each strip of coal is cut, allowing the roof to fall into the cavity behind, creating various sized blocks, collectively referred to as 'goaf' or 'gob'. Ideal strata for coal faces consist of a continuous large thickness of coal with a fairly strong roof and floor. The cost of equipping a heavy duty longwall face with an extraction height of 2.0–2.5 m is typically £7.5–10.0×10^6.

Storage facilities include horizontal bunkers and vertical staple shafts. Strata bunkers are normally constructed in specially designed, enlarged roadways and are required for long periods of time without need of future repair work. Staple shafts are relatively short vertical shafts sunk between two roadways at different horizons. Strong strata are required in which to locate the base of staple shafts and are preferred for horizontal bunkers.

The construction of all underground excavations consists of two separate operations, cutting and supporting, and the effects that the sedimentary facies may have on these operations are considered in the next section. Variations in the mire facies (which form the target coal seams), and the relationships of the mire facies to other facies (especially channels and their marginal facies), have a significant effect on coal production and are discussed in the next section.

Effects and significance of sedimentary facies on mining

Sedimentary facies, which are often characterized by certain lithologies, exercise an important control on the *in situ* geotechnical characteristics of Coal Measures rocks, which ultimately influence all deep-mining activities. These characteristics include geometry (thickness and continuity), geotechnical properties (mainly related to strength) and hydrogeological properties. The characteristics of coal seams, represented by the mire facies, are usually the prime consideration in deciding the location of areas to be mined, so there is a particular need to examine the effects of mire facies variation. Additionally, the relationship between the mire facies and others immediately above and below the seam, especially channel-related deposits, may be important in influencing the choice of location, production method and rate of extraction of coal.

The geotechnical properties of the different facies encountered during mining have major effects on both cutting and supporting mine excavations. Typical geotechnical parameters for major rock types of the southern part of the Pennine coalfields are summarized in Table 2. These are a function of both original sediment properties and later changes due to diagenesis and burial. The diagenetic and burial history within a sedimentary basin clearly varies according to stratigraphic position and location; however, the parameters shown in Table 2 are believed to be fairly typical of the coal-bearing deposits of Pennine Basin. Geotechnical

Fig. 7. Typical longwall face, showing hydraulic chocks on the left and a shearer drum to the right on which are located picks.

Table 2. *Typical geotechnical properties of Coal Measures rocks from a low rank area of the Pennine Basin*

Lithology	Descriptive categories for uniaxial compressive strength BS5930 (1981)	Uniaxial compressive strength (MPa)	Young modulus (GPa)	Angle of friction (degrees)	Indirect tensile strength (MPa)	Abrasivity
Coal	Moderately strong	15.0–50.0	2.0–4.0	30–40	5.0–10.0	—
Seat earth claystone	Moderately weak to moderately strong	10.0–40.0	5.0–15.0	15–30	—	<0.3
Seat earth siltstone	Moderately strong	30.0–50.0	10.0–20.0	25–35	10.0–18.5	0.3–1.0
Claystone	Moderately strong to strong	30.0–65.0	5.0–15.0	20–35	10.0–16.0	<0.3
Siltstone, very fine to fine	Moderately strong to strong	40.0–65.0	13.0–20.0	25–35	13.5–16.0	0.3–0.7
Siltstone, medium to coarse	Strong	50.0–70.0	13.0–23.0	25–35	13.5–18.5	0.3–1.0
Layered sandstone/siltstone	Strong	55.0–75.0	15.0–25.0	35–40	10.0–20.0	0.5–2.0
Sandstone	Strong	60.0–90.0	17.0–30.0	30–45	15.0–25.0	1.0–2.5
Sandstone, well cemented	Strong to very strong	70.0–120.0	18.0–50.0	35–50	25.0–35.0	1.5–4.0

Results are from tests carried out using standard procedures at British Coal rock testing laboratory, Technical Services Research Establishment, Bretby.

properties such as those in Table 2 are generally derived from the laboratory measurement of small samples and exclude a variety of larger scale discontinuities, including joints and bedding planes, which contribute to the *in situ* behaviour of rock masses in mine workings. Where strata change vertically or laterally, or form interlayered sequences, the interfaces of different rock types often form discontinuities, especially where hard and soft strata are juxtaposed (Moebs & Ferm 1982) or where channels are present (Weisenfluh & Ferm 1991). Burrows, roots, layers of comminuted plant debris, parting planes, listric surfaces and slickensides contribute to the geotechnical properties of the various facies that form rock masses in mine workings. The main *in situ* geotechnical properties of the various facies and their effects on deep mining are summarized in Table 3.

To determine the full response of the rocks to excavation, additional factors must also be considered, including tectonic or mining-induced discontinuities, *in situ* and induced stress, types of cutting machinery, nature of support and shape and direction of excavation (Hoek & Brown 1980). Non-circular roadways, such as arch-supported coalgates, produce stress configurations comprising relatively high induced stresses and relatively low confining stresses in the unsupported floor and shoulder of the arches. At these locations the strength of the excavated rock may be lower than the induced stress and failure may occur. Elliott (1974, 1984) has discussed the relationship between the properties of sedimentary rocks and their mining behaviour. The effects on mining of the different sedimentary facies described by Guion *et al.* (this volume) are described in the following sections.

Mire (coal)

The thickness, continuity and quality of coal seams are fundamental factors which affect longwall coal mining (Table 4, Fig. 8). These factors are controlled by both the original depositional environment and succeeding events related to diagenesis, compaction, burial and tectonism.

The geotechnical properties of coal (Table 2) make it easy to excavate with shearer drums with long picks (Fig. 7), which exploit the relatively low strength of the coal and its subvertical fracture system (cleat). Although coal has a relatively low strength, it may often be stronger than the underlying or overlying strata, and for this reason some coal may be left in the roof or floor for support (Table 4).

Seam thickness. Pre-depositional, syn-depositional and post-depositional events in the original sedimentary environment all influence seam thickness (Guion 1987b). Pre-depositional

Table 3. In situ *geotechnical properties of facies and their effects of mining*

Facies	Typical geotechnical properties	Effects on mining
Pedogenic		
Mire (mainly coal)	Moderately strong. Thinly laminated to thinly bedded. Few bedding discontinuities. Variably developed cleat	Easy to excavate. Intermediate support characteristics, often strong enough for roof and floor support of faces
Palaeosol (mainly seat earth)	Variable strength 1. Usually moderately weak to moderately strong. Thinly to very thickly bedded. Common discontinuities at all angles (listric surfaces)	Easy to excavate Poor support characteristics 1. Increased closure rates in roadways, dinting required in D-shaped roads 2. Machinery becomes 'bogged down' Increased run of face ash on faces due to cutting floor dirt Machinery becomes gummed up with clay
	2. Sometimes strong to very strong. Thinly to very thickly bedded. Occasional bedding and root-related discontinuities	More difficult to excavate 1. Increased pickwear 2. Base of extraction often moved up-section, resulting in cutting roof dirt Good support but may result in slabbing in arch-supported roads
Marine	Moderately strong to strong. Thinly laminated to thickly bedded. Common bedding discontinuities	Easy to excavate. Intermediate to poor support characteristics
Lake-fill		
Lacustrine	Moderately strong to strong. Thinly laminated to medium bedded. Common bedding discontinuities	Easy to excavate. Poor support characteristics
Distal lacustrine delta	Moderately strong to strong. Thinly laminated to thickly bedded. Occasional bedding discontinuities	Intermediate excavating and support characteristics (ideal conditions for roadways and roof of faces)
Proximal lacustrine delta	Moderately strong to very strong. Thinly to very thickly bedded. Occasional to common bedding discontinuities	Intermediate to difficult excavation characteristics—often abrasive. Intermediate to good support characteristics. High incendive temperature potential risk. Occasional increased airborne dust
Channel		
Major channel	Variable strength, but often strong to very strong. Thickly to very thickly bedded. Few bedding discontinuities Occasional syn-sedimentary faults subparallel to channel margins	Often extremely abrasive and very difficult to excavate Good support for roof and floor in sandstone channels, poorer in other lithologies and at channel margins. High incendive temperature potential. Risk of increased volumes of airborne dusts
Minor channel	Variable strength from moderately weak to very strong. Thinly to very thickly bedded. Occasional to common bedding discontinuities. Common syn-sedimentary faults subparallel to channel margins	Abrasive and often difficult to excavate. Often good support characteristics in sandstone, poorer in other lithologies and at channel margins. High incendive temperature potential risk. Risk of increased volumes of airborne dust
Near-channel		
Overbank	Moderately strong to strong. Medium to very thickly bedded. Rare bedding discontinuities. Common syn-sedimentary faults subparallel to channel margins. Common upright trunks	Variable excavation and support characteristics. Where massive, excavation and support are intermediate. Where faulted, excavation is easy and support characteristics are poor. Upright trunks difficult to support
Crevasse splay	Moderately strong to strong. Thickly laminated to medium bedded. Common bedding discontinuities at lithological interfaces	Intermediate excavating characteristics Intermediate to poor support characterstics

Terminology for geotechnical properties is based on BS 5930 (1981).

factors which are important include the topography, lithology, rate of compaction and drainage conditions of the substrate on which the peat accumulated (cf. Fig. 9). All of these are likely to influence seam thickness on a local scale (Horne *et al.* 1978*a*). They may result in an uneven base to the seam, which may cause the shearer to cut floor dirt, with a consequent increase in the run of face ash content. If the seam floor is a ganister (siliceous palaeosol), there may also be problems of pick wear, unacceptable levels of dust generation and a high incendive temperature potential. Elliott (1965) showed that seams commonly thicken above underlying abandoned channels which form swilleys, and thin over the channel

Table 4. *Effects on coal faces and reserves of variation in mire (coal) facies*

Facies-related problem	Adverse effect on faces/reserves	Solution
Regional thinning	Faces stop, reserves reduced	Change production to new areas
Local thinning	Cut roof and/or floor dirt, increase in run of face ash	Reduce height of extraction
Uneven base	Cut floor dirt, increase in run of face ash	Leave floor coal
Decrease in coal quality	Increase in run of face mineral matter including ash, S, Cl	Blend with better quality coal
Seam splitting	Increase in interleaf dirt and run of face ash, ultimately face stops	Blend with low ash coal

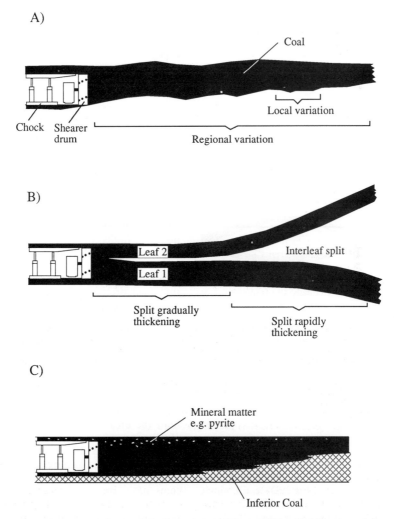

Fig. 8. Effects of lateral changes in mire (coal) facies on longwall faces. (**A**) Local thickness variation affects the working of individual faces; regional variation affects the location of faces. (**B**) Interleaf splits which thicken gradually affect the quality of the face product: rapid thickening may stop faces. (**C**) Change in mineral content affects the quality of the face product. Modified from Guion & Fulton (1993).

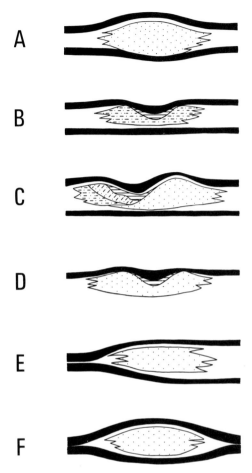

Fig. 9. Thickness variation of seams and interseam strata that accompany channels. (**A**) Interseam strata are thicker than in adjacent regions. (**B**) Interseam strata are thinner than in adjacent regions due to underlying unplugged or partially plugged channels. (**C**) Interseam strata show marked thickness variation due to complex channel-fills. (**D**) seams above the channel show thickness variation. (**E**) Abrupt lateral seam splitting on one side of a channel. (**F**) Abrupt lateral seam splitting on both sides of a channel. Reproduced with permission of Elsevier Science Ltd from Guion (1987b).

are important controls on peat deposition and influence seam thickness and quality on a regional scale (Fig. 8A and 8C) (Fulton, 1987a, b), affecting the choice of areas to be mined. The topographic elevation of the site of peat accumulation and the rate of subsidence have also been shown by Ferm & Staub (1984) to control the distribution of mineable coal bodies.

Many of the post-depositional influences on coal thickness and continuity are related to the mode of inundation and drowning of the peat mire, and the effects of subsequent channels.

Seam splitting. Seam splits occur as a consequence of the lateral increase in thickness of thin siliciclastic layers (Fig. 8B). Thus a thin bed of claystone within a seam may pass laterally within several hundreds of metres into a thick clastic interval containing sandstone (cf. Fig. 9E and 9F). The rates of increase in the thickness of seam splits are a function of the different processes which caused them.

Seam splits can often be attributed to differential rates of subsidence, which may be either the result of active tectonic influence (e.g. Fielding 1984a; Broadhurst & France 1986; Fulton & Williams 1988) or caused by differential compaction resulting in the switching of depositional systems (e.g. Elliott 1969; Fielding 1984a, c, 1986b). Careful study, e.g. relating lines of split to the positions of known faults, may reveal which of the controls was important, and whether the split is likely to be linear (possibly linked to faults) or irregular (possibly linked to compactional subsidence).

Where the thickness of the seam split is small, coal seams may be mined, but in this instance the effect of the seam split is to decrease the quality of the mineable product coming from faces through an increase in mineral matter derived from the clastic intercalation (Table 4). A lateral increase in the thickness of the split may result either in unacceptable levels of mineral matter, if the coal is worked in one lift, or a reduced thickness of extraction if only one of the split seams is thick enough to work (see later). In the former, seam splits ultimately limit the areas of workable coal (Figs 8B and 10).

Where seam splitting results in the presence of a rider coal and accompanying palaeosol close above a seam being worked, poor roof conditions often prevail. These may be caused by weak strata and a parting forming at the base of the rider coal or the low strength of the rider coal, which causes roof falls up to the rider coal (Horne *et al.* 1978a, b). The use of roof bolts for support in this situation would be particularly

margins. Other characteristics associated with swilleys include changes in the gradient and quality of the seam (cf. Fig. 9D) and compactional effects linked with the swilley margins. The latter may induce discontinuities causing roof support problems.

Palaeobotanical, edaphic, climatic and hydrological factors, as well as rates of subsidence,

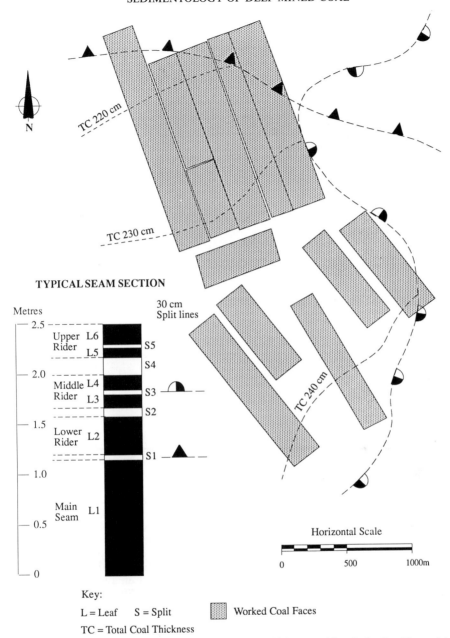

Fig. 10. Example of a mire (coal) facies map showing coal thickness and interleaf splits. The coal in the seam increases in thickness from less than 220 cm in the north to greater than 240 cm in the south. The interleaf split S1, between leaves 1 and 2, limits workings to the north, whereas the interleaf split S3, between leaves 3 and 4, limits workings to the east.

unwise, unless they are extended to suitable strata, often well above the rider coal.

Coal quality. A considerable variety of factors affect coal quality and a full discussion of these is outside the scope of this paper. Only those of sedimentary origin will be briefly discussed. Impurities that affect coal quality such as mineral matter (including quartz and clay minerals), chlorine, phosphorus, titanium and nitrogen have been outlined in Elliott (1984). The sulphur content is a particularly important

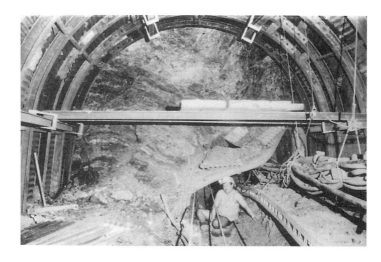

Fig. 11. Example of severe archway crush which has reduced the size of the original roadway to a small opening (bottom right). The roadway is being re-driven and new archway supports erected.

aspect of coal quality and normally occurs in the mineral pyrite. Williams and Keith (1963) relate pyrite occurrence to the availability of sulphate ions from overlying marine roof beds. Rippon (1984), however, noted sulphur reduction in a seam directly overlain by a *Lingula* band, which is in conflict with the relationships suggested by Williams & Keith (1963). Horne *et al.* (1978a, b) showed that intercalations of non-marine clastic sediment between a seam and underlying marine strata resulted in a lowered sulphur content where the coal lay beneath the clastic wedge. Alternatively, White & Birk (1989) showed that pyritic sulphur was higher in seams overlain by sandstone-filled channels. (Pyrite also forms epigenetically and is often found in cleat.) The controls on sulphur content are complex, not always clear and must be evaluated with care (Casagrande, 1987).

Palaeosol (seat earth) facies

Palaeosols are usually encountered beneath worked coal seams and as they form the floor of mine workings they are required to support mining machinery. Many palaeosols are intrinsically weak (Tables 2 and 3), especially if they are formed of clays, thus the geotechnical properties of the palaeosol underlying a seam may be critical to the success of mining operations. The discontinuities caused by roots and small polished curved ('listric') surfaces result in the low compressive strength of many palaeosols. Clay palaeosols may be further weakened by hydration water, including water used in dust suppression, causing difficult mining conditions (Table 3).

A weak palaeosol is the facies most likely to fail in roadways where induced stresses are high and confining stresses are low, causing 'floor lift' (Guion & Fulton, 1993) and archway 'crush'. The remedy for both these effects is expensive. The former usually requires re-excavation of the floor or 'dinting', whereas the latter may require re-excavation of the roof and replacement of the crown of the arch, or in more severe instances (Fig. 11), re-drivage of the road. Where weak palaeosols are present in the roof of a worked seam, for instance when the lower leaf of a split seam is being worked, bad roof conditions on the face are likely, especially if the palaeosol is clay-rich and 'listric' (Moebs & Ferm 1982). Movement of chocks on a weak floor may also be impeded by floor lift, which causes uneven floors.

Certain palaeosol lithologies are strong and very hard, especially quartz-rich seat earths (ganisters). When cut, the result is excessive pick wear and a high incendive temperature potential, which may cause frictional ignition (Table 3). 'Slabbing' may occur in roads left open for long periods of time with an unsupported floor (Fig. 12), where strong seat earths (underlain by weaker strata) fail and lift upwards into the roadway so that large rock slabs form. This effect may be observed in face roads and clearance of the slabs is made more difficult because of their high strength.

Fig. 12. Slabbing of a strong palaeosol located in the floor of a face road. Strata have moved upwards into the unsupported floor and then have failed in the centre of the road, creating a longitudinal crack. The strength of the strata has restricted further failure, creating large slabs.

Palaeosols are particularly likely to be cut on the margins of swilleys accompanying underlying abandoned, vegetated channels (Elliott 1965, 1974, 1984; Rippon & Spears 1989), where the seam thickness is reduced. The cutting of claystone palaeosols, either underlying or interbedded with coal seams, may cause clogging of moving parts in the transport system and may be difficult to remove during coal preparation.

Marine facies

Coal seams immediately underlying marine facies are not commonly worked in the Pennine coalfields because of the low quality and high sulphur content of the seams (Pearson 1974). However, workings in the Clowne seam enabled Rippon (1984) to document the progressive inundation of the seam by the Clowne Marine Band in the East Midlands.

Marine facies are generally fine-grained, fissile, carbonaceous at their base and laminated with claystone or siltstone above, and sometimes contain layers of concretionary ironstone lenses. They are fairly easy to excavate (Table 3), have low abrasivities and where massive and highly carbonaceous may have intermediate support characteristics. However, where layers or lenses of ironstone are present, discontinuities of compactional origin form at their boundaries and these may be polished. If these surfaces are present, or accompanied by fissile or laminated sequences, the support characteristics are poor. Isolated concretions, in particular, may fall from the roof (Sames & Moebs 1991). If marine facies are allowed to stand for considerable periods of time, they tend to deteriorate due to the oxidation of finely disseminated sulphides.

Lacustrine facies

Lacustrine facies commonly form the immediate roof to coal seams and the geotechnical properties of these fine-grained rocks determine the behaviour of roof strata during longwall mining. They are lithologically similar to the marine facies described previously, but overall they are usually less carbonaceous and less massive. As a result, they are easy to excavate, but have poor support characteristics (Table 3).

It is often possible to control weak lacustrine facies roofs on faces by the use of full shield supports, or by not extracting the upper part of the seam. The latter leads to a reduced loss of tonnage. In roadways, support may be given by rock bolts or cable bolts (Morris 1990) through the lacustrine facies into stronger facies above. Weak roof strata may require careful design of adequate support systems to avoid cavitation, followed by convergence of the roof in roadways (Elliott 1974, 1984). Where strata are exceptionally weak, it may be necessary to install costly support systems, or drive additional 'sacrifice' or 'following' roadways in de-stressed ground.

Distal lacustrine delta facies

Distal lacustrine delta deposits generally coarsen upwards gradually from siltstones to thickly interbedded siltstone and sandstone sequences and, in some instances, the upward coarsening continues into proximal lacustrine delta deposits. Major erosion surfaces, abrupt contacts between widely differing lithologies, slickensides and polished surfaces are rare, so that failure surfaces are unlikely to be a problem.

This facies has intermediate excavating and support characteristics and is therefore ideal for the drivage of roads. Where dominated by siltstone, it may be moderately strong to strong (Table 2) and this, together with a low abrasivity, makes these rocks easy to cut. In addition, the strength of these siltstones and the rarity of plant-rich layers and other discontinuities, allows them to provide good roof support. The paucity of discontinuities and the upward coarsening of this facies give rise to the best roof conditions for faces (compare Moebs & Ferm 1982).

Rock bolts and cable bolts from gate roads, inserted to support weaker rocks below, usually terminate in these deposits, which generally become stronger upwards. When tied together in this manner, these rocks form a strong natural bridge. Lacustrine delta deposits may be extensive laterally, with strata showing only gradual lateral change, resulting in similar mining conditions over relatively large areas.

Proximal lacustrine delta facies

This facies generally occurs in coarsening upward sequences, and usually passes upwards from distal lacustrine delta deposits. Often it lies several metres above coal seams and may not be encountered in normal workings, except where it occupies a proximal position relative to a delta system. The facies usually consists of interbedded sandstones and siltstones, with sandstones being dominant, and forms beds decimetres to metres in thickness. Small channels with basal erosion surfaces commonly occur within this facies. Minor channels are associated with this facies, especially towards the top of proximal lacustrine delta sequences.

The sandstones in this facies have high compressive strengths, although the presence of bedding contributes to its intermediate excavation characteristics. Thick beds of sandstone with high abrasivities can make cutting difficult. Laterally extensive sandstones may result in the persistence of poor cutting conditions in roadways for hundreds of metres. Excellent support characteristics result where these sandstones occur in the roof. However, minor channels with associated discontinuities, such as erosion surfaces and plant-rich layers, may reduce its support characteristics locally.

Major channel facies

Major channel deposits are likely to be readily detected in the subsurface by exploratory drilling because of their large dimensions, which may exceed 20 m in thickness and 5 km in width. Thus they can generally be avoided during deep coal mining, although the exact positions of their margins may be difficult to delineate accurately. The deposits of major channels are dominated by cross-bedded and cross-laminated sandstones (Guion *et al.*, this volume), but finer grained beds and lenses, layers of plant debris and breccias and conglomerates containing intraformational clasts and coal fragments may also be encountered.

The main effect of major channels on coal mining is to cause considerable loss of reserves, especially if the channels replace more than one seam vertically (Table 5).

The driving of roadways through sandy major channels is likely to be difficult because of their high strength and abrasivity (Table 2), especially where a carbonate cement ('cank') is present. They are likely to cause excessive airborne dust (Table 3) and have a high incendive temperature potential when cut (Elliott 1974, 1984). Problems are compounded when the fracture density is low, as large blocks, which are difficult to handle, are created during excavations. The consequences of these factors are slow drivage rates and high costs. The sandstones of this facies generally result in good support conditions. However, discontinuities within major channels, such as fine-grained layers, channel bases and weak lithologies within channel lag deposits, may give rise to locally poor support conditions.

Where sandy major channels occur above worked seams and fractures are widely spaced, excessive 'weighting' may occur on faces and there may be difficulties in caving into the goaf. Below such major channels there is a risk of wash-outs and rolls of erosional origin, 'squashouts', faults of compactional origin and slips, where sedimentary discontinuities have been emphasized by subsequent compaction. The patterns of these features are not predictable in detail, so that it is necessary for the mine design to allow for unforeseen face stoppages.

Where strong ground is required for large excavations, e.g. the base of vertical and

Table 5. *Effects on coal faces and reserves of channel facies*

Facies-related problems	Adverse effect on coal faces/reserves
Major channel	
Removal of one, or occasionally more than one, coal seam	Faces stop, major loss of reserves
Introduction of water onto face	Temporary flooding
Excessive weighting, where joints are widely spaced	Damage to chocks, catastrophic falls
Minor channel	
Channel removes a coal seam (wash-out)	Increased pickwear and incendive temperature potential risk, reduced production, face eventually stops
Partial removal of coal seam by channel; channel base approaches seam; displacement of seam by compaction faults; complex channel-fill with abundant discontinuities	Production reduced or stopped; poor roof conditions and as a result (1) cutting horizon lowered and floor, dirt and run of face ash increased; (2) reduce extraction height and coal output
Seam directly overlying channel deposits results in: (1) seam thinning over channel levees; (2) changes in gradient in swilleys; and (3) irregular seam base	Reduced coal production, increased run of face ash from cutting roof or floor dirt
Bank collapse or displacement along syn-sedimentary faults parallel to the channel margins, resulting in partial or total removal of seam	Poor roof conditions, effects as above; production reduced or stopped

horizontal bunkers, thick channel sandstones can be used to provide excellent long-term support. In attempting to predict areas of strong ground it is important to understand that the channel facies may exhibit more lateral variability than the more persistent but homogeneous proximal lacustrine delta facies.

Major channels may act as aquifers, introducing water or even oil into mine workings, although, in general, both permeability and yields of water at depth are low (Elliott 1984).

Minor channel facies

Minor channels probably contribute more to adverse mining conditions than any of the other facies of the productive Coal Measures (Tables 3 and 5). They are much more difficult to detect and predict in the subsurface than major channels because of their smaller dimensions relative to borehole spacing. The geotechnical properties of minor channels are very variable, as they may contain a wide range of lithologies, arranged in a complex manner. The strength, cutability, abrasivity and supportability depends on the particular lithologies and discontinuities encountered in the minor channels. Numerous adverse effects may accompany minor channels: (1) post-depositional erosion of seams resulting in 'wash-outs', leading to abandonment of working faces; (2) partial erosion of seams ('rock rolls'), resulting in cutting of roof strata, reducing advance rates or stopping faces; (3) cutting of sandstones, causing excessive pick wear and airborne dust, and giving rise to a high incendive temperature potential; (4) reduction of seam thickness by compaction around channel sandstones and rotated blocks from channel bank collapse in the roof or floor of seams causing 'rock rolls' or 'squash outs'; (5) roof falls generated by weak channel-fill strata, such as coal and claystone, or discontinuities arising from heterogeneous fill, inclined heterolithic stratification, slickensides and erosion surfaces, all of which may be emphasized by compaction; (6) reactivation during mining of failure surfaces accompanying bank collapse deposits; (7) displacement of seams by compaction faults or folds around sand bodies, resulting in face hindrance or stoppage; (8) changes in seam thickness and gradient above and below minor channels, and 'swilleys' formed above abandoned reaches of minor channels; (9) seam splitting accompanying some minor channels; and (10) ingress of water from relatively permeable sandstones.

A wide range of deleterious effects thus accompanies minor channels and it is therefore important that every effort is made to predict their presence ahead of mining. Production should therefore be kept away from known minor channels, or if the economic situation allows it, replacement face capacity should be made available if operating in their vicinity.

Overbank facies

Overbank deposits are usually dominated by siltstones and may form relatively good roof conditions, being moderately strong to strong (Table 3). They are likely to be relatively easy to support and cut, having low abrasivities (Table 2), although large blocks might arise during excavation as they are often massive. However, overbank deposits occur close to channel margins and may therefore be affected by small faults and fractures of syn-sedimentary and compactional origin. These features may make excavation relatively easy, but contribute to poor support characteristics.

In situ tree trunks may be locally abundant in this facies (e.g. Guion 1987*b*), especially where it lies immediately above a coal seam. These are referred to as 'potholes' in the UK and 'kettles' or 'kettlebottoms' in the USA (Horne *et al.* 1978*a*; Chase & Sames 1983; Sames & Moebs 1991; Greb 1991). These trunks, which are generally formed of large lycopsids, are a hazard in mine workings, as the trunks become wider downwards and are surrounded by a layer of vitrain, representing the outer woody material of the tree. The siltstone infill of these stumps has a tendency to detach on the weak vitrain layer and to fall downwards into mine workings (Chase & Sames 1983). Where stumps are present, roof bolts are unlikely to provide a good method of support and arch supports may be necessary.

Crevasse splay facies

Crevasse splay deposits may, under ideal circumstances, form a reasonable mining environment over areas of about 1 km^2 (Table 3) as they consist of thin sandstones that tend to be strong, interbedded with moderately strong siltstones (Table 2). The sandstone beds are relatively thin and are unlikely to form large slabs or blocks during excavation (Elliott 1984).

The interbedded nature of crevasse splay deposits means that weaknesses may occur at the boundaries between sandstone and siltstone beds, especially where mica or comminuted plant debris is abundant. Bed separation may occur and roof strata may then be difficult to support (Moebs & Ellenburger 1982).

Other hazards in mining with a roof consisting of crevasse splay deposits result from the likely proximity of minor channels and crevasse channels, which may cause wash-outs, rolls, sand bodies and other channel-associated phenomena (Guion 1984).

Evaluate and standardize data from various sources

↓

Correlate strata sequences in as much detail as possible

↓

Group strata into facies or facies associations by use of:

(1) diagnostic characteristics, e.g. lithology, sedimentary and biogenic structures, geometry
(2) isopach maps for channel facies, including interseam intervals and channel facies thickness
(3) palaeocurrent analysis

↓

Construct facies maps and cross-sections to show facies distributions and intrafacies variation, incorporating isopach maps and palaeocurrent analysis

↓

Use the facies maps and cross-sections to predict:

(1) best areas for working
(2) best horizons for roadway drivages

↓

Verify interpreted maps and cross sections against facies encountered during mining

Fig. 13. Scheme for the prediction of geological conditions ahead of mining.

Mine planning and prediction of mining conditions

Successful mine planning requires adequate forecasting of those geological factors which are likely to affect the various mining operations. Before operating coal faces, it is important to predict seam thickness, quality and continuity, as well as the properties of roof and floor strata. The prediction of rock properties is important when planning roadways, especially those which are cross-measures.

A simple scheme for the prediction of mining conditions is given in Fig. 13. Data first require evaluating and standardizing before correlation. Each sedimentary facies has a different effect on mining conditions, as discussed previously, so the next stage in the scheme is the recognition and forecasting of the distribution of facies. The prediction of mining conditions may then take place, aided by the construction of maps and cross-sections. Finally, there should be verification of both the predicted facies distributions and effects on mining conditions. Each stage of the scheme is explained in more detail below.

Data evaluation and standardization

The need to evaluate and standardize data arises from the variety and range of ages of the data sources. A less refined data set than fully cored borehole records is yielded by: (1) the use of old records such as colliery shaft sections from the older parts of the coalfield and old cored boreholes; and (2) open-hole borehole data where only cuttings and/or geophysical logs are available.

Thus the degree of sedimentological interpretation possible from either old records or open-hole data may be limited. The terms used in the past to describe lithologies, especially in older shaft sections, may be both parochial and colourful, but do not convey much geological information. Examples of the vernacular terminology that has been used in older geological records are given in Wilcockson (1950: xiii), Edwards (1951: 275) and Smith et al. (1967: 283). Although most recent boreholes are cored and geophysically logged, and have detailed and accurate descriptions, it is necessary to evaluate the terms used in both these and the older borehole descriptions and to standardize them by comparison with well-recorded adjacent sections (Fulton, 1993). Any data collected underground also need to be incorporated into any interpretation. In some instances data have been collected by geologists, but other underground data may consist of mining records, collected by non-geologically trained staff. Thus the data sets may be variable in quality. The problems of geological forecasting in mines with heterogeneous data sets have been discussed by Dreesen (1993), who has shown that some interpretation is possible even with low quality or variable quality data. Nevertheless, the quality of the sedimentological interpretation depends on the quality of the data set. Poor and widely spaced data are likely to lead only to generalized sedimentological interpretations. The most meaningful sedimentological interpretations can be made only where closely spaced data, either from underground exposures or cored boreholes, are integrated into a well-proved regional sedimentological framework.

Many current deep mine workings are in areas containing recent exploratory boreholes. Nevertheless, it is important to improve the data sets by collecting data from underground exposures, especially in regions of known or suspected geological complexity, to maximize the precision of any sedimentological interpretations.

Correlation

Accurate correlation of seams and interseam strata in each section is an obvious requirement for producing accurate sedimentological interpretation and mapping. This is often difficult using some of the older borehole and shaft records, where data were poorly recorded, and where there was poor core recovery. The properties of the coal seams are the usual basis for correlation. Other facies may also be used, such as marine bands (which may be widespread and contain diagnostic fauna) or, for more local correlation, lacustrine deposits, palaeosols, faunal markers or certain diagnostic lithological markers such as ironstone horizons. In most current deep mines, correlation is not perceived to be a major problem, although in the initial stages of exploration of a new prospect, remote from areas of known stratigraphy, correlation may be difficult and special effort is required to tie the sections into known sequences.

Recognition of facies and facies associations

The key to any sedimentological interpretation is the recognition of facies and facies associations of strata. Identification is based on a number of diagnostic characteristics, which include lithology, sedimentary structures, fossils and geometry (Guion et al., this volume). Many facies may consist of more than one lithology, but the lithologies within a facies are often arranged in a similar manner, for instance forming the coarsening-upwards sequences of the lacustrine delta facies. Strata from recorded sequences may thus be grouped together, using their diagnostic characteristics. A degree of prediction of the lithological variation within a facies may be possible, once the facies has been identified, e.g. proximal–distal relationships.

It is not always initially possible to assign all strata to a particular facies on the basis of lithological and organic characteristics alone, and the identification of the facies must be made on the basis of its relationships with other facies, its overall geometry and its directional properties. For instance, channel facies form elongate belts, which may be recognized through the plotting of isopach maps (see later section).

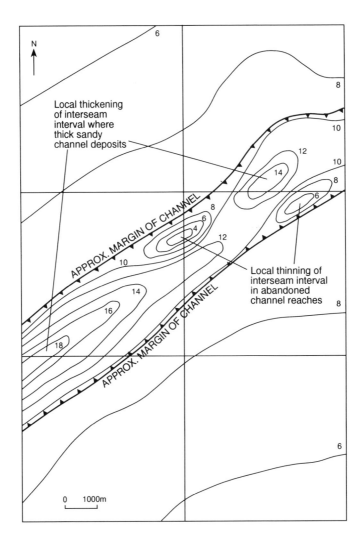

Fig. 14. Simulated isopach map showing the effects of a channel on interseam interval thickness (isopachs in metres). Where channels are present, the interseam interval is often thicker than in surrounding regions due to excess sand deposition, or thinner where there are unplugged or partially plugged channel reaches.

Palaeocurrent analysis may also aid the identification and orientation of channel facies by the recognition of characteristic flow patterns.

Prediction of facies distribution

The construction of maps and cross-sections is invaluable for predicting the distribution of sedimentary facies and usually only one seam or one interseam interval is analysed at a time. Isopach maps can be plotted to represent the thickness variation of one individual facies (e.g. a mire), groups of facies or interseam intervals. Palaeocurrent data may also be incorporated into facies maps and help to constrain geometries and establish proximal–distal relationships.

Isopach maps. Isopach maps of coal and clastic intercalations ('dirt bands') within seams are routinely plotted as a basis for calculating reserves and predicting the ash content of the mined product. Seam isopachs (Fig. 10) are also used to forecast the range of extraction heights. In addition, isopach maps may be useful in

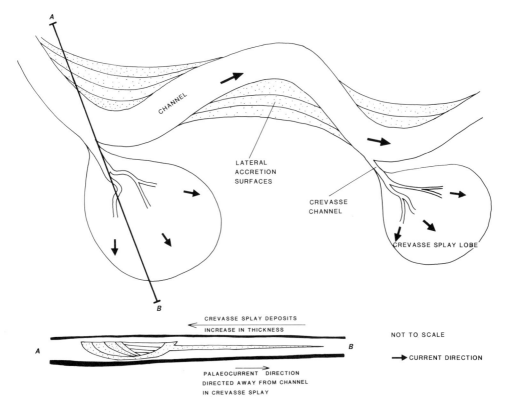

Fig. 15. Relationship between crevasse splay deposits and channels. Crevasse splays thicken towards feeding channels and show a radial palaeocurrent pattern. Careful subsurface mapping of individual crevasse splay lobes and palaeocurrents may enable positions of feeder channels to be inferred.

delineating the axes of channel belts and their margins. Guion (1987b) showed that the presence of channels may be reflected in isopach patterns in a number of ways, as illustrated in Fig. 9: (1) interseam strata are thicker than in adjacent regions—linear belts of locally thick roof commonly indicate channels that have been infilled with sandstone (Fig. 9A); (2) interseam strata are thinner than in adjacent regions—this accompanies abandoned channels that may have remained unplugged or partially plugged by clastic sediment, or that received a fine-grained infill that has been compacted resulting in 'swilleys' (Fig. 9B); (3) interseam strata show marked thickness variation as a result of complex channel-fills, where thick infills are present in some reaches and thin infills elsewhere (Fig. 9C); (4) the seam overlying the target seam is subject to thickness variations. Above channels the seam may either thin onto topographic highs or thicken into abandoned channels forming 'swilleys' (Fig. 9D); and (5) seam splitting may occur on one or both sides of a channel system (Fig. 9E and 9F).

In general, with sufficient data points, channels are accompanied by closely spaced isopachs of interseam and seam thicknesses, forming linear belts showing rapid thickness variation (Fig. 14). In regions where channels are absent, isopachs tend to be much more widely spaced.

Palaeocurrent analysis. Most palaeocurrent measurements come from observations made in mine workings. Although dipmeter tools are routinely run in many British Coal boreholes which may yield usable sedimentological directional data, the use of these geophysical logs for palaeocurrent analysis of Coal Measures rocks has not been evaluated. Palaeocurrent measurements can be used to establish the directions of sediment dispersal, and when integrated with facies maps may improve the understanding of complex facies distributions (Guion 1984, 1987a; Fielding 1984a, b, 1986a). Some of the most reliable palaeocurrent data come from rib and furrow patterns produced by trough cross-lamination, but palaeocurrent directions may also be obtained from the foreset orientation of

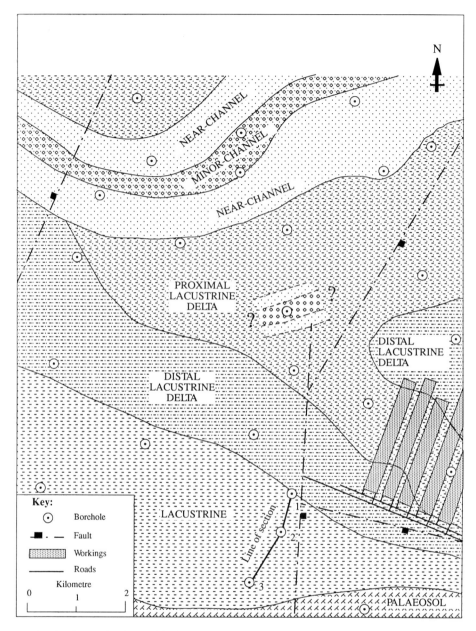

Fig. 16. Simulated example of a roof facies distribution map showing areas occupied by each facies above the Eight Feet seam and sources of data (boreholes and workings in the Eight Feet seam).

ripples, climbing ripples and cross-bedding. Orientation may also be obtained from the alignment of plant stems and bivalve traces (*Pelecypodichnus*). The crest-line orientation of oscillation ripples may also be measured, but the implications of this may not always be clear.

Sometimes, if ripple foreset orientation is taken from two-dimensional mine exposures, great accuracy is not always possible (e.g. Guion 1987*a*), but the readings are nevertheless useful. The mapping of crevasse splay thicknesses, together with their palaeocurrent pattern (Fig.

15), may enable the likely positions of minor channels to be determined (Guion 1984). Individual crevasse splay deposits thin distally away from the minor channel from which they originated and palaeocurrents fan out from the feeder crevasse channels (Fig. 15).

Sedimentary facies mapping. Maps showing the distribution of sedimentary facies are an invaluable aid in the prediction of mining conditions. The mining of coal may yield large amounts of information on the mire facies, represented by the coal seam, and those facies immediately above and below it, if geological sections are recorded in headings as mining proceeds. Most sedimentary-related problems are caused by rocks present in the lower roof and upper floor of the seam and thus the most commonly produced maps are those of these horizons (Elliott, 1984, Guion 1984, 1987*a*). Geological records of cross-measure drivages and boreholes enable cross-sections and maps to be drawn at horizons some distance above and below seams. The construction of these maps is aided by the use of conceptual models which represent the most likely geometrical arrangement of facies (Fig. 3). Facies maps are by their nature interpretative, especially where data are sparse. However, knowledge of likely facies geometries, relationships and dimensions, embodied in a realistic facies model for the strata being studied, will increase the accuracy of prediction. In this way the presence of minor channels, although not initially encountered in boreholes or workings, may be inferred by documenting the distributions of near-channel facies such as overbank deposits and crevasse splays (Fig. 3).

Optimization of face location

The primary concern when selecting an area to mine coal is often the thickness and quality of coal, both of which have been shown to be influenced by sedimentary processes. Mire (coal) facies maps showing coal isopachs and the critical thickness of interleaf splits (Fig. 10) are crucial for the prediction of the best areas to locate coal faces. Other maps useful for predicting quality parameters which relate to sedimentological factors include those showing the sulphur and ash (mineral matter) content and the thickness of particular coal lithologies, such as cannel coal.

Roof facies maps are used to forecast the effects of sedimentary facies above coal seams on coal faces. An example of how these effects may govern the choice of working areas is discussed in the following and illustrated by reference to a hypothetical roof facies map (Fig. 16). Coal faces are shown working below lacustrine delta facies in the southeast of the area of the map. In the south, a weak palaeosol roof may require too much roof coal to be left up for support to allow economic working of the seam. To the northwest, the presence of proximal lacustrine delta facies may give rise to difficulties due to the discovery of minor channel deposits in a single borehole (Fig. 16). More exploration would therefore be needed to determine the limits of this channel and its associated near-channel facies, which would be expected to have a deleterious effect on roof conditions. Much further to the north, minor channel and near-channel deposits are better defined, indicating that this area should be avoided because of the likelihood of partial seam removal and poor roof conditions. The most favourable area for working is in the west, where distal lacustrine delta and lacustrine facies may form a stable roof, especially if some roof coal is left unextracted to help control the weaker lacustrine facies.

Floor facies maps can also be used to predict conditions on faces and unsupported roadway floors in the same way as roof facies maps. Rates of roadway closure have a strong relationship with the lithology of palaeosol facies which occur beneath coal seams. It would be possible to contour rates of closure based solely on the values calculated for palaeosol lithologies present at each data point. However, rather than looking at the values in isolation, a detailed knowledge of the distribution of floor facies allows more realistic and accurate contouring.

Optimization of drivage horizons for roadways

The ideal strata for the drivage of major roadways are those which are weak enough to cut easily, yet strong enough to stand well. This combination cannot generally be found in Coal Measures rocks, so an ideal compromise is to select strata of intermediate strength, which can often be found in lacustrine delta facies.

An example of the analysis required to select the optimum horizon for a roadway will be discussed with reference to Fig. 17. This represents sections of three boreholes shown on the roof facies map (Fig. 16). The facies and lithologies in each borehole are illustrated in Fig. 17. To locate a major roadway above the Eight Feet Seam, which has a very weak floor, a 10 m thick sequence of intermediate strength strata is required. A thickness greater than the planned height of the roadway is sought, because when

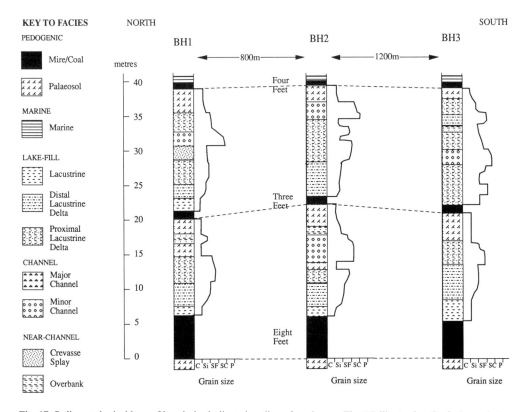

Fig. 17. Sedimentological logs of boreholes indicated on line of section on Fig. 16, illustrating the facies variation between boreholes.

driving major roads to a fixed gradient, tectonic structures such as folds or faults may alter the stratigraphic horizon of strata along the length of the road. The interval between the Eight Feet and Three Feet Seams reveals 10–15 m of strata, consisting mostly of lacustrine and lacustrine delta facies, overlain by a palaeosol. The facies and lithological attributes of these rocks, including grain size, suggest that intermediate strength strata comprise the central 10 m of this interval, except in Borehole 2, where a thin, weak palaeosol occurs in the middle of the interval. Strata of intermediate strength above this palaeosol consist of a minor channel deposit, which has cut down into the palaeosol. The channel may have only limited lateral extent, and thus a greater thickness of weaker palaeosol facies may be present beyond the margins of the channel. As a consequence, the area to each side of Borehole 2 may consist of only a 5 m thickness of intermediate strength strata, overlain by a thick palaeosol facies.

The interval between the Three Feet and Four Feet Seams, with a thicker, laterally continuous lacustrine delta facies and an absence of channels, may be a more suitable horizon for main roadway drivage. This is in spite of the greater cost involved in driving the additional distance above the Eight Feet seam.

Verification

An element often missing from many predictive schemes is verification. It is important to build confidence in a predictive scheme, and to test the validity of predictions by comparing forecasts of sedimentological interpretations and their effects on mining with the actual geology and mining conditions encountered later underground. Hence the strata encountered during mining should be recorded in roadways, faces and cored boreholes. In addition, monitoring of mining conditions is required through direct underground examination, and also less direct indicators including analysis of coal quality and measurement of face performance, rates of drivage and rates of roadway closure. This should enable continuous refinement and improvement of the prediction of sedimentological effects on mining, enabling mining staff to be convinced of the importance of sedimentological analysis in the deep mining of coal.

Conclusions

Sedimentology is of fundamental importance in all stages of deep-mined coal exploration, development and extraction. In the investigative stage, knowledge of mire (coal) facies variation and the relationship between mire and interseam facies, particularly channels, aids in the calculation of reserves or coal in place. During the feasibility stage, all facies are examined and preliminary facies maps for seams, roofs and floors drafted, so that geotechnical assessments can be made and recoverable reserves calculated.

Sedimentary facies can affect mining at the productive stage in two ways. On faces, mire (coal) facies variation influences coal thickness, continuity and quality, and is governed mainly by rates of subsidence at the time of deposition, seam splitting and influx of clastic sediment and sulphate-bearing waters. Other facies, particularly those that are channel-related, also affect coal faces, mainly by forming weak roofs and by the removal of part or all of the coal.

The geotechnical properties of facies are also important as they control the excavation and support characteristics on faces and in roadways. In general, strong facies such as channel sandstones reduce both the rate of roadway drivage and the strength of supports needed to keep the roadway open. Weak facies, such as clay palaeosols, need more expensive support systems to prevent excessive roadway closure. Distal lacustrine delta deposits, which are mainly of intermediate strength, appear to be the ideal medium for cutting and supporting roadways, and also form good roofs and floors for faces.

A scheme for the prediction of sedimentary facies and their effects on mining involves evaluation, standardization and correlation, before prediction. Isopach maps aid in forecasting seam thickness and the location of channels, whereas other contour maps, e.g. iso-sulphur and iso-ash maps, enable the prediction of seam quality. Palaeocurrent maps are useful in predicting the location and orientation of channel and near-channel facies.

Cross-sections and maps showing facies distribution and seam variation are invaluable in mine planning for optimizing the location of faces and roadways and in the choice and design of equipment to cut and support these excavations. In the current financial and operational environment, the accurate prediction of geological conditions is essential. The application of sedimentology to mine planning is likely to lead to increased productivity and a reduction in risks and costs.

This paper is dedicated to the memory of the late R. E. Elliott, former Chief Geologist of the National Coal Board, who was one of the first to recognize the applications of sedimentology in the development and extraction of deep-mined coal. We thank British Coal for access to data, and M. J. Allen, Chief Geologist, British Coal and J. Wardle, Managing Director, Technical Services and Research Executive, Bretby, for approval for publication. The authors have benefited from numerous discussions with colleagues in British Coal over many years. Financial and technical assistance from Golder Associates and Oxford Brookes University are gratefully acknowledged. Thanks are given to C. Williams for drafting many of the figures and M. Conibear for the photographs. N. S. Jones publishes with the permission of the Director of the British Geological Survey (NERC). The views expressed are those of the authors, and not necessarily those of British Coal.

References

BROADHURST, F. M. & FRANCE, A. A. 1986. Time represented by coal seams in the Coal Measures of England. *International Journal of Coal Geology*, **6**, 43–54.

BS5930 1981. *Code of Practice for Site Investigations*. British Standards Institution, London.

CASAGRANDE, D. J. 1987. Sulphur in peat and coal. *In:* SCOTT, A. C. (ed.) *Coal and Coal-bearing strata: Recent Advances*. Geological Society, London, Special Publication, **32**, 37–105.

CHASE, F. E. & SAMES, G. P. 1983. *Kettlebottoms: their Relation to Mine Roof and Support*. United States Bureau of Mines Report of Investigations 8785.

CLARKE, A. M. 1963. A contribution to the understanding of washouts, swalleys, splits and other seam variations and the amelioration of their effects on mining in south Durham. *Transactions of the Institution of Mining Engineers*, **122**, 667–706.

DREESEN, R. J. M. 1993. Seam thickness and geological hazards forecasting in deep coal mining: a feasibility study from the Campine Collieries (N-Belgium). *Bulletin de la Société Belge de Géologie*, **101**, 209–254.

EDWARDS, W. 1951. *The Concealed Coalfield of Yorkshire and Nottinghamshire*, 3rd Edn. Memoir, Geological Survey of Great Britain.

ELLIOTT, R. E. 1965. Swilleys in the Coal Measures of Nottinghamshire interpreted as palaeo-river courses. *Mercian Geologist*, **1**, 143–152.

—— 1969. Deltaic processes and episodes; the interpretation of productive Coal Measures occurring in the East Midlands, Great Britain.

Mercian Geologist, **3**, 111–135.

—— 1974. The mine geologist and risk reduction. *Mining Engineer*, **133**, 173–184.

—— 1979. Exploration in the East Midlands, United Kingdom: some procedural and interpretation principles. *In:* ARGALL, G. O. (ed.) *Coal Exploration 2. Proceedings of the 2nd International Coal Exploration Symposium, Denver*. Miller-Freeman, San Francisco, 294–312.

—— (ed.) 1984. *Procedures in Coal Mining Geology*. National Coal Board Mining Department, London.

FERM, J. C. & STAUB, J. R. 1984. Depositional controls of mineable coal bodies. *In:* RAHMANI, R. A. & FLORES, R. M. (eds) *Sedimentology of Coal and Coal-bearing Sequences*. International Association of Sedimentologists, Special Publication, **7**, 275–289.

FIELDING, C. R. 1984a. A coal depositional model for the Durham Coal Measures of N.E. England. *Journal of the Geological Society, London*, **141**, 919–931.

—— 1984b. Upper delta plain lacustrine and fluviolacustrine facies from the Westphalian of the Durham Coalfield, N.E. England. *Sedimentology*, **31**, 547–567.

—— 1984c. 'S' or 'Z' shaped coal seam splits in the Coal Measures of County Durham. *Proceedings of the Yorkshire Geological Society*, **45**, 85–89.

—— 1986a. Fluvial channel and overbank deposits from the Westphalian of the Durham Coalfield, N.E. England. *Sedimentology*, **33**, 119–140.

—— 1986b. The anatomy of a coal seam split, Durham Coalfield, Northeast England. *Geological Journal*, **21**, 45–57.

FULTON, I. M. 1987a. *The Silesian sub-system in Warwickshire, some aspects of its palynology, sedimentology and stratigraphy*. PhD Thesis, University of Aston.

—— 1987b. Genesis of the Warwickshire Thick Coal: a group of long-residence histosols. *In:* SCOTT, A. C. (ed.) *Coal and Coal-bearing Strata: Recent Advances*. Geological Society, London, Special Publication, **32**, 201–318.

—— 1993. *A Guide to the Interpretation of Facies and Recognition of Sedimentary Environments in the UK Coal Measures*. British Coal Technical Services & Research Establishment Internal Report for British Coal Opencast.

—— & WILLIAMS, H. 1988. Palaeogeographical change and controls on Namurian and Westphalian A/B sedimentation at the southern margin of the Pennine Basin, Central England. *In:* BESLY, B. M. & KELLING, G. (eds) *Sedimentation in a Synorogenic Basin Complex: the Upper Carboniferous of NW Europe*. Blackie, Glasgow, 178–199.

GREB, S. F. 1991. Roof falls and hazard prediction in Eastern Kentucky coal mines. *In:* PETERS, D. C. (ed.) *Geology in Coal Resource Utilisation*. TechBooks, Fairfax, 245–262.

GUION, P. D. 1984. Crevasse splay deposits and roofrock quality in the Threequarters Seam (Carboniferous) in the East Midlands Coalfield, U.K. *In:* RAHMANI, R. A. & FLORES, R. M. (eds) *Sedimentology of Coal and Coal-bearing Sequences*. International Association of Sedimentologists, Special Publication, **7**, 291–308.

—— 1987a. Palaeochannels in mine workings in the High Hazles Coal (Westphalian B), Nottinghamshire Coalfield, England. *Journal of the Geological Society, London*, **144**, 471–488.

—— 1987b. The influence of a palaeochannel on seam thickness in the Coal Measures of Derbyshire, England. *International Journal of Coal Geology*, **7**, 269–299.

—— 1992. Westphalian. *In:* COPE, J. C. W., INGHAM, J. K. & RAWSON, P. F. (eds) *Atlas of Palaeogeography and Lithofacies*. Geological Society, London, Memoir, **13**, 80–86.

—— & FULTON, I. M. 1993. The importance of sedimentology in deep-mined coal extraction. *Geoscientist*, **3**(2), 25–33.

——, ——, JONES, N. S. 1995. Sedimentary facies of the coal bearing Westphalian A and B north of the Wales–Brabant High. *In:* WHATELEY, M. K. G. & SPEARS, D. A. (eds) *European Coal Geology*. Geological Society, London, Special Publication, **82**, 45–77.

HOEK, E. & BROWN, E. T. 1980. *Underground Excavations in Rock*. Institution of Mining and Metallurgy, London.

HORNE, J. C., FERM, J. C., CARRUCCIO, F. T. & BAGANZ, B. P. 1978a. Depositional models in coal exploration and mine planning in the Appalachian region. *American Association of Petroleum Geologists Bulletin*, **62**, 2379–2411.

——, HOWELL, D. J., BAGANZ, B. P. & FERM, J. C. 1978b. Splay deposits as an economic factor in coal mining. *Colorado Survey Resources Series*, **4**, 89–100.

HOUSEKNECHT, D. W. & IANNACCHIONE, A. T. 1982. Anticipating facies-related coal mining problems in Hartshorne Formation, Atkoma Basin. *American Association of Petroleum Geologists Bulletin*, **66**, 923–946.

MCCABE, K. W. & PASCOE, W. 1978. *Sandstone Channels: their Influence on Roof Control in Coal Mines*. United States Mines Safety and Health Administration Report 1096.

MATHEWSON, C. C. & GOWAN, S. W. 1981. Maximize geology—minimize your exploration budget. *In:* ARGALL, G. O. (ed.) *Coal Exploration 3. Proceedings of the 3rd International Coal Exploration Symposium, Calgary*. Miller Freeman, San Francisco, 23–27.

MOEBS, N. N. 1981. *The Geologic Characters of Some Coal Wants at the Westland Mine in Southwestern Pennsylvania*. United States Bureau of Mines Report of Investigations 8555.

—— & ELLENBERGER, J. L. 1982. *Geologic Structures in Coal Mine Roof*. United States Bureau of Mines Report of Investigations 8620.

—— & FERM, J. C. 1982. *The Relation of Geology to Mine Roof Conditions in the Pocohontas No. 3 Coal Bed*. United States Bureau of Mines Information Circular 8864.

MORRIS, C. J. 1990. The influence of rock mechanics on mine design. *South Staffordshire Group of the*

Institute of Mining Engineers (available from British Coal TSRE, Ashby Road, Bretby, Burton-on-Trent, Staffs, UK.

NELSON, W. J. 1983. *Geologic Disturbances in Illinois Coal Seams*. Circular of the Illinois State Geological Survey 530.

PEARSON, H. W. 1974. Coal. *In:* RAYNER, D. H. & HEMMINGWAY, J. E. (eds) *The Geology and Mineral Resources of Yorkshire*. Yorkshire Geological Society, 309–327.

PETERS, D. C. (ed.) 1991. *Geology in Coal Resource Utilisation*. TechBooks, Fairfax.

RAMSBOTTOM, W. H. C., CALVER, M. A., EAGAR, R. M. C., HODSON, F., HOLLIDAY, D. W., STUBBLEFIELD, C. J. & WILSON, R. B. 1978. *A Correlation of Silesian Rocks in the British Isles*. Geological Society, London, Special Report, **10**.

RIPPON, J. H. 1984. The Clowne Seam, Marine Band, and overlying sediments in the Coal Measures (Westphalian B) of North Derbyshire. *Proceedings of the Yorkshire Geological Society*, **45**, 27–43.

—— & SPEARS, D. A. 1989. The sedimentology and geochemistry of the sub-Clowne cycle (Westphalian B) of northeast Derbyshire, U.K. *Proceedings of the Yorkshire Geological Society*, **47**, 181–198.

SAMES, G. P. & MOEBS, N. N. 1991. Geologic diagnosis for reducing coal mine roof failure. *In:* PETERS, D. C. (ed.) *Geology in Coal Resource Utilisation*. TechBooks, Fairfax, 203–223.

SMITH, E. G., RHYS, G. H. & EDEN, R. A. 1967. *Geology of the Country around Chesterfield, Matlock and Mansfield*. Geological Survey of the United Kingdom, Memoir.

SMITH, W. J. 1975. A study of the stratigraphical and tectonic settings of a coal seam in relationship to their effect on face performance. *Compte Rendu du 7me Congrès International de Stratigraphie et de Géologie du Carbonifère, Krefeld 1971*, Vol. 4, 239–251.

STOPPEL, D. & BLESS, M. J. M. (eds) 1981. Flozunregelmassigkeiten im Oberkarbon. *Mededelingen Rijks Geologische Dienst*, **35**, 369–332.

TAYLOR, M. 1981. Preparation and analysis of coal seam data utilising paleoenvironmental modeling, Hazard #7 coal, Eastern Kentucky. *International Journal of Coal Geology*, **1**, 213–233.

VOELKER, R. M. (ed.) 1978. *Proceedings, Coal Seam Discontinuities Symposium*. D'Appolonia, Pittsburgh.

WEISENFLUH, G. A. & FERM, J. C. 1991. Roof control in the Fireclay Coal Group, southeastern Kentucky. *Journal of Coal Quality*, **10**, 67–74.

WHITE, J. C. & BIRK, W. D. 1989. *The Influence of Sandstone Channels on Coals from the Sydney Basin: Mineralogy, Chemistry, Petrography, Rheology and Thermal Characteristics*. Report of Atlantic Coal Institute, Sydney, Nova Scotia, Canada.

WILCOCKSON, W. H. 1950. *Sections of Strata of the Coal Measures of Yorkshire*. Midland Institute of Mining Engineers, Sheffield.

WILLIAMS, E. G. & KEITH, M. L. 1963. Relationship between sulfur in coals and the occurrence of marine roof beds. *Economic Geology*, **58**, 720–729.

WILLIAMS, H. 1986. *The sedimentology of the Upper Westphalian A in the South Derbyshire Coalfield*. PhD Thesis, Oxford Polytechnic.

Sedimentary facies of the coal-bearing Westphalian A and B north of the Wales–Brabant High

PAUL D. GUION,[1] IAIN M. FULTON [2,3] & NEIL S. JONES [1,4]

[1] *Geology and Cartography Division, Oxford Brookes University, Oxford OX3 0BP, UK*
[2] *British Coal, Technical Services and Research Executive, Stanhope Bretby, Burton-upon-Trent, DE15 0QD, UK*
[3] *Present address: Golder Associates (UK) Ltd, Landmere Lane, Edwalton, Nottingham NG12 4DG, UK*
[4] *Present address: British Geological Survey, Kingsley Dunham Centre, Keyworth, Nottingham NG12 5GG, UK*

Abstract: Important coal deposits are present in Britain north of the Wales–Brabant High in the Pennine Basin, which was initiated by late Devonian to early Carboniferous extension. The formation of coals of sufficient thickness, quality and continuity to be commercially exploited was favoured during the latest Westphalian A (Langsettian) and much of the Westphalian B (Duckmantian), when deposition took place in an environment similar to an upper delta plain, with limited marine influence. During the early part of the Westphalian A, marine influence was important, with the consequence that the coals were thinner, less persistent and of inferior quality. By the late Westphalian, alluvial deposition became important, and coals of commercial quality are rare.

Accurate determination of facies, lithological attributes and geometry is an important tool in exploration and mine planning. Sedimentary facies have thus been rationalized into those that may be consistently recognizable in boreholes and mine workings. However, facies may be further subdivided when conditions allow, and intermediate facies exist, reflecting the complexities of depositional environments. The characteristics of the following facies are described here: mire, palaeosol, marine, lacustrine, lacustrine delta, major channel, minor channel, overbank and crevasse splay. Channel deposits have a number of adverse effects on mining and the range of channel-fill deposits reflects the operation of many different processes.

A model of sedimentation is presented which emphasizes that local sedimentary and autocyclic processes, local tectonics, subsidence and compaction were important in controlling the distribution of facies at a mine scale.

Carboniferous coal deposits in Western Europe occur over a wide stratigraphical range from Dinantian to Stephanian, but some of the most productive are Westphalian A and B deltaic deposits situated north of the Wales–Brabant High (Figs 1 and 2). These include the coalfields of the Pennine Basin in northern England, where future UK mining activity is likely to be concentrated (Fig. 3). Coal-bearing Westphalian rocks extend across the southern part of the North Sea where they form hydrocarbon exploration targets (Besly 1990; Leeder & Hardman 1990; Leeder *et al.* 1990; Evans *et al.* 1992; Collinson *et al.* 1993). There are extensive coal-bearing Westphalian deposits onshore in Western Europe in Belgium, the Netherlands and Germany, which show similar characteristics to those known in Britain (Van Wijhe & Bless 1974; Bless *et al.* 1980; Lorenzi *et al.* 1992).

This paper is mainly concerned with the sedimentological considerations relevant to coal exploration and development, particularly in the UK, based on coal exploration and mining data, and forms the background to the contributions by Fulton *et al.* and Jones *et al.* (this volume) on applications of sedimentology to deep and opencast mining, respectively. The main purpose of sedimentological studies of coal-bearing strata is to establish the geographical distribution of deposits, to ascertain the thickness, quantity and continuity of seams and to determine the nature of the interseam strata. The success of mining operations depends on being able to predict not only seam properties and geometries, but also those of the host sediments in which mining takes place. These

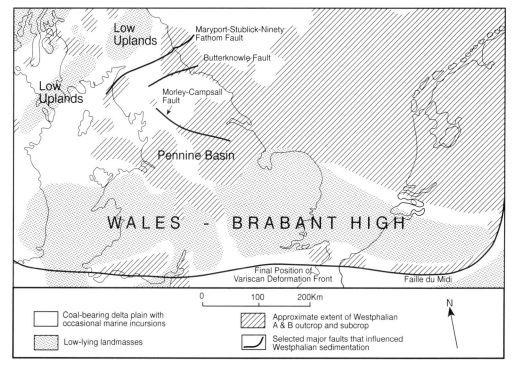

Fig. 1. Simplified palaeogeographical map of the Westphalian A and B north of the Wales–Brabant High, showing some of the major fault systems known to have influenced Westphalian sedimentation in Britain. Modified from Guion (1992).

are controlled by the sedimentary facies and hence the original depositional environment. A wide variety of factors influences the nature of coal deposits and accompanying interseam strata, at every scale from the sedimentary basin to local variation at a mine site.

The first set of factors relates to the original sedimentary environment in which peat and accompanying sediment accumulated, and includes contemporaneous tectonics and palaeogeography, climate, eustasy, bathymetry and salinity, delta switching, substrate control, subsidence and compaction; some of these have been discussed in Besly & Kelling (1988). The second set of factors, which are tectonic in origin, results in the redistribution, translocation and fragmentation of the original deposits, accompanied by burial, folding, faulting and erosion. These are beyond the scope of this paper, but nevertheless are of major importance in coal exploration and development.

The sedimentological factors relevant to exploration for Westphalian oil and gas deposits have been detailed by Besly (1990) and Collinson et al. (1993). There are some fundamental differences between oil and coal exploration and exploitation, many of which relate to the size of the target deposit. Oil reservoirs are relatively large, the well spacing is often less than one per square kilometre and the wells are uncored with the consequence that correlation between wells is not necessarily exact. In coal exploration, the target seam may be only a metre thick, and any small discontinuity in the seam can have severe consequences for its exploitation. Thus much coal exploration effort is expended in attempting to detect minor discontinuities or changes before mining. For this reason, exploration boreholes should ideally be closely spaced, and locally may exceed five per square kilometre for deep-mined coal, although 0.5–1.5 boreholes per square kilometre is typical for many coalfield areas. The coal-bearing section is normally cored, with accurate stratigraphic correlation being of paramount importance. Fortunately, the intensity of boreholes and a long experience of working now ensure that correlation is not usually a problem in established mining areas. For geological studies to be of value to the mining industry, it is

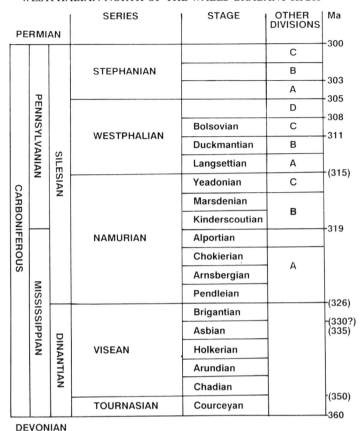

Fig. 2. Major stratigraphic divisions of the Carboniferous recognized in Western Europe. Stage names for Westphalian A–C from Owens *et al.* (1984); radiometric dates from Lippolt *et al.* (1984).

essential that the rocks are accurately described, the sedimentary facies are clearly differentiated and the distribution of sedimentary facies is understood in detail on the scale of mine workings.

Influence of palaeogeography and palaeoclimate

Westphalian coal-bearing strata were deposited north of the Wales–Brabant High in a deltaic setting within the Pennine Basin or Pennine Province in central and northern England (Wills 1951, 1956; Calver 1968a; Guion & Fielding 1988; Guion 1992) (Fig. 1).

The equatorial position of the Pennine Basin during the Westphalian (Scotese *et al.* 1979) resulted in a humid tropical climate. The influence of climate on Pennsylvanian coal development in North America, which occupied a similar palaeolatitude to Britain, has been discussed in detail in Phillips & Cecil (1985). High precipitation rates lead to the development of a largely submerged delta plain with a high water-table. This favoured the development of mainly poorly drained palaeosols and enabled peat accumulation, forming coal deposits.

In the early Westphalian, the northern limit of the Wales–Brabant High extended across Central England and the Southern North Sea into mainland Europe (Fig. 1). Embayments on the northern boundary of the landmass, for instance in Warwickshire (Fulton & Williams 1988) and in the Hereford region (Trueman 1947; Wills 1956), may have connected the Pennine Basin with sedimentary basins to the south. The exact nature of the northern boundary of the Wales–Brabant High is still not accurately delineated. Active tectonism affected the Wales–Brabant High from Duckmantian (Westphalian B) times (Besly 1988; Kelling & Collinson 1992), including block faulting and the initiation and infill of small basins by locally derived sediments. The Wales–Brabant High was probably breached through Warwickshire and Oxfordshire in the Duckman-

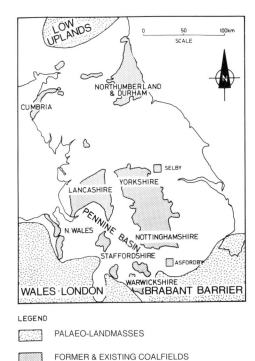

Fig. 3. Generalized late Westphalian A (Langsettian) palaeogeography of the Pennine Basin, showing recent coal exploration prospects. Reproduced from Guion & Fulton (1993).

tian, linking the Pennine Basin with sedimentary basins in southern Britain (Fulton & Williams 1988). By Westphalian D times, the topography of the Wales–Brabant High was subdued, with major connections existing between the Pennine Basin and basins to the south (Kellaway 1970; Cleal 1987; Besly 1988; Foster et al. 1989).

Exploration has shown that the Nottinghamshire Coalfield extends both east and south. It is currently being developed at a new mine at Asfordby (Davies 1978). Coal Measures have also been shown to continue south from Warwickshire into Oxfordshire (Fig. 3).

Exploration during the last two decades has proved considerable extensions to the northeast, eastern and southern margins of the Pennine coalfields (Moses 1981) (Fig. 3). A continuation of the Durham Coalfield offshore has been confirmed and the Yorkshire Coalfield has been shown to extend further north and east and is currently being worked at Selby (Stephenson 1985) (Fig. 3).

The northern limit of the Pennine Basin was formed by palaeo-landmasses in the Scottish borders, although connections with the Scottish Basin occupying the Midland Valley were present (Kirk, 1982, 1983; Guion & Fielding 1988; Guion 1992) (Figs 1 and 3).

Tectonic setting and provenance

There has been considerable debate about the development of the Carboniferous sedimentary basin north of the Wales–Brabant High, and many of the ideas have been summarized in Besly & Kelling (1988), Leeder (1988a), Besly (1990) and Fraser et al. (1990).

The development of a Carboniferous Basin north of the Wales–Brabant High is believed to have been in response to late Devonian to Dinantian extension, which resulted in the formation of a series of grabens and half-grabens (basins), separated by horsts (blocks) in northern England.

The location of some of these features and their bounding faults was in part controlled by earlier Caledonian tectonic features and granite plutons (Fraser et al. 1990). However, there is no consensus about how this extension was initiated and hypotheses have included: N–S extension to the north of a Variscan subduction zone (Leeder 1982, 1987, 1988a); E–W extension accompanying an early phase of Atlantic opening (Hazeldine 1984a, 1988); E–W dextral strike-slip related to Variscan shear (Badham 1982; Arthurton 1984); and N–S extension accompanied by dextral strike-slip (Dewey 1982). A model involving E–W extension is largely discredited, and N–S extension is the most probable mechanism, possibly accompanied by significant dextral transtension (Read 1988; Burn 1990).

Both the thickness and facies of Dinantian sediments were strongly controlled by the blocks and basins (e.g. Grayson & Oldham 1967; Gutteridge 1987; Strank 1987; Fraser et al. 1990). Although active extension decreased in the Silesian, Namurian sedimentation was still influenced by the tectonic elements established in the Dinantian, with a strongly differentiated bathymetry between the blocks and basins (e.g. Collinson 1988; Steele 1988). However, delta systems progressively infilled these inherited topographic elements from north to south throughout the Namurian (Besly 1990).

By the Westphalian, thermal sag was the dominant subsidence mechanism (Leeder 1982), and subsidence rates were much more uniform and water depths shallow, with the depocentre of the Pennine Basin being situated in the Lancashire–Staffordshire region of northern England (Calver 1968a; Guion & Fielding 1988). Nevertheless, some of the major faults inherited from the Dinantian continued to have an effect on Westphalian sedimentation, particularly in the Langsettian (Guion & Fielding 1988).

The source of much of the sediment in the Pennine Basin is generally believed to have been to the north of Britain in Westphalian A and B times (Leeder 1988b), although palaeocurrents are variable due to local tectonic influences and sediment dispersal patterns (Guion & Fielding 1988). The Wales–Brabant High appears to have been a topographically subdued feature during the Westphalian, and only provided a small amount of sediment to the Pennine Basin. However, sediment source areas have been claimed to the west in the lower part of the Westphalian A (Chisholm 1990) and to the east of Britain in the Westphalian B (Turner & O'Mara 1993).

The effects of the Variscan orogeny to the south of the Wales–Brabant High began to be manifested by mid-Duckmantian (Westphalian B) times (Besly 1988), resulting in uplift, erosion and local sediment sources in the southern part of the basin, and the development of an unconformity at or near the base of the Westphalian D, the 'Simon Unconformity' recognized both in the English Midlands (Besly 1988) and in the Southern North Sea Basin (Tubb et al. 1986). The rising Variscan Mountains to the south of the Wales–Brabant High provided sandy lithic detritus to the north, in part by the erosion of earlier Westphalian deposits (Gayer & Pesek 1992), forming the Pennant Sandstones, which reached South Wales and other basins to the south of the Wales–Brabant High during the Bolsovian (Westphalian C) (Besly 1988, 1990). The Wales–Brabant High effectively prevented this southerly-derived detritus from reaching the southern parts of the Pennine Basin until the Westphalian D, via Oxfordshire (Besly 1988, 1990; Foster et al. 1989; Guion 1992).

Major depositional environments and distribution of productive seams

The Westphalian of Britain north of the Wales–Brabant High was initially deposited in a marine-influenced deltaic setting, but by the late Westphalian, the marine influence had decreased and the rocks are believed to have been deposited in an alluvial environment.

The lowest Westphalian A, up to just above the top of the Lenisulcata Chronozone and just beneath the Kilburn Coal of the East Pennine Coalfields, was subject to repeated marine flooding events, alternating with lowstands, and shows some similarities to a lower delta plain environment (Fielding 1984a, b; Fulton & Williams 1988; Guion & Fielding 1988) (Fig. 4). Thus several laterally extensive horizons with marine or brackish faunas (Calver 1968a, b; Ramsbottom et al. 1978) form widespread maximum flooding surfaces of possible glacio-eustatic origin across the delta plain (Leeder 1988a; Collier et al. 1990). Episodes of deposition of fluvially dominated shallow water delta systems took place within this setting. Throughout the lowermost Westphalian A, there appears to have been a transition from coarse-grained, low sinuosity major channel systems to finer grained, more sinuous major channel systems (Fielding 1986) (Fig. 4). This transition accompanied an overall gradual reduction in gradients and energy levels, and the diminution of marine influence, resulting in a change from conditions similar to a lower delta plain to those comparable with an upper delta plain (Guion & Fielding 1988) (Fig. 4).

Economic coal seams are rare in the lowermost Westphalian A, where high eustatic stands and horizons containing marine and brackish faunas were important. Repeated marine transgressions may have inhibited the formation of extensive peat mires in a number of ways. Rapid eustatic rises would drown any growing vegetation and increase the water depths so that peat accumulation could not keep pace. The accompanying saline groundwater conditions may also have reduced the development of arboreal swamp vegetation (Frazier & Osanik 1969; Donaldson et al. 1970; Fielding 1984a).

From the mid-Westphalian A (Langsettian) to upper Westphalian B (Duckmantian) deposition took place in an environment similar to an upper delta plain, and most economic seams are concentrated in this interval (Fielding 1984a, b; Guion 1984, 1987a; Fulton & Williams 1988; Guion & Fielding 1988) (Fig. 4). The end of marine transgressions across the delta plain had the result that the transition to 'upper delta plain' conditions took place virtually simultaneously across the Pennine Basin at about the level of the Kilburn Coal and its equivalents (Fulton & Williams 1988) (Fig. 4). Thus from mid-Westphalian A to upper Westphalian B times, marine influence as a result of eustatic changes was minimal, groundwater conditions were fresh and rates of sedimentation and subsidence were at an optimum for the development of extensive peat mires, ultimately leading to the repeated formation of widespread coal deposits (Fig. 4). The Vanderbeckei (Clay Cross) Marine Band, which marks the Westphalian A–

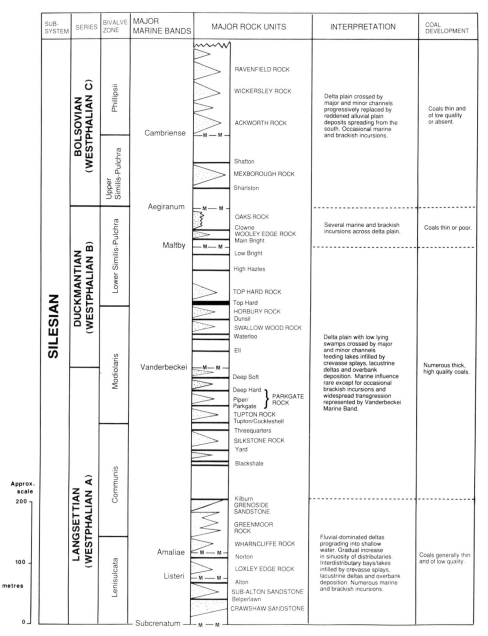

Fig. 4. Generalized stratigraphy of the Westphalian A–C (Langsettian–Bolsovian) of the East Pennine coalfields showing major coals and sandstones, with interpretations of overall depositional environments and potential for coal formation. Reproduced from Guion & Fulton (1993).

B (Langsettian–Duckmantian) boundary, represents the only significant marine incursion of eustatic origin (Ramsbottom 1979) across the low-lying upper delta plain (Fig. 4).

Several marine and brackish horizons occur towards the top of the Westphalian B (Calver 1968a, 1973; Ramsbottom et al. 1978) (Fig. 4). The Aegiranum (Mansfield) Marine Band is thick, laterally extensive and contains a fully marine fauna (Calver 1968a). It is succeeded by a thickened progradational sequence and, like the Vanderbeckei Marine Band, represents an important eustatic flooding event (Duff & Walton 1962). The other marine bands are geographically restricted in their occurrence and contain impoverished faunas (Calver 1968a, 1973). Thus it is likely that these horizons represent short-lived marine to brackish incursions across an 'upper delta plain', rather than the re-establishment of lower delta plain conditions as indicated by Fielding (1984a, b).

Three marine bands are also present in the lowermost part of the Westphalian C (Bolsovian), but these are generally restricted to the more central parts of the Pennine Basin (Calver 1968a, 1973). These horizons are believed to represent temporary incursions into the more rapidly subsiding central part of the basin. The Cambriense (Top) Marine Band (Fig. 4) is the highest marine horizon containing faunas in the British Westphalian, and occurs at a higher horizon than marine deposits generally recorded from elsewhere in northwest Europe (Calver 1968a, 1973). However, marine faunas of Westphalian D age have been recorded by Tate & Dobson (1989) in an oil well to the west of Ireland.

Red bed facies of the Etruria Formation are present in the Westphalian C (Bolsovian) of the southern part of the Pennine Basin (Besly 1988), and extend downwards into the Westphalian A and B immediately adjacent to the northern margin of the Wales-Brabant High. These represent well-drained alluvial environments that coexisted with deltaic deposition further north in the Pennine Basin. The alluvial red bed facies appears to have spread progressively northwards throughout the Westphalian C (Edwards 1951, fig 36), gradually replacing poorly drained deltaic environments (Fig. 4), possibly as a consequence of tectonic uplift heralding the main Variscan movements.

Economic coals are present at the base of the Westphalian C, but above the Cambriense Marine Band coals of economic thickness are generally rare, except in the Oxfordshire region, where the Wales-Brabant High had been breached, and where a number of relatively thick seams occur in the Westphalian D (Dunham & Poole 1974; Ramsbottom et al. 1978). The overall paucity of economic seams north of the Wales-Brabant High at this time is attributed to the well-drained conditions that led to the development of red beds and which inhibited the development and preservation of thick peat accumulations in this alluvial, basin margin setting.

Major facies of the productive Coal Measures

The first step in recognizing sedimentary facies, and hence depositional environments, is the detailed description of rocks exposed in mine workings and borehole cores; British Coal has adopted procedures for recording geological data (Elliott 1984; Fulton 1992).

The mid-Westphalian A (Langsettian) to upper Westphalian B (Duckmantian) 'upper delta plain' deposits, which contain the main productive coal seams of central and northern England (Fig. 4), have been extensively studied (e.g. Elliott, 1968, 1969; Scott 1978, 1984; Haszeldine 1983a, b, 1984b; Fielding 1984a, b, 1986; Guion, 1984, 1987a, b; Fulton 1987a; Jones 1992). Much of this work has been reviewed and the depositional setting discussed by Guion & Fielding (1988). Each worker has recognized and named different sedimentary facies, although there is broad agreement on the overall depositional environment. Some of the facies erected by Fielding (1984a, b, 1986) and Guion and Fielding (1988) have proved to be difficult to recognize consistently in mine workings and borehole cores, although it is acknowledged that they may be recognizable in good three-dimensional exposures such as opencast sites.

Thus we have attempted to simplify and rationalize the facies into those that we consider to be consistently recognizable with the more limited data obtainable from cores and deep mine workings. The characteristics and geometry of each of these major facies, which represents a different depositional subenvironment of the upper delta plain, are summarized in Fig. 5 and Table 1. The facies have been grouped into the following facies associations: pedogenic, which includes coals and seat earths; marine; lake-fill; channel; and near channel.

In some instances, lithology may be diagnostic of a depositional environment, but in other instances the lithology alone may not be diagnostic (e.g. minor channels) and other features such as the nature of the basal contact, internal organization, thickness, lateral extent and relationship to other rock bodies may be important in interpreting the deposits. Thus the facies described do not necessarily correspond simply to particular lithologies, but rather it is the organization of several lithologies that enables facies to be recognized.

Certain of the facies that contain diagnostic lithologies or fossils are laterally extensive, and may generally be correlated between sections or

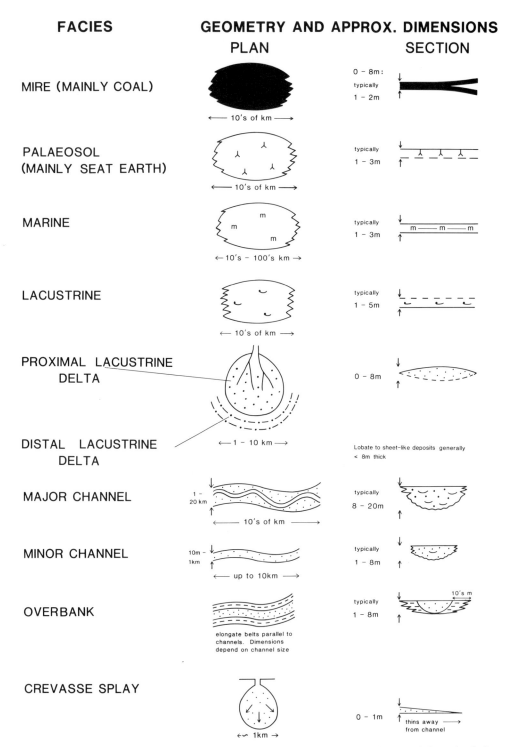

Fig. 5. Schematic representation of the geometry and approximate vertical and lateral extent of the major facies of coal-bearing Westphalian (mid-Langsettian–upper Duckmantian) of the Pennine Basin.

Table 1. *Facies associations and characteristics of the major facies of the Westphalian (mid-Langsettian to upper Duckmantian) coal-bearing upper delta plain deposits of the Pennine Basin.*

Facies associations and facies	Lithology	Sedimentary structures	Geometry	Fossils
Pedogenic				
Mire (mainly coal)	Coal, impure coal	Lamination, banding	Extensive sheets greater than 0.1 m thick; may split or die out laterally	Plant remains
Palaeosol (mainly seat earth)	Grey to white, brown/cream or red claystone, siltstone or sandstone depending on substrate and drainage conditions. Thin impure coals	Irregular laminations disturbed by rootlets, or lamination totally destroyed. Common mottling. Abundant polished ('listric') surfaces, siderite nodules, sphaerosiderite or iron oxides, depending on drainage conditions	Generally extensive sheets	*In situ* rootlets, *Stigmaria*, *Calamites* roots in some palaeosols
Marine	Dark grey to black fissile carbonaceous claystone	Well-developed thin lamination, sometimes massive	Extensive sheets	Marine goniatites, bivalves, brachiopods, gastropods, crinoids, bryozoa, forams, *Lingula*, *Orbiculoidea*, plant fragments, trace fossils
Lake fill				
Lacustrine	Medium grey to black claystone or siltstone, carbonaceous mudstone, cannel, boghead, coal	Flat lamination, rare sandy laminae and scours	Sheet-like	Non-marine bivalves, fish, ostracods, *Spirorbis*, plant fragments, trace fossils
Lacustrine delta (a) distal	Siltstones, interlaminated siltstone/sandstone forming upward-coarsening sequences	Flat lamination, current- and wave-ripple cross-lamination, climbing ripples, backflow ripples, ripple form sets, soft sediment deformation	Sheet-like to lobate deposits occur distally and beneath proximal lacustrine delta deposits, generally shows upward coarsening above lacustrine deposits	Trace fossils, plant debris, rare non-marine bivalves
(b) Proximal	Sandstone, interlaminated sandstone/siltstone, forming upward-coarsening sequences	Current-ripple cross-lamination, cross-bedding at tops of sequences, climbing ripples, wave-ripple cross-lamination, occasional trough-like scour surfaces, flat lamination, ripple form sets, backflow ripples, soft sediment deformation	Lobate to sheet-like deposits up to about 10 km across and 8 m thick, with gradational or sharp bases, forming upward-coarsening sequences above distal lacustrine delta	Plant debris, trace fossils
Channel				
Major channel	Thick erosively based sandstone bodies, often with horizons of breccia of conglomerate. Subordinate siltstone, claystone	Erosion surfaces, trough and planar cross-bedding, inclined heterolithic surfaces, ripple and cross-lamination, soft sediment deformation	Elongate belts, typically 1–20 km wide, 10s of km long, greater than 8 m thick	Plant fragments and debris, often abundant, rare trace fossils
Minor channel	Sandstone, siltstone, claystone, typically heterolithic, breccia, conglomerate; very variable	Erosion surfaces, ripple cross-lamination, trough and planar cross-bedding inclined heterolithic surfaces, flat lamination, soft sediment deformation, bank collapse deposits	Elongate belts typically up to 1 km wide, several km long, up to 8 m thick	Plant fragments and debris, often abundant, rare trace fossils
Near-channel				
Overbank	Siltstone, claystone; sometimes interbedded sandstone	Massive or weakly laminated siltstone, passing distally into laminated claystone. Occasionally interbedded siltstone/sandstone with low dips away from adjacent channel	Elongate belts adjacent and parallel to channel margins	Abundant plant leaves and stems, *in situ* trees
Crevasse splay	Sharp based sandstones interbedded with siltstones or claystones	Current-ripple cross-lamination, flat lamination, cross-bedding, wave rippled tops	Lobate deposits adjacent to channels. Individual splays generally < 1 m thick, often vertically stacked. Thin distally away from channel	Plant debris. *Pelecypodichnus* escape shafts

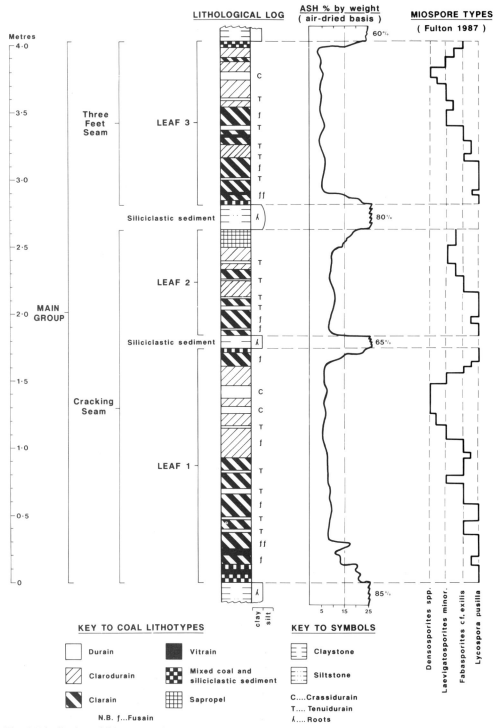

Fig. 6. Distribution of lithotypes, ash content and miospores in a group of coal leaves. Leaf 1 represents a full cycle. Leaves 2 and 3 are dulling-upwards sequences, succeeded in leaf 2 by a sapropelic coal (e.g. cannel) and in leaf 3 by a mixture of coal and siliciclastic sediment. Tenuidurains are more often found in the centre of coal leaves, whereas crassidurains are found towards the top. Ash content is often low in the centre and high at the top and base. Leaf 1 has a full cycle of miospore types, leaf 2 a partial cycle and leaf 3 has a partial succession from *Lycospora pusilla* to *Laevigatosporites minor*.

boreholes. These include mire (coal), palaeosol (seat earth), marine and lacustrine facies (Fig. 5). These facies can therefore be relatively easily mapped in the subsurface, although their properties may vary laterally. In contrast, channels and near-channel facies are more difficult to locate in the subsurface because of their relatively restricted dimensions and wide range of lithological characteristics (Fig. 5, Table 1).

Examples of intermediate facies exist, reflecting evolutionary changes in the depositional environment—for example, crevasse splay deposits grading upwards into lacustrine delta deposits as a result of the gradual enlargement of a breach in a channel bank, causing the initial splay system to develop into a lacustrine delta. Many of the facies may also be subdivided on the basis of detailed studies and an appreciation of three-dimensional relationships with other facies. Thus different types of minor channels and minor distributaries, may be differentiated (e.g. Guion 1984; Fielding 1986; Guion & Fielding 1988).

Hence, although a framework of major facies is suggested here, it is important to acknowledge the real complexity of sedimentary systems by adopting a flexible approach to facies recognition. Where useful, subfacies may therefore be recognized to enable detailed environmental interpretation and correlation. The main characteristics of each of the facies are described in the following sections.

Mire (coal) facies

Description. These deposits are dominated by stratified humic coal, forming leaves of coal which are often grouped into seams (Fig. 6), and as such are one of the most easily recognized Coal Measures facies. In geophysical logs, coal is characterized by a low gamma ray response, a low density and a high interval transit time on sonic logs (Dison & Whitworth 1985). Coal has been distinguished from siliciclastic sediment by the National Coal Board (1972) on the basis of the ash content after combustion. Coal should not exceed 40% by weight air-dried (b.w.a-d.). Coal seams are made up of one or more leaves of coal, separated by beds of siliciclastic sediment or 'dirt'. Leaves are coherent units of coal containing no beds of siliciclastic sediment, and have been divided into low ash (less than 10% b.w.a-d.) and high ash (10–40% b.w.a-d.) coals by Fulton (1987*a*,*b*), or clean (less than 15% b.w.a-d.) and inferior (15–40% b.w.a-d.) by the National Coal Board (1972). Most economic seams contain leaves with a low ash content, although the leaves may contain beds of high ash coal, usually at the base or top (Fig. 6).

Sulphur, which is an undesirable impurity, occurs in coal in several forms. Pyrite is commonly found disseminated in cubes or framboids; as larger nodules and lenses parallel to bedding; and in veins and cleat. Organic sulphur occurs as sulphides and disulphides, thiols and thiophenes intimately incorporated in the coal. Elemental sulphur and sulphates, if present at all, usually occur in extremely small amounts. Chou (1990) showed that in a seam from Illinois, the ratio of organic to pyritic plus sulphate sulphur in most lithotypes is generally in the range 1:1 to 2:1, which is believed to be comparable with the coals of the Pennine Basin. Fusain, however, was shown to contain more pyritic plus sulphate sulphur than organic sulphur. Total sulphur values were also shown to be lower in the lithotype durain, which agrees with the study of Fulton (1987*a*) on the Warwickshire Thick Coal. A comprehensive review of sulphur in peat and coal can be found in Casagrande (1987).

Two major types of bituminous coal are generally recognized: humic coal, with a low hydrogen content, and sapropelic coal, with a high hydrogen content. Sapropelic coals (cannels and boghead coals) are considered to have originated in lakes, rather than mires (Elliott 1965; Moore 1968), are volumetrically unimportant, thin and lenticular and commonly occur at the base and tops of coal seams (Fig. 6, leaf 2); they are included in the lacustrine facies.

The macroscopic description of the internal layering of humic coal may be achieved by reference to four lithotypes: vitrain, clarain, fusain and durain (Stach *et al.* 1982). Additional subdivisions have been used, including the lithotype clarodurain (Cady 1942), and mixed coal and siliciclastic sediment (often referred to as 'coal and dirt' or 'dirty coal' in drillers' logs). Further refinement is possible by recognizing additional combinations of these lithotypes (Fulton 1987*a*). Other systems of macroscopic coal description exist (summarized in Davis 1978 and Bustin *et al.* 1985) based mainly on the amount of bright and dull coal, similar to the system used by British Coal.

Mire (coal), as defined as a facies in this paper, is restricted to leaves greater than 0.1 m thick, usually of low ash content, which may be stacked to form even thicker coal seams (Fig. 6). The leaves of coal are usually separated by palaeosol (seat earth). The basal few centimetres of these leaves are commonly formed from mixed coal and siliciclastic sediment (Fig. 6).

The siliciclastic sediment may be finely disseminated, in which instance it resembles durain, so that the lithotype may be detected by density measurement and checked by geophysical density logs or by laboratory ash analysis. It may also be concentrated in laminae and lenses. Above the base, leaves are usually dominated by clarain, with common vitrain, fusain and occasional thin beds of tenuidurain (i.e. durain containing thin-walled spores) (Fig. 6). Towards the centre of a leaf, the lithotype clarodurain often takes the place of clarain, and this may continue to the top, or the lithotype content of the top may mirror that in the basal part of the leaf (Fig. 6). Alternatively, in a thicker, more fully developed leaf, beds of crassidurain (containing thick-walled spores) decimetres thick, may be present in its centre or top (Fig. 6).

The microscopic content of coals may be described using maceral groups, macerals and microlithotypes (Stach *et al.* 1982; McCabe 1984; Scott 1987), but this topic is beyond the scope of this paper. On a microscopic scale, a variety of plant tissue is visible. Several studies have been published on British coals on both miospores (Smith 1957, 1962; Smith & Butterworth 1967) and megaspores (Scott 1978; Scott & King 1981), but only fairly recently has detailed work been carried out (Bartram 1987; Fulton 1987*a,b*). These show that spore assemblages can be recognized which repeatedly occur in a similar order. There appears to be a correlation between miospore assemblages and both lithotype (Fulton 1987*a*) and microlithotype (Smith 1964), in that Westphalian coals that are rich in clarain/clarite are associated with a *Lycospora pusilla* (Lycospore) assemblage, tenuidurain with a *Fabasporites* cf. *exilis* (Fabaspore) assemblage, clarodurain/clarodurite with a *Laevigatosporites minor* (Transition) assemblage and crassidurain/crassidurite (micrinite) with *Densosporites* spp. (Densospore) assemblage (Fig. 6).

The mire facies forms sheet-like deposits with leaves which rarely exceed 3 m in thickness and usually range from 0.1 to 2.0 m (Fig. 5, Table 1). In most cases, they are underlain by palaeosol (seat earth) facies, and in some instances are interbedded with thin palaeosols to form seams which may occasionally exceed 8 m in thickness. Thinner leaves of coal may also be overlain by palaeosol facies, but they are more usually overlain abruptly by lacustrine facies.

Interpretation. Coal is a floriclastic deposit, and both palaeobotanical coal-ball studies (Phillips & DiMichele 1981; Phillips *et al.* 1985) and palynological studies (Smith 1962; Smith & Butterworth 1967; Scott 1978, 1979; Phillips & Peppers 1984) have indicated that most Westphalian humic coals are composed of lycophyte remains which formed *in situ* peat deposits.

The term peat has been used to describe widely differing organic-rich deposits (Clymo 1983), but the system most easily applied to coal is based on the ash content after combustion (reviewed in Kearns & Davison 1983). In most classifications, an upper limit of 25% ash (dry weight) distinguished peats from mucks, thus the use of the term peat to describe all coal precursors (National Coal Board 1972) is rather generous. In modern environments, peat is classified as an organic soil or histosol (FAO/UNESCO 1974; Soil Survey Staff 1975) so coal can be regarded as a special type of palaeosol.

Interpretation of Westphalian histosols is dependent on comparison with present day histosols developing in a tropical climate similar to that envisaged for the Westphalian (Rowley *et al.* 1985; Fulton 1987*b*). The habitat in which peat accumulates is termed a mire (Gore 1983).

Tropical histosols can be divided into two types: ombrotrophic (Du Reitz 1954) and rheotrophic (Moore & Bellamy 1974). Anderson (1964*a*) recognized similar divisions in the histosols of Sarawak and Brunei which, using modern terminology (Moore 1987), can be described as ombrotrophic bog forests (peat swamp) and rheotrophic swamp forests (freshwater swamp). The former are oligotrophic, fed by rainwater and raised above the influence of groundwater, and have pH values lower than 4.0, ash contents less than 25%, approximately level surfaces and peat or muck soils. Anderson (1964*a*) recognized an evolutionary sequence of habitats which involved the transition from rheotrophic swamp to ombrotrophic bog. Moore (1987) suggested the use of the term 'mire complex' where both habitats are present in the same hydrological unit. Anderson & Muller (1975), from a study of plants and their pollen, identified six floral communities which occurred in a catena across the bog surface. These were believed to be controlled by a change in edaphic conditions (Anderson 1983) reflecting increasing infertility. Increases in the magnitude of water-table fluctuation may also have been important, and Anderson & Muller (1975) refer to their communities as a hydrosere.

The evaluation of ancient coal deposits must take into account their original mode of formation and their subsequent alteration through diagenesis and coalification. One diagnostic feature of the two habitats which may be preserved is the ash content, although even this

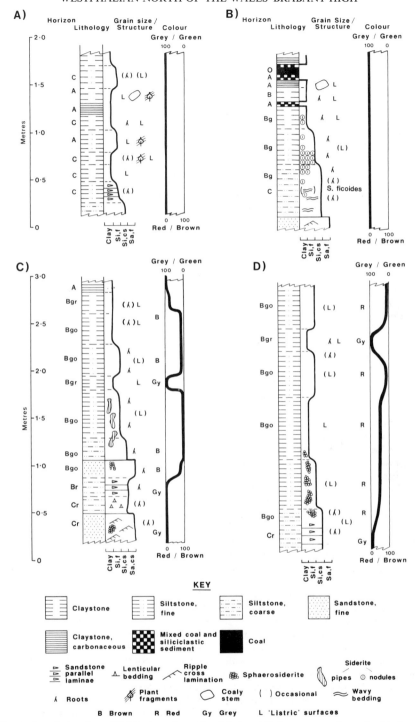

Fig. 7. Examples of the three siliciclastic palaeosol subfacies. (**A**) Immature palaeosol with few roots and stratification. (**B**) Poorly drained palaeosol with decreasing grain size and increasing number of roots upwards, topped by a thin leaf of coal. (**C**) Partially drained palaeosol, interbedded brown and grey with siderite pipes and abundant roots. (**D**) Partially drained palaeosol, interbedded red and grey with sphaerosiderite and fewer preserved roots.

may have undergone reduction during coalification (Cecil et al. 1979). Thus care must be exercised when attempting to interpret original communities from seam properties.

The upward transition from siliciclastic palaeosol to coal (Fig. 6) is interpreted as accompanying the change from rheotrophic swamp to ombrotrophic bog, similar to that observed by Anderson (1964a). This occurs when there is both a low rate of subsidence and siliciclastic sediment input, so that the amount of floriclastic sediment rapidly increases. In thick leaves of coal, this change usually occurs in the basal few centimetres corresponding to the change from mixed coal and siliciclastic sediment to other coal lithotypes (e.g. leaves 1 and 3, Fig. 6). Following Moore (1987), this may be interpreted as the product of a mire complex.

Clarain-dominated coal, which commonly occurs as the next lithotype upwards (Fig. 6), indicates a continued anoxic environment such as may be found in the catotelm (Ingram 1978) of mires. Vitrain is formed from 'woody' plant material (Teichmüller & Teichmüller 1982), presumably lycophyte periderm, and it is possible that during the wet season, water may have been above the peat surface. The lack of siliciclastic sediment in most thick coal leaves may be attributed to a combination of syndepositional factors. These include luxuriant vegetation acting as a sediment filter (Robertson 1952), reducing the water velocity and sediment-carrying power; clay flocculation by acid swamp water (Staub & Cohen 1979); the distance from a siliciclastic sediment source; and a rapid rise of the bog above groundwater levels. The Lycospore assemblage associated with clarain probably corresponded to a pioneering community of lycophytes (Fig. 6, leaves 1–3) from which the peat was derived. Fusain, which often occurs near the base of coal leaves, can form by the burning of plant material, and may indicate that the pioneering communities were prone to lightning (cf. Anderson 1964b) and subsequent fires.

Drier conditions may have been responsible for the formation of thin beds of tenuidurain caused by the oxidation of plant material and the associated Fabaspore assemblage. When these, together with other edaphic changes, become more persistent, clarodurain is formed, and there is an accompanying shift from the Fabaspore assemblage to the *Laevigatosporites* assemblage (Fig. 6, leaves 1–3). Miospores of the Denospore assemblage are believed to originate from the climax vegetation. Edaphic changes, the most important of which was probably an increased length of oxidation brought about by increase in thickness of the acrotelm or surface aerobic layer, are also believed to have been responsible for the production of crassidurain associated with the Denospore assemblage.

An increase in the rate of subsidence, if moderate, would overcome the rate of peat production, and the edaphic conditions would cause a reversal of the previous trends (Fig. 6). If the subsidence were rapid, the bog flora would drown at its climax stage and be overlain by lacustrine deposits.

Clymo (1987) suggested that the thickness of modern ombrotrophic peat deposits is proportional to the rate of addition of new material, minus that which has decayed, and must therefore be limited. His calculations for peat indicate a maximum thickness of 7–15 m. Assuming a compaction ratio of 13.5:1 (Elliott 1985a), this would give a typical coal leaf thickness of only 0.5–1.1 m. Although the calculations of Clymo (1987) may not be wholly appropriate for Carboniferous peats, the commonly observed maximum leaf thickness suggests that there may be limits to the thickness of peat accumulation.

Palaeosol (seat earth) facies

Description. Palaeosols or seat earths are commonly fine-grained siliciclastic rocks in which rhizophores (root-bearing axes) or roots (rootlets), or evidence of their activity, are preserved. *Stigmaria ficoides* is the most common axis, with roots or rootlets arranged spirally upon it, thus when *in situ*, roots show a variety of orientations.

Most palaeosols are clay or silt grade, although pedogenesis may have caused the evolution of any of the facies of the Coal Measures into a palaeosol. Some palaeosols consist of sandy sediment and where they are well cemented and highly siliceous are referred to as 'ganister'. In some instances sandy palaeosols or ganisters have been mapped on the flanks of underlying abandoned channels or 'swilleys' (Rippon & Spears 1989).

Certain palaeosols may display a variability of colour not normally found in other Coal Measures facies and are sometimes mottled. Inorganic sedimentary structures are usually poorly preserved, giving rise to a massive appearance in some instances, and the parent material of many palaeosols exhibits upward fining. Palaeosols of clay grade commonly contain abundant irregularly curved polished surfaces which are generally referred to as 'listric' surfaces by British Coal geologists.

Siderite concretions are common and exhibit a variety of forms including rounded, elongate and irregular. In many instances they can be shown to have formed around rootlets. Sphaerosiderite occurs in some palaeosols and may be dispersed or aggregated into nodules or patches. Organic-rich examples of siliciclastic palaeosols grade into coal, which may be distinguished on the basis of ash content.

Some palaeosols show vertical changes in the characteristics mentioned above and can be differentiated into a number of horizons to form profiles. Birkeland (1974, table 1.1) recognized six master soil horizons in modern soils based on process, with subdivisions defined using laboratory measurements of the chemical and physical properties. Special features are denoted by suffixes. Although exact equivalents of these horizons are impossible to define in palaeosols, it is feasible in some instances to use this horizon nomenclature, with the addition of 'r' and 'o' suffixes for reduced or oxidized layers, as suggested by the Working Group (1971) (Fig. 7).

Three siliciclastic palaeosol subfacies are recognized based on Fulton (1987a): (1) immature palaeosol subfacies; (2) poorly drained palaeosol subfacies; and (3) partly drained palaeosol subfacies. However, intergradational types occur and it is not always easy to designate a particular deposit to a specific subfacies.

Immature palaeosols (Fig 7A) usually consist of pedogenic sequences of thin grey interbeds of siliciclastic palaeosol and either non-palaeosol or sporadically rooted lithologies. The main attributes of such a palaeosol are the occasional to common *in situ* carbonaceous root traces. Non-palaeosol lithologies often contain foliage of pteridosperms and *Calamites* stems. Each interbed is usually thin (less than 0.5 m) and deposits of the whole subfacies rarely exceed 2 m in thickness. The lateral extent of this subfacies is variable and it may be underlain and overlain by a wide range of other facies.

Poorly drained palaeosols (Fig. 7B) are also grey in colour, and are sometimes weakly mottled with millimetre-size irregular paler and darker blotches. The palaeosol sequences may form thicker profiles than immature palaeosols and can include coal. They often fine upwards from coarser lithologies containing a few carbonaceous roots to finer lithologies of silt or clay grade with abundant roots, which may pass upwards into a thin coal seam. The coarser lithologies at the base often contain *Stigmaria* rhizophores and siderite nodules, although the nodules may also be spread throughout the profile. Sphaerosiderite, consisting of radial iron carbonate spheres up to about 2 mm in diameter, may also be found, often in the coarser lithologies. Thin coals at the top of the profiles often contain siliciclastic sediment, fusain and clarain, and are rarely more than a few centimetres thick. Individual profiles may attain up to 4 m in thickness, but they are sometimes repeated vertically to form stacked palaeosol profiles up to 10 m thick. The poorly drained palaeosols are the most common palaeosol type in the Pennine coalfields and form sheet-like deposits, which are most often underlain by lacustrine delta and crevasse splay facies, and overlain by other siliciclastic palaeosol subfacies, mire (coal) or lacustrine deposits.

Partly drained palaeosols, which are relatively rare, are formed from similar parent material to the other palaeosol subfacies, but in contrast contain a range of colours and lack coal. Two variants occur: those coloured brown or cream, which may be uniformly coloured or grey mottled (Fig. 7C); and those with a variety of colours including yellow, red, green and purple (Fig. 7D). The latter often become progressively redder upwards, but abruptly revert to grey when overlain by grey lacustrine or palaeosol facies. Coloration often occurs as subvertical elongate mottles, blotches or streaks. Root traces are rarely preserved, especially in clay lithologies where listric surfaces are abundant. Sphaerosiderite is fairly common in this subfacies, usually occurring towards the base, but often extending towards the centre of the profile. Siderite nodules are also common, and sometimes elongate, subvertical, slightly irregular pipes of siderite up to 2 cm in diameter and 15 cm long are developed in the centre of the profile. Profiles are typically 1–3 m thick, which may repeat to form stacked profiles many metres in thickness. This subfacies is often underlain by other palaeosol subfacies, mire (coal) and lacustrine delta deposits. It is overlain most often by other palaeosol subfacies, mire (coal) and lacustrine mudstone deposits.

Interpretation. Siliciclastic palaeosols (seat earths) of the Pennine Basin have received little attention until recently. Elliott (1968) recognized grey and brown varieties, whereas Guion (1978) suggested a tripartite division by the addition of immature seat earth. This tripartite division was retained by Fulton (1987a), who stressed the importance of syndepositional drainage conditions by including drainage classes similar to those of Young (1976, p39) in their names. Guion & Fielding (1988) and Besly & Fielding (1989) also recognized well-drained podzolic palaeosols, but these are believed to be rare in the main productive Coal Measures of the Pennine Basin.

Palaeosols may be developed in rocks with a wide range of original grain size, structure and fabric, thus the original substrate is an important factor influencing the characteristics of the palaeosol ultimately produced, as well as the conditions in which pedogenesis took place. Listric surfaces, which occur mainly in clay grade palaeosols, may result from a pedogenic fabric, or may also originate from the compaction of organic-rich clays.

Immature palaeosols are believed to have formed in a more active environment of deposition than the other palaeosol subfacies. Interbeds lacking root penetration are of subaqueous origin, formed by the episodic influx of sediment into standing water, the depth of which may have been beyond that capable of allowing *in situ* plant colonization. As water bodies became progressively filled, they became shallow enough to support plant growth, resulting in pedogenesis and the formation of a palaeosol. Where subsidence periodically exceeded sediment influx, the residence time of individual palaeosols was short and immature palaeosols developed.

Organic-rich mudstones can be regarded as pedogenic A horizons, whereas the barely modified rooted lithologies can be regarded as C horizons (Fig. 7A). Modern equivalents of the immature palaeosol subfacies may be found in interdistributary areas of the Sepik River of New Guinea, which are subject to periodic influxes of siliciclastic sediment (Pandago and Kabuk Land Systems) (Haantjens *et al.* 1968), or the marsh environment of Coleman & Prior (1980). Similar modern soils may be classified as fluvents (Soil Survey Staff 1975), weakly developed riverine or lacustrine soils (D'Hoore 1964), fluvisols (FAO/UNESCO 1974) and alluvial soils (Duchaufour 1982).

The immature palaeosol subfacies is similar to the floodbasin facies, and possibly the seat earths and coals in the marginal lake/delta top facies of Scott (1978). The alluvial palaeosol of Guion & Fielding (1989) would also appear to be equivalent.

Poorly drained palaeosols commonly exhibit upward-fining which may reflect the diminution of sediment supply and grade with time. It is believed that a greater length of pedogenic residence time compared with immature palaeosols allowed the formation of a better developed profile. Coal can be interpreted as a well-developed pedogenic O horizon, and sometimes a dark organic mudstone A horizon may be recognized towards the top of the profile (Fig. 7B). Below these, the thoroughly rooted zone can be interpreted as a pedoturbated B horizon. Its grey colour indicates that any iron present is in the ferrous state, which, together with an abundance of organic matter, attests to reducing conditions during formation, with the water-table at or above the sediment surface. Episodes of influx of siliciclastic sediment, followed by long periods of pedogenesis, may lead to the formation of stacked palaeosol profiles.

It is likely that the poorly drained palaeosol subfacies developed either in areas more distant from active deposition than the immature palaeosol facies, or in areas undergoing lower rates of subsidence. Poorly drained swamp deposits recognized by Coleman (1966) in the Atchafalaya Basin are comparable, as are those with 'permanent' vegetation, producing alluvial black clays and peats in the Sepik River of New Guinea (Sanai and Pora Land Systems) Haantjens *et al.* 1968). The siliciclastic sediments in these types of soils are likely to be classified as aquents or aquepts (Soil Survey Staff 1975), gleysols (FAO/UNESCO 1974), hydromorphic soils (D'Hoore 1964) and gley soils (Duchaufour 1982).

The poorly drained palaeosol subfacies appears to be the equivalent of the swamp association of Scott (1978), who did not distinguish between thick low ash coal deposits of mire complexes and thinner high ash coals that occur within poorly drained palaeosol sequences. It also appears to be similar to the passive lake margin facies of Fielding (1984*a*, *b*), although coals are excluded from his facies. This subfacies may also be compared with the gleysols of Guion & Fielding (1988) and hydromorphic palaeosols of Besly & Fielding (1989), which both, however, exclude thin coals.

Partly drained palaeosols commonly show vertical variation in their profiles (Fig. 7C and 7D), with the lowest grey horizon showing little pedoturbation being interpreted as a Cr horizon, the grey well-rooted beds above as a Br horizon, the oxidized sediment above this as a Bgo horizon and the rare, organic-rich layer above representing an A horizon (Fig. 7C).

The striking colour of partly drained palaeosols can be attributed to oxidation caused by periodic water-table lowering. Van Wallenburg (1973) showed that brown mottles begin above the mean high water-table and sometimes the change from brown to brownish grey matrix indicates the position of this water-table. The brown colour is likely to be due to hydrated iron oxide, which has become stable by virtue of insufficient organic matter to cause reduction, and a lack of time for complete oxidation to hematite. The formation of sphaerosiderite and siderite pipes is probably due to an upward movement of ferrous ions by the capillary rise of water and their subsequent combination with carbonate ions released by the decay of organic matter.

The changes in reddening of the coloured variant of this subfacies may result from an alternation between poorly drained and imperfectly drained conditions. Periods of oxidation longer than that required to form the brown variant would result initially in the formation of red streaks and mottles, due to oxygenated water passing along pathways created by roots. Progressive oxidation would lead to an increase in reddening upwards. Reversions to anoxic conditions would result in gleying of the sediment around the roots, causing grey haloes.

Improved drainage conditions are favoured by a combination of pre-existing topographic elevation and water-table lowering. This subfacies appears to occur in areas undergoing low rates of subsidence or relative uplift, such as found in more elevated areas towards basin margins (Besly 1988), where local water-table lowering has taken place (Rippon & Spears 1989), or in better drained areas overlying former channels and their associated deposits (e.g. Guion 1987a).

There is a paucity of descriptions of modern environments exactly equivalent to the partly drained palaeosol subfacies, but alluvial soils with 1–2 m changes in the water-table give rise to brown soils (Young 1976) and the coarser sediments of the Nagam Land System in the Sepik River area (Haantjens et al. 1968) produce brown and ochreous soils. These soil types are not likely to be separated from other palaeosol subfacies in classifications of modern soils.

Elliott (1968) recognized similar brown and red Westphalian seat earths, Guion & Fielding (1988) described partly drained or temporarily drained gleysols, whereas Besly & Fielding (1989) described partly drained palaeosols which include a semi-gley palaeosol with mottled brown, cream or red upper horizons.

Marine facies

Description. This facies consists of a medium grey to black carbonaceous claystone to siltstone and contains fossils indicative of marine conditions. The claystones often exhibit a well-developed fissility, which is enhanced by weathering, and sometimes thin dark and pale laminations, which Spears (1969) attributed to seasonal sedimentation events. Typical fauna includes goniatites such as *Anthracoceras* and *Anthracoceratites*, bivalves such as *Dunbarella*

and *Posidonia*, the inarticulate brachiopods *Lingula* and *Orbiculoidea*, and a variety of microfossils including foraminifera and conodonts. Some of the marine horizons, particularly the Aegiranum (Mansfield) Marine Band (Fig. 4) contain rich benthonic faunas (Calver 1968a), but in many, the faunas are restricted and impoverished, especially in the marginal areas of the basin (Fulton & Williams 1988). Calver (1968a) showed systematic lateral and vertical changes in the faunas of each marine horizon, which he related to salinity control.

The marine facies generally shows a higher level of radioactivity than other mudstones, which often enables their ready detection on gamma ray logs and may help to differentiate them from non-marine mudstones (Whittaker et al. 1985). In most instances, this is probably due to a high proportion of illitic clays containing potassium, but some instances have been described where the radioactivity results from uranium (Knowles 1964; Spears 1964; Rippon 1984). Leeder et al. (1990) have urged caution in the use of gamma ray logs to differentiate marine bands and have suggested that analysis of the carbon to sulphur ratios, as discussed by Berner & Raiswell (1984), may be a useful tool for the detection of marine mudstones, especially offshore where the stratigraphy is not well established.

Marine mudstones generally form thin horizons, a metre or so in thickness, but are known to attain 9 m in the case of the Aegiranum (Mansfield) Marine Band (Calver 1968a). Certain of the marine mudstone horizons, such as the Vanderbeckei and Subcrenatum Marine Bands (Fig. 4) extend from the English coalfields into Europe (Calver 1968a; Ramsbottom et al. 1978), which indicates the establishment of marine conditions over hundreds of thousands of square kilometres, whereas others are more restricted in extent.

Interpretation. There have been many studies of the faunas, palaeoecology, petrology and geochemistry of Westphalian marine mudstones (e.g. Clift & Leek 1943; Edwards & Stubblefield 1948; Calver 1968a,b, 1969, 1973; Taylor 1971; Ashby & Pearson 1979; Spears & Sezgin 1985; Dobson & Kinghorn 1987; Leeder et al. 1990) and reference should be made to these works for detailed discussions of the depositional environment. It has been established that many of the marine mudstones contain distinct faunal assemblages which correspond to stages of increasing salinity to a maximum, followed by reversion to non-marine conditions (Calver 1968a). No convincing evidence has been presented that tidal activity accompanied these marine incursions into the Pennine Basin during the Westphalian. The lateral continuity and extensive occurrence, together with the presence of expanded cycles above them (Duff & Walton 1962) suggests significant eustatic rises across the delta plain in the case of the Subcrenatum, Vanderbeckei and Aegiranum Marine Bands: the effects of eustasy were probably less significant in some of the other Westphalian marine bands, which are more localized in extent.

Lacustrine facies

Description. This facies generally consists of black to medium grey claystone or siltstone and also includes the volumetrically unimportant sapropelic cannel and boghead coals. The claystones may be carbonaceous, possess well-developed fissility and grade into sapropelic coal. The darkness of the mudstone is usually related to the amount of carbonaceous material. Thin, medium and dark grey coloured laminations are present in some mudstones. Iron carbonate beds, patches, nodules or lenses are often present, in many instances nucleated around fossil remains. Lacustrine mudstones are characterized by non-marine fossils, especially bivalves such as *Carbonicola, Anthracosia, Anthraconaia* and *Naiadites*, which may be locally abundant. The bivalve fossils show various states of preservation ranging from compressed shells to uncrushed shells preserved in carbonate concretions. In many instances, the two valves are preserved in an articulated state. Less common fossils include the ostracods *Geisina* and *Carbonita*, and scales and spines of fish such as *Rhabdoderma, Rhizodopsis* and *Rhadinichthys*, which are present particularly in carbonaceous claystones. The small serpulid *Spirorbis* is sometimes present epifaunally on bivalve shells or plant stems.

Bioturbation is sometimes present, but in many instances the traces are too indistinct to assign to a particular ichnogenus. However, trace fossils including *Planolites, Arenicolites, Cochlichnus* and *Gyrochorte* have been recorded (Elliott 1968, 1985*b*; Guion 1978; Eagar *et al.* 1985; Pollard 1988). Plant fragments, usually consisting of compressed stems and trunks, are sometimes present and often consist of *Lepidodendron* where lacustrine deposits immediately overlie coal seams.

Lacustrine facies generally occur as sheets, which range from a few centimetres to a few metres in thickness (Fig. 5). In many instances, lacustrine facies immediately overlie coal seams and exhibit upward-coarsening, passing upwards into distal lacustrine delta deposits with which they are transitional. Particular horizons may be traced laterally for tens of kilometres (Fig. 5). Where the mudstones contain a distinctive fauna, such as abundant bivalves of the *Carbonicola cristagalli–C. rhomboidalis* group above the Cockleshell Seam of the East Midlands, this may be useful for local correlation purposes (Smith *et al.* 1967; Elliott 1984).

Interpretation. This facies is comparable in part with the 'faunal mudstones' of Elliott (1968, 1969), prodelta lake clays of Scott (1978), floodbasin lake deposits of Guion (1984) and the prodelta/lacustrine mudstones of Guion (1987*a*). It also includes the anoxic lake floor and outer minor delta/overbank claystones of Fielding (1984*a, b*) and the suboxic lake/bay floor of Guion & Fielding (1988). Black mudstones and rhythmite muds were considered by Haszeldine (1984*b*) to be lake floor deposits. The mudstones were deposited from suspension in a low energy lake floor environment where sedimentation rates were low. The paucity of current-formed structures, the articulated nature of bivalve shells and the undamaged nature of other fossils indicates minimum current activity (Broadhurst 1964). The high organic content, presence of lamination and fissility and lack of bioturbation and benthonic organisms in some of the lacustrine deposits suggests that conditions were inimical to benthos as a result of poor oxygenation of the substrate (Demaison & Moore 1980).

The presence of benthonic bivalves which closely resemble modern freshwater unionids is generally interpreted as indicating non-marine conditions (Trueman 1946; Eager 1960; Calver 1968*b*). However, 'non-marine' bivalves have been recorded on some occasions in the same bed as marine fossils (Calver 1968*b*), suggesting that they may have been able to tolerate saline conditions in some circumstances. Several studies have been carried out on the palaeoecology of the faunas, including Broadhurst (1964) and Eagar (1960, 1974, 1987). Pollard (1969) considered that the darker, more fissile claystones reflected the slowest rates of sedimentation and that the common upward-coarsening of many lacustrine mudstones is accompanied by changes in fauna and mineralogy which reflect increasing rates of sedimentation. Haszeldine (1984*b*) suggested that rhythmically bedded dark and pale mudstones formed as a result of turbid underflows which produced laminations by dynamic sorting mechanisms.

The ultimate source of the sediment compris-

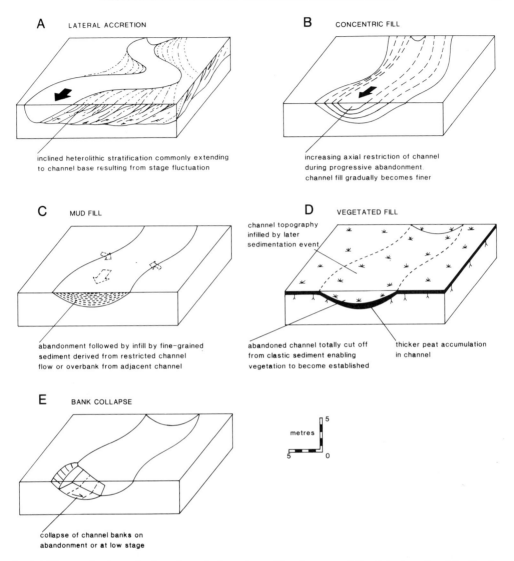

Fig. 8. Filling mechanisms of minor channels from the coal-bearing upper delta plain deposits of the Pennine Basin. (**A**) Lateral accretion, forming inclined heterolithic surfaces. (**B**) Concentric fill, formed by decreasing flow through the channel. (**C**) Mud-fill of an abandoned channel derived from restricted flows or overbank flows from adjacent channels. (**D**) Vegetated-fill: colonization of abandoned channel by vegetation. (**E**) Bank collapse, resulting in translation of bank material into channel.

ing the lacustrine facies is difficult to determine, but it may either have been derived as prodelta deposits of lacustrine deltas, or from overbank flows from nearby channels, there being insufficient criteria in many instances to differentiate between the two modes of origin.

Distal lacustrine delta facies

Description. Distal lacustrine delta deposits form sheet-like to lobate bodies, which generally underlie, or occur down-palaeocurrent from the proximal lacustrine delta facies, with which it is gradational (Figs 6, 10). In many instances they pass upwards from the lacustrine facies, forming upward-coarsening sequences. This facies is dominated by parallel-laminated siltstones and interbedded or interlaminated siltstones and sandstones (Table 1), with sandstone comprising less than 50% of the total. Thin sandstone beds with current and wave-ripple cross-lamination, wavy and lenticular bedding with both

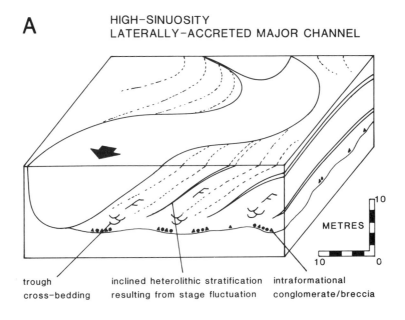

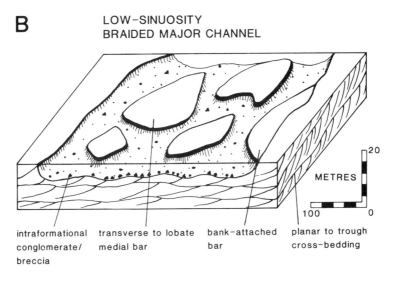

Fig. 9. Simplified representation of types of major channels from the coal-bearing upper delta plain deposits of the Pennine Basin. (**A**) High sinuosity laterally accreted major channel. (**B**) Low sinuosity braided major channel.

current and wave-formed ripples may be present, especially towards the top. Laterally continuous drapes of siltstone are sometimes present in some of the sandier beds. Climbing ripples (ripple drift), developed in siltstone with sandstone laminae picking out the structures, are common in this facies, and occasional backflow or regressive ripples, termed 'train drift' by Elliott (1968), may be present. Comminuted plant debris and sideritic patches are often concentrated on foresets, but larger plant fragments are rare. Soft sediment deformation is sometimes developed, especially within sandstones.

Bioturbation is common and trace fossils include *Planolites*, *Arenicolites* and *Pelecypodichnus*. Occasional non-marine bivalves have also been recorded.

Proximal lacustrine delta facies

Description. Proximal lacustrine delta deposits are dominated by sandstones and interlaminated and interbedded sandstones and siltstones (Table 1), and contain more than 50% sand. The proximal lacustrine delta deposits usually form parts of upward-coarsening sequences up to about 8 m thick and a few kilometres across (Fig. 5), which lie above and proximal to distal lacustrine delta deposits. Current-ripple cross-lamination is the dominant sedimentary structure, usually showing unidirectional palaeocurrents with low to medium dispersion. In many instances the cross-lamination takes the form of climbing ripples (ripple drift), often with comminuted plant debris and sideritic patches concentrated on the foresets. Backflow ripples are sometimes present on ripple foresets. Other sedimentary structures include low amplitude irregular three-dimensional ripples with wavelengths of between 5 and 20 cm, which was termed 'complex interlamination' by Clarke (1963) and 'complex' stratification by Guion (1978). Wave-ripple cross-lamination is sometimes present at certain horizons. Flat, parallel lamination, wave or current-ripple formsets and wavy bedding may also be present, especially in the less sandy parts of the sequence. Horizons showing soft sediment deformation, particularly convolute bedding or load casts, are often present.

Contacts between beds tend to be gradational, but abrupt and erosive contacts are occasionally present. Erosion surfaces generally have a relief up to about a metre or so and define small channels. Brecciation and intraformational clasts accompany some of these small channel bases.

Fossils include occasional plant stems and trucks and trace fossils such as *Pelecypodichnus* (bivalve resting traces) and small inclined to vertical burrows referred to as 'colonial burrows' by Elliott (1968), which may be compared with *Trichichnus*.

Interpretation of distal and proximal lacustrine delta facies. Lacustrine delta deposits have been widely recognized from the Westphalian of northern England. Elliott (1969) assigned his layered sand–siltstone and rippled sandstone facies to inner distributary mouth bar and outer distributary facies, respectively. Scott (1978) recognized deposits representing the outermost portion of a river distributary mouth bar entering a small lake (facies 1B) and a proximal distributary mouth bar sequence (facies 2A). Guion (1978) differentiated lacustrine delta front (distributary mouth bar) deposits and distal siltstones, and later Guion (1987a) described the characteristics of proximal and distal deposits of a lacustrine delta complex. Haszeldine (1984b) also recognized lacustrine delta mouth bar deposits. Fielding (1984a, b) and Guion & Fielding (1988) included lacustrine delta deposits within medial crevasse splay/minor delta and distal crevasse splay/minor delta facies. Thus detailed interpretations of lacustrine delta deposits are available in the above papers.

Lacustrine deltas form where a river enters a lake, often with little density contrast between inflowing water and the lake, resulting in homopycnal flow (Bates 1953). This condition results in friction-dominated flow (Wright 1977), accompanied by rapid sediment dumping, resulting in the formation of lacustrine delta mouth bars, and is responsible for the abundance of climbing ripples in lacustrine delta deposits. Hyperpycnal flow (Bates 1953) may also take place, where denser water enters a less dense lake, possibly due to a high sediment load. This may result in density underflows, as described from modern lakes. Haszeldine (1984b) and Fielding (1984b) considered this to be an important depositional mechanism in Westphalian lacustrine delta sequences.

There is often evidence of discharge fluctuation, which has resulted in the deposition of sandy and silty interbeds, and alternations of climbing ripples resulting from rapid sedimentation, with wave-formed structures formed by reworking during low discharge periods. However, discharge variation was not generally as episodic as that responsible for crevasse splays. Thus the deposition of individual sandy crevasse splay deposits took place only at high river stage, whereas lacustrine delta mouths received sediment throughout their history, even though the discharge may have varied.

Lacustrine delta deposits commonly form overall coarsening-upward sequences, which result from progradation. The small channels present in the upper parts of the proximal lacustrine delta facies represent distal extensions of distributary channels which bifurcated at the delta front. These channels would be expected to be rapidly cut and filled as the delta prograded and new channels formed.

At any locality the sequence may be terminated at any point in the succession, and be followed by other facies, especially palaeosol and mire deposits, which formed when the delta was abandoned. In some instances more than one upward-coarsening sequence is present within an interseam interval, indicating more than one episode of delta progradation into some lakes. Several mechanisms may be possible, including the following: (1) rises in lake level, drowning a delta and causing renewal of delta progradation; (2) abandonment of a delta or part of a delta, due to upstream avulsion, followed by growth of another delta lobe into the lake; (3) progradation of several deltas or parts of deltas into a lake from different directions at different times, giving overlapping lobes; (4) major variation in discharge of feeding channels; and (5) episodes of rapid subsidence controlling delta building.

In most instances it is believed that episodes of delta abandonment, or the progradation of more than one delta into the lake, were responsible for the multiple coarsening-upwards sequences.

Crevasse splays sometimes pass upwards into lacustrine delta facies, which suggests that crevasse systems evolved into lacustrine deltas, as crevasses enlarged and diverted more of the flow from distributaries. Deposits showing intermediate characteristics of crevasse splays and lacustrine deltas are also developed.

Major channel facies

Description. Major channel deposits are dominated by sandstones which form belts tens of kilometres long, which may be typically 1–10 km wide and 8–20 m thick, although they may occasionally exceed these dimensions (Figs 5, 9). Little detailed information has been published on the dimensions of major channel belts: Fielding (1986) considered that major channel deposits were generally less than 5 km wide, but he also recorded some channel belts exceeding 10 km in width. Some of the major channel deposits form important surface features and outcrops, and thus have been individually named by the Geological Survey, e.g. the Silkstone Rock of the East Pennines (Eden *et al.* 1957) and the Seaton Sluice Sandstone of Northumberland (Land 1974). Where major channel deposits are present, this may result in the absence of one or more coal seams that are developed elsewhere in the coalfield.

The major channel deposits possess basal erosion surfaces, which may have a relief of several metres. The basal erosion surfaces are often overlain by beds of breccia or conglomerate, several centimetres thick, which include intraformational clasts of claystone or siltstone, which sometimes exceed 5 cm in diameter. Bank collapse deposits, consisting of disrupted blocks of bank material showing anomalous dips, are present in some major channels (cf. Fig. 8E). Other clasts present in major channel deposits include sideritic clasts derived from the erosion of bank material, large masses of coal resulting from the erosion of peat and coaly clasts of plant stems and trunks. The presence of clasts comprised of siderite root nodules in some major channel deposits indicates that they have eroded through coal horizons into the underlying seat earth. Extraformational clasts are generally rare, but they have been recorded in some of the coarser channel deposits, especially in the northern part of the Pennine Basin (e.g. Land 1974).

The major channel deposits are generally fine-grained, although medium to coarse sandstones are present in the more northerly parts of the Pennine Coalfields. Fielding (1986) concluded that major distributaries were responsible for the transport of sediment from the north towards the depocentre of the Pennine Basin in the south, and that a down-palaeocurrent decrease in grain size would be expected.

Beds of mudstone tend to be subordinate in the major channel deposits, but fine-grained beds, often rich in mica and comminuted carbonaceous debris, are sometimes present. Thin drapes of siltstone or claystone on the tops of abandoned bedforms, and clay plugs infilling abandoned channels, have been recorded, which form effective permeability barriers.

Sedimentary structures include cosets of trough and planar cross-bedding with sets up to about 2 m in thickness. Inclined heterolithic surfaces (Thomas *et al.* 1987), representing lateral accretion surfaces of high sinuosity rivers, have been observed, but do not appear to have been common. Stratification surfaces inclined down-palaeocurrent, representing downstream-accreting bar forms, appear to be more common (Haszeldine 1983*a, b*; Jones 1992). Current-ripple cross-lamination, with foresets picked out by mica and comminuted plant debris, is also a common sedimentary structure. In some instances the sandstone may appear to be structureless. Wave-formed structures tend to be rare, but wave ripples developed on the tops of abandoned bedforms have been recorded (Jones 1992).

Internal erosion surfaces and bounding surfaces which may have complex organization

(Haszeldine 1983a, b; Fielding 1986; Jones 1992) are present within major channels. Fossils and trace fossils, other than plant remains, are rare in the major channel facies.

Interpretation. This facies corresponds in part to the 'wash out sandstone' facies of Elliott (1968, 1969) and the major distributary channel facies of Fielding (1984a, 1986) and Guion & Fielding (1988). The characteristics of major channels have been discussed in detail by Fielding (1986) and thus a full interpretation will not be given here. Both high and low sinuosity major channels have been recognized from the Westphalian A and B of the Pennine Basin (Fig. 9A and B). The deposits of high sinuosity laterally-accreted major channels may be recognized by the presence of inclined heterolithic surfaces and palaeocurrent patterns with a high dispersion (Fig. 9A). Low sinuosity channels are dominated by planar to trough cross-bedded sandstones, often indicating downstream-accreting barforms with a low palaeocurrent dispersion (Fig. 9B). Haszeldine (1981, 1983a, b) showed that in Westphalian B channels in northeast England, deposition of medial or bank-attached bars took place in large, low sinuosity channels. Fielding (1986) and Guion & Fielding (1988) proposed that in the early Westphalian A, a gradual transition took place from broad, coarse-grained, low sinuosity channel belts to more sinuous, narrow, finer-grained channels, which accompanied the evolution from lower delta plain to upper delta plain conditions.

The controls on the sinuosity of channels present in the upper delta plain of mid-Westphalian A to late Westphalian B are not yet clear, although low sinuosity channels appear to be more abundant in the more proximal northerly part of the basin (Haszeldine 1981, 1983a, b; Jones 1992). Fielding (1986) considered that stage fluctuation of channels took place and that the sinuosity of channels decreased during the flood stage. Field evidence from major channels provides evidence for discharge variation; thus mudstone drapes on abandoned bedforms, sometimes with wave-reworked tops, are the products of low stage flow.

Although the deposits of the major channels form relatively thick, wide belts, the dimensions of the active channel were considerably smaller than the deposit ultimately formed by the channel shifting across the delta plain. Typical bankfull depths of 10–12 m were suggested by Fielding (1986).

Minor channel facies

Description. Minor channel deposits contain a wide range of lithological fills which have a variety of modes of origin (Guion 1984, 1987a, b; Williams 1986; Jones 1992; Fig. 8). For this reason, lithology is not a good criterion for recognizing the presence of minor channels. Typical dimensions range from about 10 m to 1000 m in width, and 1–8 m in thickness (Fig. 5, Table 1), but examples outside this range also exist. Minor channels have been mapped for several kilometres in the subsurface, but in some instances they have been shown to die out and give way distally to other facies such as lacustrine delta or crevasse splay deposits (Guion 1978, 1984; Fielding 1984a, b, 1986; Guion & Fielding 1988).

Minor channels may result in the erosional removal of underlying coal, causing wash-outs, and have numerous other deleterious effects on mining. The dimensions of minor channels are small compared with the spacing of exploratory boreholes for deep-mined coal, and their detection in the subsurface is difficult, especially in the investigative phase of exploration.

Minor channel bases are marked by erosion surfaces, which are often picked out by concretionary siderite. Intraformational breccias or conglomerates are common, and may include claystone, siltstone and sideritic clasts. Masses of coal, plant stems and trunks are often present in minor channel fills. Many of the other lithologies present in minor channels depend on the mechanism of channel-filling. Sandstone beds are usually fine-grained, and may be massive, ripple cross-laminated or cross-bedded. In some instances the sandstones may be rich in comminuted plant debris and mica. Inclined heterolithic surfaces (Thomas *et al.* 1987), which are typically sandy at the base, becoming silty upwards, are common in some minor channels. Fills of siltstones and claystones are also present, in some instances occupying the full thickness of the minor channel. The siltstones may be massive or laminated, and sometimes contain indistinct patches of sandstone which may have been introduced into the channel by bank sloughing (Guion 1987a). These fine-grained channel-fills often contain abundant plant stems and fronds and sideritic concretions. *In situ* coal, cannel and carbonaceous mudstones may be present, in particular occupying abandoned channels and forming 'swilleys' (Elliott 1965).

Beds showing anomalous dips, contortion and soft sediment deformation are relatively common in minor channels. These have a variety of modes of origin, including channel bank collapse, post-depositional adjustment of sediment and compaction (Guion 1987a).

Fossil fauna are generally absent from minor channels, but locally trace fossils may be present, representing colonization of channels during low stage or on abandonment.

Interpretation. This facies may be included in part with the 'wash out sandstone' facies of Elliott (1968, 1969) and the distributary channel and crevasse channel of Guion (1984, 1987a, b). A number of different types of channel recognized by Fielding (1984a, b, 1986) and Guion & Fielding (1988) have been included in minor channel facies. These include proximal major crevasse splay channel, minor distributary channel, minor crevasse channel and distal feeder channel facies. The recognition of these channel types is dependent on a good understanding of three-dimensional relationships, which is generally not achievable with widely spaced boreholes or deep mine data. Thus it is considered to be more useful to group the various channel types into one facies. It must nevertheless be acknowledged that the organization of minor channels is hierarchical, and that sediment and water were distributed by a system of channels of decreasing magnitude. Detailed subsurface mapping (e.g. Guion 1984) may enable the recognition of various types of channel on a local scale.

The operation of a variety of different processes may be recognized during the history of minor channels. Each channel has an individual history and thus the nature of minor channel fills is variable. Guion (1984, 1987a, b), Fielding (1986), Williams (1986) and Jones (1992) have recognized several phases of channel-filling during the active life of the channel and during abandonment, as proposed by Meckel (1972). Thus it is useful to attempt to differentiate the deposits represented by each of the following stages of channel evolution: (1) active channel, high stage; (2) active channel, falling and low stage; and (3) abandoned channel.

High stage processes include bedform and bar-form migration, channel widening and bank collapse. Thus at high stage, the transport and deposition of a sandy bedload is important and channel bases would be covered by dunes, ripples and compound bar-forms.

During falling stage, flow is reduced and altered, and this, together with wave activity, may result in the modification of structures formed at high stage (Collinson 1970). Lateral accretion may occur on the sides of pre-existing structures within the channel, small scours may be cut into bar-forms, and erosion of the bar crest may form reactivation surfaces (Jones 1992). As the stage drops further, bedload transport is reduced, scours may be filled and suspended sediment deposition occurs in standing water, draping earlier structures. Wave activity may also modify both the lee and stoss sides of earlier structures, forming symmetrically crested ripples. Colonization by organisms may also take place under low energy conditions at low stage, and trace fossils such as *Skolithos*, *Planolites* and *Pelecypodichnus* may be preserved.

Channels can be abandoned by a variety of mechanisms, including avulsion, chute cut-off and neck cut-off (Allen 1965). Channel abandonment is indicated by the deposition of fine-grained sediment or organic material, forming a mud plug, concentric fills or inclined heterolithic surfaces which gradually become finer as the flow through the channel reduces.

Many of the minor channels had a history in which active processes alternated with periods of reduced flow, resulting in complex channel-fills (e.g. Guion, 1984, 1987a).

The sinuosity of minor channels varies from highly sinuous to straight. Some of the sinuous channels underwent appreciable lateral migration, generating inclined heterolithic surfaces (Fig. 8A), whereas others appear to have undergone little shifting of course throughout their life. The inclined heterolithic surfaces were deposited as laterally accreted point bars during the active phase of the channel (Fig. 8A). The heterolithic nature of the lateral accretion surfaces is a product of stage fluctuation, each couplet recording the passing of a flood event. In some channels the lateral accretion surfaces are predominantly silty, which indicates that these channels carried a very fine-grained load. Sand deposits also occur in various positions in channels, indicating bedload deposition on mid-channel and side-attached bars during the high stage activity of the channel.

Much of the fill of minor channels represents falling and low stage or abandoned channel processes, and a variety of types of fill have been recognized. Concentrically filled channels (Fig. 8B) are formed by the progressive abandonment of a reach of channel. The dimensions of the channel and the grain size of the load gradually decreases with time, resulting in the concentric fill, with the deposits becoming finer inwards and upwards. The final fill of the channels is often organic-rich claystones, deposited from suspension when the channel was completely abandoned. Channel fills of this type have been described by Wing (1984), Hopkins (1985) and

Guion (1987a).

Mud-filled minor channels (Fig. 8C) are the result of the rapid abandonment of a reach of channel or a channel system, and may be compared with the clay plugs described by Fisk (1947). The infill of channels of this type ranges from siltstone to organic-rich claystone, with drifted plant remains such as *Calamites*, *Sphenophyllum*, *Neuropteris* and *Cordaites* being common. According to Hopkins (1985), mud-filled minor channels result from rapid channel abandonment, followed by the deposition of mud from suspension, either from sluggish flows in the channel, or derived by overbank floods from an adjacent channel. The fill of mud-filled channels is often difficult to differentiate from overbank material, and this type of channel may easily be missed in the subsurface.

Some abandoned channels were subsequently starved of clastic sediment, enabling vegetation to become established both within the channel and on its banks, resulting in vegetated abandoned channels (Fig. 8D). Evidence for the presence of vegetation is now represented by coal, which generally thickens into the channel and thins onto the channel margins, resulting in 'swilleys'. These 'swilleys' in which coal seams thicken and pass downwards into the abandoned channels have been described by Raistrick and Marshall (1939), Clarke (1963), Elliott (1965, 1984) and Rippon & Spears (1989). Some swilleys have been traced in mine workings for several kilometres, suggesting that an episode of avulsion and rapid abandonment took place, followed by a prolonged period of sediment starvation, enabling peat to accumulate. The presence of these extensive swilleys suggests that the abandonment of large areas may have sometimes taken place. In these instances the depositional system is believed to have built up close to base level, with the consequence that channels became sluggish and inefficient. Any gradient advantage developed elsewhere would result in an avulsion event, with consequent clastic sediment starvation and the accumulation of peat mires on the abandoned areas in the abandoned channels. The subsequent deposition of clastic sediment over vegetated abandoned minor channels would be accompanied by differential compaction of the thickened peat in the abandoned channel, resulting in a thicker mass of sediment over the position of the abandoned channel.

Bank collapse is another mechanism which contributed to the fill of minor channels (Fig. 8E) and has been documented by Guion (1987a, b). Bank collapse deposits consist of blocks of sediment, often with irregular or slickensided boundaries, which exhibit anomalous dips, brecciation and soft sediment deformation. They consist of channel bank material that collapsed into a channel as a result of bank failure. Bank collapse is favoured where discharge in a channel decreases after a flood. However, the preservation of the collapsed material is most likely to occur if the bank collapse is followed by low stage flow or channel abandonment.

Overbank facies

Description. Overbank deposits characteristically consist of massive pale grey siltstones (Table 1), although poorly developed parallel lamination and thin beds of sandstone may be present. Where the overbank deposits consist of interbedded siltstones and sandstones, they usually show low dips away from the associated channel. Overbank deposits occur adjacent to channel margins (Fig. 5) and may sometimes occur interbedded or interdigitated with channel or crevasse splay deposits. They typically form narrow wedge-shaped elongate belts, usually only tens or hundreds of metres wide, parallel to the channel margins, which decrease in thickness and become finer away from the channel (Fig. 5), often giving way laterally to laminated claystones containing plant leaves and stems (Guion 1987b). In some instances the wedge of overbank sediment and its associated channel system occupy a seam split (e.g. Guion 1987b). The siltstones adjacent to channel margins usually contain isolated stems and fronds of plants, which may be flat-lying or oblique to bedding, such as *Neuropteris*, *Calamites* or *Cordaites*, which are commonly preserved in sideritic patches or nodules. Plants, particularly *Calamites*, sometimes occur in growth position, rooted in these deposits. Trace fossils, which may be horizontal or oblique, are occasionally present. Soft sediment microfaults are also often a feature of this facies.

Where overbank deposits occur above a coal seam, vertical *in situ* trunks, generally infilled with siltstone, may be present, up to several metres high. The trees are rooted in the coal seam or its immediate roof and pass upwards into the overlying overbank deposits (Guion 1978, 1987b). The tree trunks sometimes extend through the entire thickness of the overbank deposits into the overlying facies and may be infilled by sediment derived from overlying deposits, e.g. sandstone.

Interpretation. This facies corresponds to the massive siltstone facies of Elliott (1968, 1969), overbank flood deposits of Guion (1978, 1984, 1987a, b) and the siltstone-dominated overbank deposits of Fielding (1984c, b, 1986) and Guion & Fielding (1988). The coarse-grained overbank facies of Fielding (1984a, b, 1986) is not defined as a separate facies in this paper, but corresponds to the interstratified sandstone and siltstone overbank deposits that dip away from the channel.

The overbank facies has been interpreted as the product of the subaqueous deposition of silt and clay from suspension by episodes of sheet flow of sediment over the banks of channels during high river stage. The occasional preservation of plants in growth position indicates periods of colonization between sedimentation events. There is little evidence of the emergence of these deposits above the water-table. Rapid sedimentation, especially close to the channel margins, is indicated by the presence of the *in situ* tree trunks (Teichmüller 1955; Elliott 1968, 1985; Broadhurst & Loring 1970).

Crevasse splay facies

Description. Crevasse splay deposits consist of laterally continuous beds of very fine or fine sandstone, often containing comminuted plant debris, which are generally between a few centimetres and 1 m in thickness, but may exceptionally exceed 2 m (Fig. 5, Table 1). The individual sand beds sometimes become finer or siltier upwards. The sand beds are interbedded with siltstones or claystones which may contain lamination, wave or current ripples. The bases of the sandstone beds are generally abrupt, but may be locally erosive. Occasionally, the tops of the sandstone beds show signs of reworking into wave ripples. The main sedimentary structure in the sandstone beds is trough cross-lamination, although cross-bedding may be present at the base of the thicker beds.

At a given locality, a series of crevasse splay sandstones may be stacked vertically and, in many instances, the thickness of individual sandstone beds increases upwards. Detailed examination of the relationships of the sandstone beds can often reveal a pattern of lateral offset stacking (Guion 1984; Williams 1986).

Crevasse splay deposits occur adjacent to channels, forming fan-shaped or lobate bodies up to about 1 km across (Fig. 5) and exhibiting a palaeocurrent pattern that radiates away from the source. Proximal to distal thinning is evident, with individual sandstone beds gradually thinning to zero, or becoming silty distally (Guion 1984). Proximal crevasse splay sandstones are more likely to show erosive bases and intra-formational conglomerates and plant stems and trunks may be present. Trace fossils such as *Planolites* occasionally occur on the bases of sandstone beds, and *Pelecypodichnus*, representing bivalve escape shafts which traverse sandstone beds, are sometimes present, indicating rapid sedimentation (Guion 1984).

Interpretation. Crevasse splay deposits have not been widely recognized from productive Coal Measures of the Pennine coalfields, although they are a relatively common and distinctive facies. Guion (1978, 1984), Williams (1986) and Jones (1992) differentiated crevasse splays as a separate facies, whereas Fielding (1984a, b) and Guion & Fielding (1988) included crevasse splays in medial crevasse splay/minor delta and distal crevasse splay/minor delta facies. An account of the characteristics of Westphalian crevasse splay deposits was given by Guion (1984).

Crevasse splays are formed when water and sediment leave a main channel through a crevasse channel, which is created by breaching of the channel margin. Individual beds of sand represent episodic depositional events from sediment-laden floodwater at high river stage. Each sand bed is the product of a decelerating flow, and the upward-fining of some beds may reflect the waning flow during deposition. The upward thickening of individual crevasse splays within a package may indicate progradation of the splay system into an interchannel area, or progressive enlargement of the feeder crevasse channel. The lateral offset stacking of crevasse splay sands is believed to be caused by individual splays forming slight topographical highs, which controlled the position of deposition of successive splay lobes. The siltstones or claystones interbedded with the crevasse splay sands may be the product of either fine material carried out of the main channel during the flood event, or ambient sedimentation in the lake into which the splays flowed.

Wave reworked tops and wave ripples in siltstones interbedded with the crevasse splay sandstones indicate that wave activity was important in the bodies of water into which the crevasse splays flowed.

Crevasse splays may be traced proximally into crevasse channels (Fielding 1984a, b, 1986; Guion 1984), which may have operated over relatively long periods. Although Fielding described various types of crevasse channels, it is unlikely that these are sufficiently distinctive to be readily recognized without establishing relationships with both the main channel and the associated crevasse splay system. The

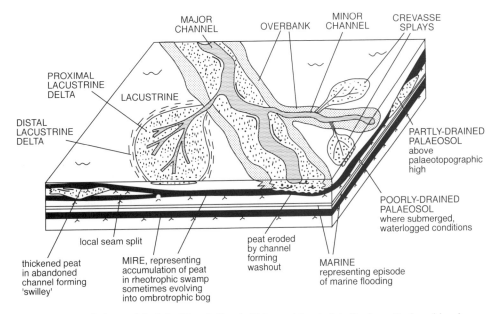

Fig. 10. Schematic facies model of the Westphalian (mid-Langsettian to late Duckmantian) coal-bearing upper delta plain environment, showing relationships of major facies.

differentiation of crevasse splay deposits from lacustrine delta deposits should, however, be possible in many instances. The deposition of crevasse splays from episodic decelerating flows would be expected to generate a series of sharp-based sandstones which may show upward fining, interbedded with siltstones or claystones. Lacustrine delta mouth bars, on the other hand, are generally formed by more continuous flow, and form upward-coarsening sequences resulting from the progradation of the delta system. However, as Fielding (1984b) pointed out, facies showing mixed or intermediate characteristics of crevasse splays and 'minor mouth bars' may be developed.

Organization of facies

The organization of the major facies of the mid-Westphalian A to late Westphalian B 'upper delta plain' depositional environment is summarized in Fig. 10, which is a simplification and synthesis of facies models previously developed by Fulton (1987a), Jones (1992) and Guion & Fulton (1993). Major channels of variable sinuosity were the main pathways of sediment dispersal across the delta plain. These channels fed a hierarchy of minor channels, lacustrine deltas and crevasse splays, depositing sediment in shallow lakes formed as a result of locally enhanced subsidence rates, which were induced in part by the compaction of underlying peat and mud (Guion 1978, 1984; Fielding 1984a, 1986; Guion & Fielding 1988). Infilling of the lakes by sediment and the abandonment of channel systems enabled conditions to be established in which palaeosols (seat earths) and mires could form, leading to coal development. Botanical, edaphic, climatic and hydrological factors were undoubtedly important controls on peat formation and hence seam thickness (Fulton 1987b). However, the immediately pre-existing topography and lithology were also important in influencing seam thickness. Thus on a local scale, seams commonly thicken above underlying abandoned channels and thin above palaeotopographic highs such as channel margins, causing swilleys. On a more regional scale, abandoned major channel deposits may create relatively well-drained, elevated areas, with the result that peat formation starts later above these compared with the surrounding areas, and thus coal seams are thinner above abandoned channel belts. In other instances, however, thicker coals formed above relatively elevated areas and exhibit lateral deterioration into the flanking regions, which were presumed to have been subject to more waterlogged conditions.

Seam splitting is attributed to differential subsidence, which may be the result of either local tectonic influence (e.g. Fulton & Williams 1988) or differential compaction resulting in the switching of depositional systems by avulsion

(Guion & Fielding 1988). A study of the relationship between seam split lines and known faults may enable the controlling mechanism to be identified, although the relationships are often subtle.

The 'upper delta plain' environment in which most of the economic coals formed was generally remote from marine influence, with the exception of the marine incursion represented by the Vanderbeckei (Clay Cross) Marine Band (Fig. 4). The patterns of rapid lateral and vertical variation in both seams and interseam strata indicate that local factors such as avulsion events, gradient advantages, rates of compactionally induced subsidence and local tectonics were more influential on sedimentation than external base level controls. The influence of base level, representing an interplay between tectonic subsidence, eustasy and sediment supply, must not be dismissed completely (see Flint et al., Read, this volume), but its effects are difficult to distinguish at a mine scale, where local controls on sedimentation were much more important in controlling the distribution and properties of the various facies.

Although the sediment was originally of mainly northerly provenance, palaeocurrents are highly variable due to their control by local sedimentary patterns (Guion & Fielding 1988).

The range of depositional environments which occurs from mid-Westphalian B times was similar across most of the Pennine Basin. The overall succession is appreciably thicker near the basin depocentre in the north Staffordshire–Lancashire region, where the subsidence rates were highest (Guion & Fielding 1988). However, the distributive system was able to keep the sedimentary basin topped up throughout, regularly producing telmatic/terrestrial conditions, resulting in the formation of peat mires. Only shallow water depths were attained, and lake-fill sequences ('cycles') were deposited, which, according to Duff & Walton (1962), show a mean thickness of about 25 ft (7.6 m). Duff & Walton (1964) also showed that in the east Pennine coalfields, the number of cycles showed a strong correlation with overall thickness and thus favoured autocyclic controls on sedimentation.

One of the major controls on peat formation is believed to be the optimum rates of overall subsidence, such that the accumulation of peat could keep pace with subsidence (Fulton 1987b), and seams sufficiently thick to be considered economically viable were generated. Some of the seams, such as the Top Hard group and its equivalents, are laterally continuous over hundreds or thousands of square kilometres, although complex splitting patterns of its leaves are common (Elliott 1968). Optimum subsidence rates were prolonged at the margins of the Pennine Basin (e.g. in Staffordshire and Warwickshire), enabling the formation of thick multi-leaf coals, such as the Warwickshire Thick Coal (Wills 1956; Butterworth 1964; Fulton 1987b; Fulton & Williams 1988).

Conclusions

An understanding of the overall depositional environment and palaeogeography, together with their controls has allowed potential new reserves to be identified, and conversely new exploration has enhanced our understanding of the sedimentology and palaeogeography of the deposits.

Studies of the Westphalian A and B coal-bearing strata of the Pennine Basin have enabled a number of sedimentary facies to be differentiated, which should be recognizable with a reasonable degree of consistency in boreholes and mine workings. The processes that resulted in the deposition of these facies, together with their geometries and interrelationships, are beginning to be understood, but there is undoubted scope for an improvement in our knowledge of this area.

We thank British Coal for access to data, and approval for publication by M. J. Allen, Chief Geologist, British Coal Corporation. Numerous people within British Coal, including geologists and mining staff, have facilitated access to records and colliery workings. N. S. Jones publishes with the permission of the Director of the British Geological Survey (NERC). Oxford Brookes University provided financial and technical assistance to enable the completion of this paper. J. Rippon, C. Watkins and R. Brown are thanked for suggested improvements to the manuscript. The views expressed are those of the authors, and not necessarily those of British Coal.

References

ALLEN, J. R. L. 1965. A review of the origin and characteristics of recent alluvial sediments. *Sedimentology*, **5**, 91–180.

ANDERSON, J. A. R. 1964a. The structure and development of the peat swamps of Sarawak and Brunei. *Journal of Tropical Geography*, **18**, 7–16.

—— 1964b. Observations on climatic damage in peat swamp forest in Sarawak. *Commonwealth Foresty Review*, **43**, 145–158.

—— 1983 Tropical peat swamps of Western Malesia. *In:* GORE, A. J. P. (ed.) *Ecosystems of the World*, Vol. 4B. *Mires: Swamp, Bog, Fen and Moor*,

Regional Studies. Elsevier, Amsterdam, 181–199.

—— & MULLER, J. 1975. Palynological study of a Holocene peat and a Miocene coal deposit from NW Borneo. *Review of Palaeobotany and Palynology,* **19,** 291–351.

ARTHURTON, R. S. 1984. The Ribblesdale fold belt, N.W. England—a Dinantian/early Namurian dextral shear zone. *In:* HUTTON, D. & SANDERSON, D. J. (eds) *Variscan Tectonics of the North Atlantic Region.* Geological Society, London, Special Publication, **14,** 131–138.

ASHBY, D. A. & PEARSON, M. J. 1979. Mineral distribution in sediments associated with Alton Marine Band near Penistone, South Yorkshire. *In:* MORTLAND, M. M. & FARMER, V. C. (eds) *International Clay Conference 1978,* Developments in Sedimentology, Elsevier, Amsterdam, **27,** 311–321.

BADHAM, J. P. N. 1982. Strike-slip orogens—an explanation for the Hercynides. *Journal of the Geological Society, London,* **139,** 493–504.

BARTRAM, K. M. 1987. Lycopod succession in coals: an example from the Low Barnsley Seam (Westphalian B), Yorkshire, England. *In:* SCOTT, A. C. (ed.) *Coal and Coal-bearing Strata: Recent Advances.* Geological Society, London, Special Publication, **32,** 187–199.

BATES, C. C. 1953. Rational theory of delta formation. *American Association of Petroleum Geologists Bulletin,* **37,** 2119–2162.

BERNER, R. A. & RAISWELL, R. 1984. C/S method for distinguishing freshwater from marine sedimentary rocks. *Geology,* **12,** 365–358.

BESLY, B. M. 1988. Palaeogeographic implications of late Westphalian to early Permian red-beds, Central England. *In:* BESLY, B. M. & KELLING, G. (eds) *Sedimentation in a Synorogenic Basin Complex: the Upper Carboniferous of NW Europe.* Blackie, Glasgow, 200–221.

—— 1990. Carboniferous. *In:* GLENNIE, K. W. (ed.) *Introduction to the Petroleum Geology of the North Sea, 3rd Edn.* Blackwell Scientific, Oxford, 90–119.

—— & FIELDING, C. R. 1989. Palaeosols in Westphalian coal-bearing and red-bed sequences, central and northern England. *Palaeogeography, Palaeoclimatology, Palaeoecology,* **70,** 303–330.

—— & KELLING, G. (eds) 1988. *Sedimentation in a Synorogenic Basin Complex: the Upper Carboniferous of NW Europe.* Blackie, Glasgow.

BIRKELAND, P. W. 1974. *Pedology, Weathering and Geomorphological Research.* Oxford University Press, Oxford.

BLESS, M. J. M., BOUCKAERT, J. & 8 others. 1980. Pre-Permian sedimentation in NW Europe. Pre-Permian depositional environments around the Brabant Massif in Belgium, the Netherlands and Germany. *Sedimentary Geology,* **27,** 1–81.

BROADHURST, F. M. 1964. Some aspects of the palaeoecology of non-marine faunas and rates of sedimentation in the Lancashire Coal Measures. *American Journal of Science,* **262,** 858–869.

—— & LORING D. H. 1970. Rates of sedimentation in the Upper Carboniferous of Britain. *Lethaia,* **3,** 1–9.

BURN, M. J. 1990. *Controls on the geometry of Carboniferous deltas: east Fife, U.K. and east Kentucky, U.S.A.* PhD Thesis, Oxford Polytechnic.

BUSTIN, R. M., CAMERON, A. R., GRIEVE, D. A. & KALKREUTH, W. D. 1985. *Coal Petrology: its Principles, Methods and Applications (Revised Edition).* Geological Society of Canada, Short Course Notes, **3.**

BUTTERWORTH, M. A. 1964. The distribution of *Densosporites sphaerotriangularis* in the Westphalian B of the Pennine Coalfields in England. *Fortschritte in der Geologie von Rheinland und Westfalen,* **12,** 317–330.

CADY, G. H. 1942. Modern concepts of the physical constitution of coal. *Journal of Geology,* **50,** 337–356.

CALVER, M. A. 1968a. Distribution of marine faunas in Northern England and adjoining areas. *Proceedings of the Yorkshire Geological Society,* **37,** 1–72.

—— 1968b. Coal Measures invertebrate faunas. *In:* MURCHISON, D. G. & WESTOLL, T. S. (eds) *Coal and Coal-bearing Strata.* Oliver & Boyd, Edinburgh, 147–177.

—— 1969. Westphalian of Britain. *Compte Rendu du 6me Congrès International de Stratigraphie et de Géologie du Carbonifère, Sheffield 1967,* **1,** 233–254.

—— 1973. Marine faunas of the Westphalian B and C of the British Pennine Coalfields. *Compte Rendu du 7me Congrès International de Stratigraphie et de Géologie du Carbonifère, Krefeld 1971,* **2,** 253–265.

CASAGRANDE, D. J. 1987. Sulphur in peat and coal. *In:* SCOTT, A. C. (ed.) *Coal and Coal-bearing Strata: Recent Advances.* Geological Society, London, Special Publication, **32,** 87–105.

CECIL, C. G., STANTON, R. W., DULONG, F. R. & RENTON, J. J. 1979. Some geologic factors controlling mineral matter in coal. *In:* DONALDSON, A. C., PRESLEY, M. M. & RENTON, J. J. (eds) *Carboniferous, Short Course and Guidebook.* American Association of Petroleum Geologists, Field Seminar, West Virginia Geological and Economic Survey Bulletin, **B-37-3,** 43–56.

CHISHOLM, J. I. 1990. The Upper Band–Better Bed sequence (Lower Coal Measures, Westphalian A) in the central and south Pennine area of England. *Geological Magazine,* **127,** 55–74.

CHOU, C-L. 1990. Geochemistry of sulfur in coal. *In:* ORR, W. L. & WHITE, C. M. (eds) *Geochemistry of Sulfur in Fossil Fuels.* ACS Symposium Series, **249,** 30–52.

CLARKE, A. M. 1963. A contribution to the understanding of washouts, swalleys, splits and other seam variations and the amelioration of their effects on mining in south Durham. *Transactions of the Institution of Mining Engineers,* **122,** 667–706.

CLEAL, C. J. 1987. Macrofloral biostratigraphy of the Newent Coalfield, Gloucestershire. *Geological Journal,* **22,** 207–217.

CLIFT, S. G. & LEEK, J. E. 1943. Marine strata in the Notts. and Debys. Coalfield with particular reference to the Clay Cross Marine Band. *Transactions of the Institution of Mining Surveyors*, **23**, 98–104.

CLYMO, R. S. 1983. Peat. *In:* GORE, A. J. P. (ed.) *Ecosystems of the World*, Vol. 4A. *Mires: Swamp, Bog, Fen and Moor, General Studies.* Elsevier, Amsterdam, 159–224.

—— 1987. Rainwater-fed peat as a precursor of coal. *In:* SCOTT, A. C. (ed.) *Coal and Coal-bearing Strata: Recent Advances.* Geological Society, London, Special Publication, **32**, 17–23.

COLEMAN, J. M. 1966. Ecological changes in a massive freshwater clay sequence. *Transactions of the Gulf Coast Association of Geological Societies*, **16**, 159–164.

—— & PRIOR, D. B. 1980. *Deltaic Sand Bodies.* American Association of Petroleum Geologists, Course Note Series, **15**.

COLLIER, R. E. Ll., LEEDER, M. R. & MAYNARD, J. R. 1990. Transgressions and regressions: a model for the influence of tectonic subsidence, deposition and eustacy, with application to Quaternary and Carboniferous examples. *Geological Magazine*, **127**, 117–128.

COLLINSON, J. D. 1970. Bedforms of the Tana River, Norway. *Geografiska Annaler*, **52**, 31–56.

—— 1988. Controls on Namurian sedimentation in the Central Province basins of northern England. *In:* BESLY, B. M. & KELLING, G. (eds) *Sedimentation in a Synorogenic Basin Complex: the Upper Carboniferous of NW Europe.* Blackie, Glasgow, 85–101.

——, JONES, C. M., BLACKBOURN, G. A., BESLY, B. M., ARCHARD, G. M. & McMAHON, A. H. 1993. Carboniferous depositional systems of the Southern North Sea. *In:* PARKER, J. R. (ed.) *Petroleum Geology of Northwest Europe: Proceedings of the 4th Conference.* Geological Society, London, 677–687.

D'HOORE, J. 1964. *Soil Map of Africa, 1:5,000,000.* Explanatory Monograph. Commission for Technical Cooperation in Africa, Publication, **93**, Lagos.

DAVIS, A. 1978. Compromise in coal seam description. *In:* DUTCHER, R. R. (ed.) *Field Description of Coal.* American Society for Testing and Materials, Special Technical Publication, **661**, 33–40.

DEMAISON, G. J. & MOORE, G. T. 1980. Anoxic environments and oil source bed genesis. *American Association of Petroleum Geologists Bulletin*. **64**, 1179–1209.

DEWEY, J. F. 1982. Plate tectonics and the evolution of the British Isles. *Journal of the Geological Society, London*, **139**, 371–414.

DISON, I. & WHITWORTH, K. R. 1985. Downhole logging: extracting information from boreholes. *Colliery Guardian*, **233**, 400–408

DOBSON, J. & KINGHORN, R. R. F. 1987. Dispersed sedimentary organic matter in Coal Measures horizons, East Midlands, UK. *Journal of Petroleum Geology*, **10**, 453–474.

DONALDSON, A. C., MARTIN, R. H. & KANES, W. H. 1970. Holocene Guadalupe Delta of Texas. *In:* MORGAN, J. P. (ed.) *Deltaic Sedimentation Modern and Ancient.* Society of Economic Paleontologists and Mineralogists, Special Publication, **15**, 107–137.

DUCHAUFOUR, P. 1982. *Pedology*, Allen and Unwin, London.

DUFF, P. McL. D. & WALTON, E. K. 1962. Statistical basis for cyclothems: a quantitative study of the sedimentary succession in the East Pennine Coalfield. *Sedimentology*, **1**, 235–255.

—— & —— 1964. Trend surface analysis of sedimentary features of the Modiolaris Zone, East Pennine Coalfield, England. *In:* VAN STRAATEN, L. M. J. U. (ed.) *Deltaic and Shallow Marine Deposits.* Developments in Sedimentology, **1**, Elsevier, Amsterdam, 114–122.

DUNHAM, K. C. & POOLE, E. G. 1974. The Oxfordshire coalfield. *Journal of the Geological Society, London*, **130**, 387–391.

DU RIETZ, E. 1954. Die Mineral Bodenwasserziegergrenze als Grundlage einer Naturlichen Zweigleiderung der Nord und Mitteleuropaischen Moore. *Vegetatio*, **5/6**, 571–585.

EAGAR, R. M. C. 1960. A summary of the results of recent work on the palaeoecology of Carboniferous non-marine lamellibranchs. *Compte Rendu du 4me Congrès International de Stratigraphie et de Géologie du Carbonifère, Heerlen 1958*, **1**, 137–149.

—— 1974. Shape of shell of *Carbonicola* in relation to burrowing. *Lethaia*, **7**, 219–238.

—— 1987. The shape of the Upper Carboniferous non-marine bivalve *Anthraconaia* in relation to the organic carbon content of the host sediment. *Transactions of the Royal Society of Edinburgh: Earth Sciences*, **78**, 177–195.

——, BAINS, J. G., COLLINSON, J. D., HARDY, P. G., OKOLO, S. A. & POLLARD, J. E. 1985. Trace fossil assemblages and their occurrence in Silesian (mid-Carboniferous) deltaic sediments in the Central Pennine Basin, England. *In:* CURRAN, H. A. (ed.) *Biogenic Structures: their Use in Interpreting Depositional Environments.* Society of Economic Paleontologists and Mineralogists, Special Publication, **35**, 99–149.

EDEN, R. A., STEVENSON, I. P. & EDWARDS, W. 1957. *Geology of the Country around Sheffield*, Memoir, Geological Survey of the United Kingdom.

EDWARDS, W. 1951. *The Concealed Coalfield of Yorkshire and Nottinghamshire*, 3rd Edn. Memoir, Geological Survey of Great Britain.

—— & STUBBLEFIELD, C. J. 1948. Marine bands and other faunal marker-horizons in relation to the sedimentary cycles of the Middle Coal Measures of Nottinghamshire and Debyshire. *Quarterly Journal of the Geological Society of London*, **103**, 209–260.

ELLIOTT, R. E. 1965. Swilleys in the Coal Measures of Nottinghamshire interpreted as palaeo-river courses. *Mercian Geologist*, **1**, 133–142.

—— 1968. Facies, sedimentation successions and cyclothems in productive Coal Measures in the East Midlands, Great Britain. *Mercian Geologist*,

2, 351–372.

—— 1969. Deltaic processes and episodes: the interpretation of productive Coal Measures occurring in the East Midlands, Great Britain. *Mercian Geologist*, **3**, 111–135.

—— (ed.) 1984. *Procedures in Coal Mining Geology*. National Coal Board Mining Department, London.

—— 1985a. Quantification of peat to coal compaction stages based especially on phenomena in the East Pennine Coalfield, England. *Proceedings of the Yorkshire Geological Society*, **45**, 163–172.

—— 1985b. An interpretation of the trace fossil *Cochlichnus kochi* (Ludwig) from the East Pennine Coalfield of Great Britain. *Proceedings of the Yorkshire Geological Society*, **45**, 183–187.

EVANS, D. J., MENEILLY, A. & BROWN, G. 1992. Seismic facies analysis of Westphalian sequences of the southern North Sea. *Marine and Petroleum Geology*, **9**, 578–589.

FAO/UNESCO 1974. *Unesco Soil Map of the World, 1:5,000,000*, Vol. 1. Legend, Ch. 3. The Soil Units. UNESCO, Paris.

FIELDING, C. R. 1984a. A coal deposition model for the Durham Coal Measures of N.E. England. *Journal of the Geological Society, London*, **141**, 919–931.

—— 1984b. Upper delta plain lacustrine and fluviolacustrine facies from the Westphalian of the Durham Coalfield, N.E. England. *Sedimentology*, **31**, 547–567.

—— 1986. Fluvial channel and overbank deposits from the Westphalian of the Durham Coalfield, N.E. England. *Sedimentology*, **33**, 119–140.

FISK, H. N. 1947. *Fine Grained Alluvial Deposits and Their Effects on Mississippi River Activity*. Mississippi River Commission, Vicksburg.

FLINT, S., AITKEN, J. & HAMPSON, G. 1995. The application of sequence stratigraphy to coal-bearing coastal plain successions: implications for the UK Coal Measures. *In:* WHATELEY, M. K. G. & SPEARS, D. A. (eds) *European Coal Geology*. Geological Society, London, Special Publication, **82**, 1–16.

FOSTER, D., HOLLIDAY, D. W., JONES, C. M., OWENS, B. & WELSH, A. 1989. The concealed rocks of Berkshire and South Oxfordshire. *Proceedings of the Geologists' Association*, **100**, 395–407.

FRASER, A. J., NASH, D. F., STEELE, R. P. & EBDON, C. C. 1990. A regional assessment of the intra-Carboniferous play of Northern England. *In:* BROOKES, J. (ed.) *Classic Petroleum Provinces*. Geological Society, London, Special Publication, **50**, 417–440.

FRAZIER, D. E. & OSANIK, A. 1969. Recent peat deposits—Louisiana coastal plain. *In:* DAPPLES, E. C. & HOPKINS, M. E. (eds) *Environment of Coal Deposition*. Geological Society of America, Special Paper, **114**, 63–85.

FULTON, I. M. 1987a. *The Silesian sub-system in Warwickshire, some aspects of its palynology, sedimentology and stratigraphy*. PhD Thesis, University of Aston.

—— 1987b. Genesis of the Warwickshire Thick Coal: a group of long-residence histosols. *In:* SCOTT, A. C. (ed.) *Coal and Coal-bearing Strata: Recent Advances*. Geological Society, London, Special Publication, **32**, 201–218.

—— 1992. *Field Manual of Sedimentological Logging of Coal Measures*. British Coal, Technical and Research Executive, Bretby.

—— & WILLIAMS, H. 1988. Palaeogeographical change and controls on Namurian and Westphalian A/B sedimentation at the southern margin of the Pennine Basin, Central England. *In:* BESLY, B. M. & KELLING, G. (eds) *Sedimentation in a Synorogenic Basin Complex: the Upper Carboniferous of NW Europe*. Blackie, Glasgow, 178–199.

——, GUION, P. D. & JONES, N. S. 1995. The application of sedimentology to deep-mined coal. *In:* WHATELEY, M. K. G. & SPEARS, D. A. (eds) *European Coal Geology*. Geological Society, London, Special Publication, **82**, 17–43.

GAYER R. A. & PESEK, J. 1992. Cannibalisation of Coal Measures in the South Wales Coalfield—significance for foreland basin evolution. *Proceedings of the Ussher Society*, **8**, 44–49.

GORE, A. J. P. (ed.) 1983. *Ecosystems of the World*, Vols 4A, B. Mires: Swamp, Bog, Fen and Moor, General and Regional Studies. Elsevier, Amsterdam.

GRAYSON, R. F. & OLDHAM, L. 1987. A new structural framework for the northern British Dinantian as a basis for oil, gas and mineral exploration. *In:* MILLER J., ADAMS, A. E. & WRIGHT, V. P. (eds) *European Dinantian Environments*. Wiley, Chichester, 33–60.

GUION, P. D. 1978. *Sedimentation of interstream strata and some relationships with coal seams in the East Midlands Coalfield*. PhD Thesis, City of London Polytechnic.

—— 1984. Crevasse splay deposits and roof-rock quality in the Threequarters Seam (Carboniferous) in the East Midlands Coalfield, U.K. *In:* RAHMANI, R. A. & FLORES, R. M. (eds) *Sedimentology of Coal and Coal-bearing Sequences*. International Association of Sedimentologists, Special Publication, **7**, 291–308.

—— 1987a. Palaeochannels in mine workings in the High Hazles Coal (Westphalian B), Nottinghamshire Coalfield, England. *Journal of the Geological Society, London*, **144**, 471–488.

—— 1987b. The influence of a palaeochannel on seam thickness in the Coal Measures of Derbyshire, England. *International Journal of Coal Geology*, **7**, 269–299.

—— 1992. Westphalian. *In:* COPE, J. C. W., INGHAM, J. & RAWSON, P. F. (eds) *Atlas of Palaeogeography and Lithofacies*. Geological Society, London, Memoir, **13**, 80–86.

—— & FIELDING, C. R. 1988. Westphalian A and B sedimentation in the Pennine Basin, U.K. *In:* BESLY, B. M. & KELLING, G. (eds) *Sedimentation in a Synorogenic Basin Complex: the Upper Carboniferous of NW Europe*. Blackie, Glasgow, 153–177.

—— & FULTON, I. M. 1993. The importance of sedimentology in deep-mined coal extraction.

Geoscientist, **3**(2), 25–33.

GUTTERIDGE, P. 1987. Dinantian sedimentation and the basement structure of the Derbyshire dome. *Geological Journal*, **22**, 25–41.

HAANTJENS, H. A. (ed.) 1968. *Lands of the Wewak–Lower Sepik area, Territory of Papua and New Guinea*. CSIRO Australia Land Research Series, **22**.

HASZELDINE, R. S. 1981. *Westphalian B coalfield sedimentology and its regional setting*. PhD Thesis, University of Strathclyde.

—— 1983a. Fluvial bars reconstructed from a deep, straight channel, Upper Carboniferous Coalfield of Northeast England. *Journal of Sedimentary Petrology*, **53**, 1233–1247.

—— 1983b. Descending tabular cross-bed sets and bounding surfaces from a fluvial channel in the Upper Carboniferous coalfield of north-east England. *In:* COLLINSON, J. D. & LEWIN, R. S. (eds) *Modern and Ancient Fluvial Systems*. International Association of Sedimentologists, Special Publication, **6**, 449–456.

—— 1984a. Carboniferous North Atlantic palaeogeography: stratigraphic evidence for rifting, not megashear or subduction. *Geological Magazine*, **121**, 442–463.

—— 1984b. Muddy deltas in freshwater lakes, and tectonism in the Upper Carboniferous coalfield of N.E. England. *Sedimentology*, **31**, 811–822.

—— 1988. Crustal lineaments in the British Isles: their relationship to Carboniferous basins. *In:* BESLY, B. M. & KELLING, G. (eds) *Sedimentation in a Synorogenic Basin Complex: the upper Carboniferous of NW Europe*. Blackie, Glasgow, 53–68.

HOPKINS, J. C. 1985. Channel-fill deposits formed by aggradation in deeply-scoured superimposed distributaries of the Lower Kootenai Formation (Cretaceous). *Journal of Sedimentary Petrology*, **55**, 42–55.

INGRAM, H. A. P. 1978. Soil layers in mires: function and terminology. *Journal of Soil Science*, **29**, 224–227.

JONES, N. S. 1992. *Sedimentology of Westphalian coal-bearing strata and applications to opencast coal mining, West Cumbrian Coalfield, U.K.* PhD Thesis, Oxford Brookes University.

JONES, N. S., GUION, P. D. & FULTON, I. M. 1995. Sedimentology and its applications with the UK opencast coal mining industry. *In:* WHATELEY, M. K. G. & SPEARS, D. A. (eds) *European Coal Geology*. Geological Society, London, Special Publication, **82**, 115–136.

KEARNS, F. L. & DAVISON, A. T. 1983. Field classification system of organic-rich sediments. *In:* RAYMOND, R. & ANDREJKO, M. J. (eds) *Mineral Matter in Peat Workshop, Los Alamos*. Los Alamos National Laboratory, 147–158.

KELLAWAY, G. A. 1970. The Upper Coal Measures of South-west England compared with those of South Wales and the Southern Midlands. *Compte Rendu du 6me Congrès International de Stratigraphie et de Géologie du Carbonifère, Sheffield 1967*, **3**, 1040–1056.

KELLING, G. & COLLINSON, J. D. 1992. Silesian. *In:* DUFF, P. McL. D. & SMITH, A. J. (eds) *Geology of England and Wales*. Geological Society, London, 239–273.

KIRK, M. 1982. Palaeoenvironments and geography in the Westphalian A and B coalfields in Scotland. *Open Earth*, **17**, S36–S37.

—— 1983. *Sedimentology and palaeogeography of the Westphalian A and B coalfields in Scotland*. PhD Thesis, University of Strathclyde.

KNOWLES, B. 1964. The radioactive content of the Coal Measures sediments in the Yorkshire–Derbyshire Coalfield. *Proceedings of the Yorkshire Geological Society*, **34**, 413–450.

LAND, D. H. 1974. *Geology of the Tynemouth District*. Memoir, Geological Survey of the United Kingdom.

LEEDER, M. R. 1982. Upper Palaeozoic basins of the British Isles—Caledonide inheritance versus Hercynian plate margin processes. *Journal of the Geological Society, London*, **139**, 479–491.

—— 1987. Tectonic and palaeogeographic models for lower Carboniferous Europe. *In:* MILLER, J., ADAMS, A. E. & WRIGHT, V. P. (eds) *European Dinantian Environments*. Wiley, Chichester, 1–20.

—— 1988a. Recent developments in Carboniferous geology: a critical review with implications for the British Isles and N.W. Europe. *Proceedings of the Geologists' Association*, **99**, 73–100.

—— 1988b. Devono-Carboniferous river systems and sediment dispersal from the orogenic belts and cratons of NW Europe. *In:* HARRIS, A. L. & FETTES, D. J. (eds) *The Caledonian–Appalachian Orogen*. Geological Society, London, Special Publication, **38**, 549–558.

—— & HARDMAN, M. 1990. Carboniferous geology of the Southern North Sea Basin and controls on hydrocarbon prospectivity. *In:* HARDMAN, R. F. P. & BROOKES, J. (ed.) *Tectonic Events Responsible for Britain's Oil and Gas Reserves*. Geological Society, London, Special Publication, **55**, 87–105.

——, RAISWELL, R., AL-BIATTY, H., McMAHON, A. & HARDMAN, M. 1990. Carboniferous stratigraphy, sedimentation and correlation of well 48/3-3 in the southern North Sea Basin: integrated use of palynology, natural gamma/sonic logs and carbon/sulphur geochemistry. *Journal of the Geological Society, London*, **147**, 287–300.

LIPPOLT, H. J., HESS, J. C. & BURGER, K. 1984. Isotopische alter pyroklastischen Sanidinen aus Kaolin-Kohlentonsteinen als Korrelationsmarken für das mitteleuropäisch Oberkarbon. *Fortschritte in der Geologie von Rheinland und Westfalen*, **32**, 119–150.

LORENZI, G., BOSSIROY, D. & DREESEN, R. 1992. *Les Mineraux Argileux au Service des Corrélations Stratigraphiques des Formations Houillères du Carbonifère*. Rapport Final, Recherche Technique Charbon, Commission des Communautés Européenes.

MECKEL, L. D. 1972. Anatomy of distributary channel-fill deposits in recent mud deltas. *American Association of Petroleum Geologists Bulletin*, **56**, 659.

McCABE, P. J. 1984. Depositional environments of

coal and coal-bearing strata. *In:* RAHMANI, R. A. & FLORES, R. M. (eds) *Sedimentology of Coal and Coal-bearing Sequences.* International Association of Sedimentologists, Special Publication, **7**, 13–42.

MOORE, L. R. 1968. Cannel coals, bogheads and oil shales. *In:* MURCHISON, D. G. & WESTOLL, T. S. (eds) *Coal and Coal-bearing Strata.* Oliver & Boyd, Edinburgh, 20–29.

—— 1987. Ecological and hydrological aspects of peat formation. *In:* SCOTT, A. C. (ed.) *Coal and Coal-bearing Strata: Recent Advances.* Geological Society, London, Special Publication, **32**, 7–15.

—— & BELLAMY, D. J. 1974. *Peatlands.* Paul Elek, London.

MOSES, K. 1981. Britain's coal resources and reserves—the current position. *In: The Watt Committee on Energy. Assessment of Energy Resources.* Report, **9**, HMSO, London, 40–49.

NATIONAL COAL BOARD (NCB) 1972. *Procedure for the Assessment of Reserves.* NCB PI 1972/4, 1–8.

OWENS, B., RILEY, N. J. & CALVER, M. A. 1984. Boundary stratotypes and the new stage names for the lower and middle Westphalian sequences in Britain. *Compte Rendu du 10me Congrès International de Stratigraphie et de Géologie du Carbonifère, Madrid 1983*, **4**, 461–472.

PHILLIPS, T. L. & CECIL, C. B. (eds) 1985. Paleoclimatic controls on coal resources of the Pennsylvanian system of North America. *International Journal of Coal Geology*, **5**, Parts 1, 2.

—— & DIMICHELE, W. A. 1981. Palaeoecology of Mid Pennsylvanian age coal swamps in Southern Illinois Herrin Coal Member at Sahara Mine No. 6. *In:* NICHOLAS, K. J. (ed.) *Paleobotany, Paleoecology and Evolution*, Vol. 1. Praeger, New York, 205–255.

—— & PEPPERS, R. A. 1984. Changing patterns of Pennsylvanian coal-swamp vegetation and implications of climatic control on coal occurrence. *International Journal of Coal Geology*, **3**, 205–255.

——, —— & DIMICHELE, W. A. 1985. Stratigraphic and interregional changes in Pennsylvanian coal swamp vegetation: environmental inferences. *International Journal of Coal Geology*, **5**, 43–109.

POLLARD, J. E. 1969. Three ostracod-mussel bands in the Coal Measures (Westphalian) of Northumberland and Durham. *Proceedings of the Yorkshire Geological Society*, **37**, 239–276.

—— 1988. Trace fossils in coal-bearing sequences. *Journal of the Geological Society, London*, **145**, 339–350.

RAISTRICK, A. & MARSHALL, C. E. 1939. *The Nature and Origin of Coal and Coal Seams.* English Universities Press, London.

RAMSBOTTOM, W. H. C. 1979. Rates of transgression and regression in the Carboniferous of N.W. Europe. *Journal of the Geological Society, London*, **136**, 147–154.

——, CALVER, M. A., EAGAR, R. M. C., HODSON, F., HOLLIDAY, D. W., STUBBLEFIELD, C. J. & WILSON, R. B. 1978. *A Correlation of Silesian Rocks in the British Isles.* Geological Society of London, Special Report, **10**.

READ, W. A. 1995. Sequence stratigraphy and lithofacies geometry in an early Namurian coal-bearing succession in central Scotland. *In:* WHATELEY, M. K. G. & SPEARS, D. A. (eds) *European Coal Geology.* Geological Society, London, Special Publication, **82**, 285–297.

RIPPON, J. H. 1984. The Clowne Seam, Marine Band, and overlying sediments in the Coal Measures (Westphalian B) of North Derbyshire. *Proceedings of the Yorkshire Geological Society*, **45**, 27–43.

—— & SPEARS, D. A. 1989. The sedimentology and geochemistry of the sub-Clowne cycle (Westphalian B) of north-east Derbyshire, U.K. *Proceedings of the Yorkshire Geological Society*, **47**, 181–198.

ROBERTSON, T. 1952. Plant control in rhythmic sedimentation. *Compte Rendu du 3me Congrès International de Stratigraphie et de Géologie du Carbonifère, Heerlen 1951*, **2**, 515–521.

ROWLEY, D. B., RAYMOND, A., PARRISH, J. T., LOTTES, A. L., SCOTESE, C. R. & ZIEGLER, A. M. 1985. Carboniferous paleogeographic, phytogeographic and paleoclimatic reconstructions. *International Journal of Coal Geology*, **5**, 7–42.

SCOTESE, C. R., BAMBACH, R. K., BARTON, C., VAN DER VOO, R. & ZIEGLER, A. M. 1979. Palaeozoic base maps. *Journal of Geology*, **87**, 217–277.

SCOTT, A. C. 1978. Sedimentological and ecological control of Westphalian B plant assemblages from West Yorkshire. *Proceedings of the Yorkshire Geological Society*, **41**, 461–508.

—— 1979. The ecology of Coal Measures floras from northern Britain. *Proceedings of the Geologists' Association*, **90**, 97–106.

—— 1984. Studies on the sedimentology, palaeontology and palaeoecology of the Middle Coal Measures (Westphalian B, Upper Carboniferous) at Swillington, Yorkshire. Part 1. Introduction. *Transactions of the Leeds Geological Association*, **10**, 1–16.

—— 1987. Coal and coal-bearing strata: recent advances and future prospects. *In:* SCOTT, A. C. (ed.) *Coal and Coal-bearing Strata: Recent Advances.* Geological Society, London, Special Publication, **32**, 1–6.

—— & KING, G. R. 1981. Megaspores and coal facies: an example from the Westphalian A of Leicestershire, England. *Review of Palaeobotany and Palynology*, **34**, 107–113.

SMITH, A. H. V. 1957. The sequence of microspore assemblages associated with the occurrence of crassidurite in coal seams of Yorkshire. *Geological Magazine*, **94**, 345–363.

—— 1962. The palaeontology of Carboniferous peats based on the miospores and petrography of bituminous coals. *Proceedings of the Yorkshire Geological Society*, **33**, 423–474.

—— 1964. Zur Petrologie und Palynologie der Kohlenflöze des Karbons und ihrer Begleitschichten. *Fortschritte in der Geologie von Rheinland und Westfalen*, **12**, 285–302.

—— & BUTTERWORTH, M. 1967. *Miospores of the Carboniferous of Great Britain.* Special Papers in Palaeontology, **1**, Palaeontological Association,

London.

SMITH, E. G., RHYS, G. H. & EDEN, R. A. 1967. *Geology of the Country around Chesterfield, Matlock and Mansfield*, Memoir, Geological Survey of the United Kingdom.

SOIL SURVEY STAFF 1975. *Soil Taxonomy*. Agriculture Handbook 436, US Department of Agriculture, Washington.

SPEARS, D. A. 1964. The radioactivity of the Mansfield Marine Band, Yorkshire. *Geochimica et Cosmochimica Acta*, **28**, 673–681.

—— 1969. A laminated marine shale of Carboniferous age from Yorkshire, England. *Journal of Sedimentary Petrology*, **39**, 106–112.

—— & SEZGIN, H. I. 1985. Mineralogy and geochemistry of the Subcrenatum Marine Band and associated coal-bearing sediments, South Yorkshire. *Journal of Sedimentary Petrology*, **55**, 570–578.

STACH, E., MACKOWSKY, M-TH., TEICHMÜLLER, M., TAYLOR, G. H., CHANDRA, D. & TEICHMÜLLER, R. 1982. *Stach's Textbook of Coal Petrology, 3rd Edn*. Gebruder Borntraeger, Berlin.

STAUB, J. R. & COHEN, A. D. 1979. The Snuggedy Swamp of Carolina: a backbarrier coal forming environment. *Journal of Sedimentary Petrology*, **49**, 133–144.

STEELE, R. P. 1988. The Namurian sedimentary history of the Gainsborough Trough. *In:* BESLY, B. M. & KELLING, G. (eds) *Sedimentation in a Synorogenic Basin Complex: the Upper Carboniferous of NW Europe*. Blackie, Glasgow, 102–113.

STEPHENSON, P. H. 1985. UK exploration—Selby Coalfield. *Colliery Guardian*, **233**, 374–378.

STRANK, A. R. E. 1987. The stratigraphy and structure of Dinantian strata in the East Midlands, U.K. *In:* MILLER, J., ADAMS, A. E. & WRIGHT, V. P. (eds) *European Dinantian Environments*. Wiley, Chichester, 157–175.

TATE, M. P. & DOBSON, M. R. 1989. Pre-Mesozoic geology of the western and north-western Irish continental shelf. *Journal of the Geological Society, London*, **146**, 229–240.

TAYLOR, R. K. 1971. The petrography of the Mansfield Marine Band cyclothem at Tinsley Park, Sheffield. *Proceedings of the Yorkshire Geological Society*, **38**, 299–328.

TEICHMÜLLER, M. & TEICHMÜLLER, R. 1982. The geological basis of coal formation. *In:* STACH, E., MACKOWSKY, M-TH., TEICHMÜLLER, M., TAYLOR, G. H., CHANDRA, D. & TEICHMÜLLER, R. 1982. *Stach's Textbook of Coal Petrology, 3rd Edn*. Gebruder Borntraeger, Berlin, 5–86.

TEICHMÜLLER, R. 1955. Uber Kustenmoore der Gegenwart und die Moore des Ruhrkarbons. *Geologisches Jahrbuch*, **71**, 197–220.

THOMAS, R. G., SMITH, D. G., WOOD, J. M., VISSER, J., CALVERLEY-RANGE, E. A. & KOSTER, E. H. 1987. Inclined heterolithic stratification, description, interpretation and significance. *Sedimentary Geology*, **53**, 123–179.

TRUEMAN, A. E. 1946. Stratigraphical problems in the Coal Measures of Europe and North America. *Quarterly Journal of the Geological Society of London*, **102**, xlix–xciii.

—— 1947. Stratigraphical problems in the Coal Measures of Great Britain. *Quarterly Journal of the Geological Society of London*, **103**, lxv–civ.

TUBB, S. R., SOULSBY, A. & LAWRENCE, S. R. 1986. Palaeozoic prospects on the northern flanks of the London-Brabant Massif. *In:* BROOKS, J., GOFF, J. C. & VAN HOORN, B. (eds) *Habitat of Palaeozoic Gas in N.W. Europe*. Geological Society, London, Special Publication, **23**, 55–72.

TURNER, B. R. & O'MARA, P. T. 1993. Fluvial channel types in the Coal Measures of northeast England: implications for hydrocarbon exploration in the southern North Sea. *Keynote Addresses and Abstracts, 5th International Conference on Fluvial Sedimentology*. University of Queensland, Brisbane, 135.

VAN WALLENBURG, C. 1973. Hydromorphic soil characteristics in alluvial soils in connection with soil drainage. *In:* SCHLICHTING, E. & SCHWERTMANN, V. (eds) *Pseudogley and Gley. Transactions of Commissions V and VI of the International Society of Soil Science*. Verlag Chemie, Nova Scotia, 393–493.

VAN WIJHE, D. H. & BLESS, M. J. 1974. The Westphalian of the Netherlands with special reference to miospore assemblages. *Geologie en Mijnbouw*, **53** 295–328.

WHITTAKER, A., HOLLIDAY, D. W. & PENN, I. E. 1985. *Geophysical Logs in British Stratigraphy*. Geological Society, London, Special Report, **18**.

WILLIAMS, H. 1986. *The sedimentology of the Upper Westphalian A in the South Derbyshire Coalfield*. PhD Thesis, Oxford Polytechnic.

WILLS, L. J. 1951. *A Palaeogeographical Atlas of the British Isles and Adjacent Parts of Europe*. Blackie, Glasgow.

—— 1956. *Concealed Coalfields*. Blackie, Glasgow.

WING, S. L. 1984. Relation of paleovegetation to geometry and cyclicity of some fluvial carbonaceous deposits. *Journal of Sedimentary Petrology*, **54**, 52–66.

WORKING GROUP ON THE ORIGIN AND NATURE OF PALAEOSOLS 1971. Criteria for the recognition and classification of palaeosols. *In:* YAALON, D. H. (ed.) *Paleopedology*. International Society of Soil Science and Israel University Press, Jerusalem, 153–158.

WRIGHT, L. D. 1977. Sediment transport and deposition at river mouths: a synthesis. *American Association of Petroleum Geologists Bulletin*, **88**, 857–868.

YOUNG, A. 1976. *Tropical Soils and Soil Survey*. Cambridge University Press, Cambridge.

Discrimination of regionally extensive coals in the Upper Carboniferous of the Pennine Basin, UK using high resolution sequence stratigraphic concepts

GARY HAMPSON

Department of Earth Sciences, University of Liverpool, PO Box 147, Liverpool L69 3BX, UK

Abstract: During the late Carboniferous, the Pennine Basin, UK was a thermally subsiding, intracratonic basin with little active tectonism. The sedimentary fill of the basin consists of coarsening-upward coal-bearing deltaic cyclothems bounded by widespread faunal-concentrate condensed horizons. The Yeadonian (Namurian G_1) Rough Rock Group comprises three such cyclothems that have been studied at outcrop and in core using detailed sedimentological logging and then applying high resolution sequence stratigraphical concepts. Key surfaces and systems tracts are recognized and a sequence stratigraphic framework is constructed using these criteria and incorporating basin biostratigraphy and palaeogeography.

Two regionally extensive coals are present in the Rough Rock Group. They formed under relatively long-lived, basin-wide conditions of (1) clastic sediment starvation and (2) rising water-table, creating accommodation space for peats to accumulate and be preserved in mires. These conditions are characteristic of deposition in transgressive systems tracts and in the context of the Rough Rock group sequence stratigraphic framework these coals are identified as up-dip equivalents of initial flooding surfaces. This example raises wider implications for the discrimination of other regionally extensive coals in the upper Carboniferous strata of the Pennine basin using high resolution sequence stratigraphic concepts.

The Upper Carboniferous Millstone grit and Coal Measures of the Pennine Basin, UK contain many regionally extensive coals. The youngest Millstone Grit strata belong to the coal-bearing Rough Rock Group, which represents a transition from typical Millstone Grit strata to typical Coal Measures strata. The Rough Rock Group is defined by the *Gastrioceras cancellatum* marine band at its base and the *Gastrioceras subcrenatum* marine band at its top, and represents deposition during the Yeadonian (Namurian G_1) stage. The Rough Rock Group is extensively exposed on both limbs of the Pennine anticline (Fig. 1a) in quarries, natural crags and streams. Data from the East Midlands Oilfield (Fig. 1a) in the form of core samples taken at 1 ft (0.3048 m) intervals and original loggers' notes describe sections through the Rough Rock Group. One continuous cored section in this area is taken from the IGS borehole at Nether Heage [GR SK 360 512]. Elsewhere scattered information is available in the form of borehole and coal-mining records, which are summarized in the relevant British Geological Survey memoirs (Wright *et al.* 1927; Wray *et al.* 1930; Bromehead *et al.* 1933; Mitchell *et al.* 1947; Edwards *et al.* 1950; Stephens *et al.* 1953; Eden *et al.* 1957; Earp *et al.* 1961; Price *et al.* 1963; Edwards 1967; Smith *et al.* 1967; Evans *et al.* 1968; Stevenson & Gaunt 1971; Smith *et al.* 1973; Frost & Smart 1979; Aitkenhead *et al.* 1985; Chisholm *et al.* 1988).

In recent years high resolution sequence stratigraphic concepts have been successfully applied to several successions of coal-bearing strata at outcrop (Arditto 1987; Shanley & McCabe 1991; McCabe 1993; Aitken & Flint 1994; Flint *et al.* this volume). These studies have integrated existing concepts regarding environments of coal deposition into a sequence stratigraphical framework.

The aims of this paper are (1) to document two regionally extensive coals in the Rough Rock group (2) provide a facies analysis and interpretation of the group based on detailed sedimentological logging of the available sections and (3) to use these interpretations to construct a self-consistent high resolution sequence stratigraphic framework in which regionally

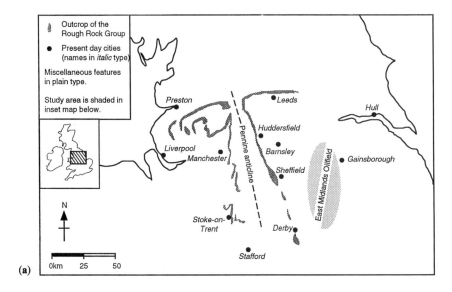

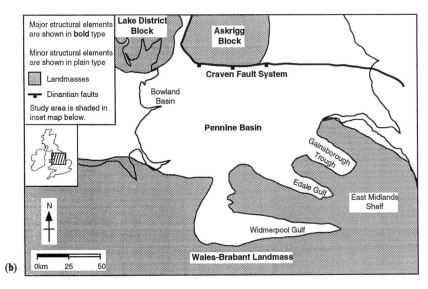

Fig. 1. (a) Simplified outcrop map of the Rough Rock Group showing the location of the study area. (b) Simplified map of the Namurian palaeogeography of the Pennine Basin and adjacent areas. After Collinson (1988).

extensive coals may be discriminated, applying the concepts of Posamentier et al. (1988), Posamentier & Vail (1988), Van Wagoner et al. (1990) and Mitchum & Van Wagoner (1991).

Geological setting

The Pennine Basin of northern England was one of a series of linked, intracratonic basins initiated during the Dinantian by extensional rifting. The basin was bounded by the Craven Fault System to the north and the Wales–Brabant land mass to the south (Fig. 1b). Rifting stopped in the Dinantian and during the Namurian and early Westphalian the basin underwent thermal subsidence with very little active tectonism. Within the study area active tectonism was largely confined to the East Midlands Shelf (Edwards 1967; Smith et al. 1973) (Fig. 1b). The basin was infilled by a series of delta systems in the lower and middle Namurian, each of which advanced progres-

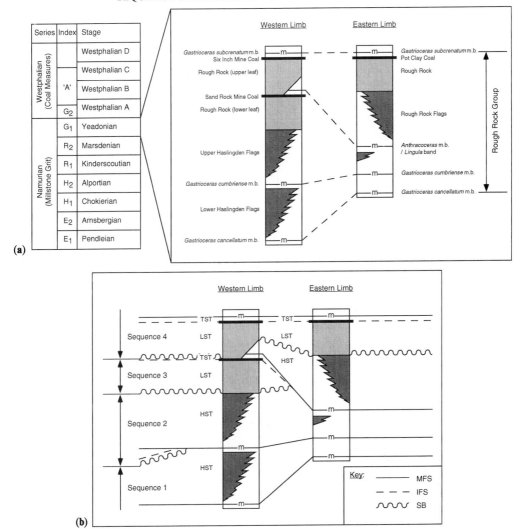

Fig. 2. (a) Schematic stratigraphy of the Rough Rock Group along the northern margin of the Pennine Basin. The stratigraphic columns represent areas on the western and eastern limbs of the Pennine anticline. No vertical scale is implied. (b) Schematic version of the author's sequence stratigraphic framework for the Rough Rock Group along the northern basin margin. Abbreviations: LST, lowstand systems tract; TST, transgressive systems tract; HST, highstand systems tract; SB, sequence boundary; IFS, initial flooding surface; and MFS, maximum flooding surface.

sively further south into the basin (Collinson 1988; Collinson *et al.* 1977; Jones 1980; Martinsen 1993). The spatial arrangement of these delta systems was strongly influenced by the inherited 'basin and block' bathymetry. Net sedimentation rates were very high and by the late Namurian the inherited bathymetry was almost completely infilled and deposition took place in a shallow depression with gently dipping ramp margins. The Rough Rock Group and overlying Coal Measures were deposited in this setting.

The basin-fill includes basin-wide faunal-concentrate condensed horizons referred to as marine bands and *Lingula* bands by previous workers (Bisat 1924; Calver 1968; Wignall 1987). These horizons bound cyclothems, or units of cyclic sedimentation, within the basin-fill. Many marine bands bear distinctive goniatite faunas which provide a biostratigraphic framework with an average time resolution of *c*. 180 ka during the Namurian (Holdsworth & Collinson 1988). Throughout the late Carboniferous the Pennine Basin lay within a broad non-seasonal,

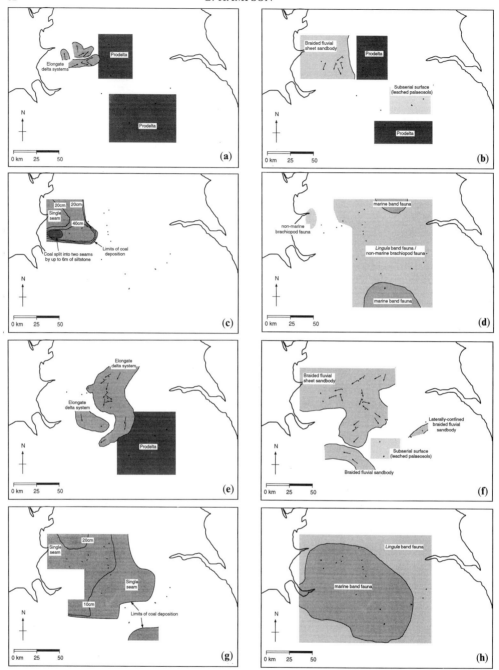

Fig. 3. Palaeogeographic reconstructions, with mean palaeocurrent directions, of the Pennine Basin are shown at the time of deposition of the (**a**) Upper Haslingden Flags (highstand systems tract, sequence 2); (**b**) lower leaf of the Rough Rock (lowstand systems tract, sequence 3); (**c**) Sand Rock Mine Coal (initial flooding surface, sequence 3); (**d**) *Anthracoceras* marine band (maximum flooding surface, sequence 3); (**e**) Rough Rock Flags (highstand systems tract, sequence 3); (**f**) upper leaf of the Rough Rock (lowstand systems tract, sequence 4); (**g**) Six Inch Mine Coal/Pot Clay Coal (initial flooding surface, sequence 4); (**h**) *Gastrioceras subcrenatum* marine band (maximum flooding surface, sequence 4). Locations of logged sections are also shown. Additional data were obtained from the relevant BGS memoirs, Bristow (1988), Calver (1968), Collinson & Banks (1975), Smith (1967) and Steele (1988).

humid, equatorial climatic belt. However, the late Carboniferous was a time of glaciation with extensive polar ice-caps which are inferred to have behaved in an analogous way to their Quaternary counterparts. Retreat and advance of the ice-caps resulted in high magnitude glacio-eustatic sea level fluctuations at several frequencies. Previous workers have interpreted the cyclic nature of sedimentation to have reflected these glacio-eustatic sea level fluctuations (Ramsbottom 1977; Collinson 1988; Holdsworth & Collinson 1988). High frequency fluctuations in particular appear to have had amplified magnitudes (Collier 1990; Maynard & Leeder 1992), suggesting that a strongly enhanced, high order sequence stratigraphic signal may be detected in the strata.

The Rough Rock Group contains regionally extensive sandstone sheets (the Rough Rock) overlying overall coarsening-upward siltstone and sandstone bodies (the Lower Haslingden Flags, Upper Haslingden Flags and Rough Rock Flags). Previous workers have studied the sedimentology of these stratigraphic units (Collinson & Banks 1975; Miller 1986; Bristow 1988; Bristow & Meyers 1989), whereas the most recent research has started to apply the concepts of sequence stratigraphy to the succession (Read 1991; Church 1992; Maynard 1992). The generalized stratigraphy of the group is summarized in Fig. 2.

Coal stratigraphy in the Rough Rock Group

On the western limb of the Pennine anticline the Rough Rock Group contains two regionally extensive coals: the Sand Rock Mine Coal and the Six Inch Mine Coal (Fig. 2). On the eastern limb only the Pot Clay Coal (also known locally as the Cottingley Crow Coal and the Thin Coal; Stephens et al. 1953) is regionally extensive, but there is at least one locally confined unnamed coal. Exposures of the coals are rare and most information about them comes from coal-mining records. The Sand Rock Mine Coal and Six Inch Mine Coal have both been mined in the Lancashire Coalfield and the Pot Clay Coal has been exploited locally where exposed.

The Sand Rock Mine Coal occurs as a single seam up to 50 cm thick throughout most of its extent (Fig. 3c). South of Preston the seam thickens considerably, locally reaching a maximum thickness of 1.5 m, and splits. Here the two leaves of the seam are separated by 'argillaceous measures up to 21 feet thick' over a lateral extent of about 7 km (Price et al. 1963: 23; fig. 6). The Sand Rock Mine Coal directly overlies the lower leaf of the Rough Rock and is directly overlain by a condensed shale horizon bearing a non-marine bivalve fauna (Price et al. 1963: 29; Smith 1967: 107), which the author correlates with an *Anthracoceras*-bearing shale (Fig. 2a, *Anthracoceras* marine band) reported to lie further east by Edwards et al. (1950) and Stephens et al. (1953: 65). Throughout much of the study area this marine band equivalent is cut out by the upper leaf of the Rough Rock, which thus directly overlies the Sand Rock Mine Coal (Fig. 2). The area extent of the Sand Rock Mine Coal is at least 2600 km^2.

The Six Inch Mine Coal occurs as a single, thin seam, locally up to to 35 cm thick (Fig. 3g). The Pot Clay Coal occurs as a single seam of poor quality coal ('cannel') or coaly shale up to 20 cm thick which thins to the south (Wray et al. 1930; Edwards et al. 1950; Stephens et al. 1953; Eden et al. 1957; Frost & Smart, 1979; Aitken et al. 1985) (Fig. 3g). The Six Inch Mine Coal and Pot Clay Coal both lie at the same stratigraphic horizon, directly overlying the Rough Rock and directly underlying the *G. subcrenatum* marine band (Fig. 2). These coals have long been correlated as belonging to the same seam (Wray et al. 1930; Stephens et al. 1953; Ramsbottom et al. 1978). The areal extent of the combined Six Inch Mine/Pot Clay Coal is at least 8400 km^2.

One unnamed coal is exposed at the top of the Rough Rock Flags in the section at Harden Clough (SE 147 042). This coal is of poor quality and occurs as a single seam 5 cm thick. No other coals at this stratigraphic level have been documented previously, and hence this coal is inferred to be locally confined.

Facies analysis

Three units of linked sedimentation termed cyclothems have been recognized by previous workers in the Rough Rock Group. Each is bounded by either a marine band or a *Lingula* band, which are taken to approximate to time lines. Each cyclothem is therefore assumed to represent approximately the same time interval across the study area, despite lateral changes in lithological character.

The Rough Rock group has been characterized in terms of lithofacies on the basis of lithology, sedimentary structures, bed geometry, bed contacts and fossil content. Lithofacies have been grouped into three associations and the salient features of these associations and their component facies are summarized in Table 1. Many of the lithofacies have been described by

Table 1. *Characteristics of Rough Rock Group lithofacies*

Facies	Interpretation	Lithology	Sedimentary structures	Geometry	Basal contact	Fossils
A1	Marine faunal-concentrate condensed horizon	Very dark grey fissile shale	Flat lamination, pyrite	Sheet	Sharp	Goniatites, pectinoid bivalves, fish
A2	Brackish water faunal-concentrate condensed horizon	Very dark grey fissile shale	Flat lamination	Sheet	Sharp	Non-marine bivalves, restricted-marine brachiopods, fish
A3	Prodelta	Siltstone	Flat lamination, siderite-cemented nodules and horizons, bioturbation	?	Gradational	*Planolites*
B1	Distal delta front	Interlaminated and thinly interbedded siltstone and fine-grained sandstone	Wave-ripple and current ripple cross-lamination in sandstone beds, siderite-cemented nodules, flat lamination and bioturbation in siltstones	Lobate (? and other)	Gradational	*Planolites*
B2	Medial/lateral delta front	Fine-grained sandstone with siltstone interbeds	Wave ripple and current ripple cross-lamination in sandstone beds, bioturbation	Lobate (? and other)	Gradational	Plant debris, *Planolites*, *Pelecepodychmus*, escape burrows
B3	Proximal delta front (distributary mouth bar)	Fine-grained sandstone	Flat bedding with internal flat lamination, rare trough cross-bedding, primary current lineation, wave ripples, current ripples, bioturbation, growth faults	Lobate	Gradational	Plant debris, *Pelecepodychmus*
C1	Distributary channel	Medium-grained sandstone	Trough and tabular cross-bedding, micaceous partings	Single-storey channel	Erosive	Plant debris
C2	Crevasse splay derived interdistributary sediments	Interbedded siltstone and medium to fine grained sandstone	Trough cross-bedding and current ripple cross-lamination in sandstone beds, flat lamination in siltstones, bioturbation	Sheet	Sharp	Plant debris, *Pelecepodychmus*
C3	?Interdistributary bay	Shale	Flat lamination, siderite-cemented nodules and horizons	?	Gradational	
C4	Major braided fluvial systems	Medium to very coarse grained sandstone	Sets of tabular and trough cross-beds, some with basal intraformational conglomerate lags, rare lateral accretion surfaces, current ripple cross-lamination	Multi-storey, multilateral sheets and laterally restricted bodies	Strongly erosive	Plant debris
C5	Leached palaeosols	Altered substrate	Bleached upper horizon with variable silica enrichment, mottles, glaebules, vertical tubules (some with carbonaceous linings), carbonaceous streaks, clay-enriched lower horizon with slickensides, red and green nodules	Sheet	Irregular	Rootlets, plant debris
C6	Peat swamp	Coal and carbonaceous shale	Flat lamination	Sheet	Sharp	Plant debris

previous workers and readers are advised to consult the relevant references for detailed facies descriptions.

All cyclothems commence with a shale containing abundant marine or marginal-marine body fossils (marine band or *Lingula* band, respectively) at its base (association A), pass upwards into progressively coarser grained clastic rocks (association B) and culminate in fluviatile and coal-bearing facies (association C). This profile represents the progradation of a land-attached depositional system into the basin. The overall coarsening-upwards profile and internal facies composition of the cyclothems implies a deltaic origin.

Facies associations A and B

Description. Facies A1 and A2 both consist of dark grey non-bioturbated, flat-laminated, fissile shales with abundant body fossils. In facies A1 the body fossils constitute a marine nektonic fauna comprising goniatites, fish and pectinoid bivalves, with pyrite as a common constituent either as finely disseminated particles or centimetre-scale spheroidal nodules. In facies A2 the body fossils constitute a brackish water fauna with a benthic component comprising marginal marine brachiopods (*Lingula*) or non-marine bivalves (*Carbonicola, Anthraconaia, Naiadites*) (Smith 1967; Calver 1968; Wignall 1987) and a nektonic component comprising fish. Facies A1 and A2 occur in regionally extensive bands up to 1 m thick, within which there is a lateral transition from facies A1 (marine band) in the centre of the basin to facies A2 (*Lingula* band) towards the basin margins (Calver 1968; Wignall 1987) (Figs. 3d and 3h). Both facies A1 and A2 are gradationally overlain by facies A3 and sharply overlie facies of associations B (delta front) and C (delta plain) across a discrete bioturbated surfaces which notably contains *Rhizocorallium*.

Facies A1 and A2 are generally overlain by gradational vertical facies successions comprising facies A3, B1, B2 and B3. Within each succession there is an overall coarsening-upward grain size profile, with pronounced interbedding of siltstones and sandstones internally, and an increasing dominance of current-generated structures over storm- and wave-generated structures upward through the succession. Laterally there are similar facies transitions within these successions. With the exception of plant debris, body fossils are absent from these successions. Tidally generated structures are also absent. Collinson & Banks (1975) provide a detailed description of one such succession, the Upper Haslingden Flags, and the characteristics of individual facies are summarized in Table 1.

Interpretation. From their fine grain size, fossil content and vertical facies context, facies A1 and A2 are interpreted as faunal-concentrate condensed horizons deposited in quiet water environments starved of coarse clastic sediment. Within each faunal-concentrate condensed horizon the lateral transition from facies A1 to facies A2 is interpreted to reflect a palaeogeographical transition from a zone with an anoxic substrate and oxygenated waters, capable of sustaining only a planktonic fauna, to a zone with both an oxygenated substrate and oxygenated waters, capable of sustaining both planktonic and benthic faunas (Calver 1968; Wignall 1987). This reflects a change from deeper, poorly circulated, marine environments (facies A1) to shallower, marginal marine environments (facies A2) (Calver 1968; Wignall 1987). There are four faunal-concentrate condensed horizons in the Rough Rock Group, three of which contain distinctive thick-shelled goniatite faunas (the *G. cancellatum, G. cumbriense* and *G. subcrenatum* marine bands) and one containing a more ubiquitous thin-shelled goniatite fauna (an *Anthracoceras* marine band) (Fig. 2). The sharp bioturbated basal contact of facies A1 and A2 with facies of associations B and C is interpreted to represent an abrupt suppression of clastic sediment supply synchronous with a rapid transgression.

From their lithology, sedimentary structures and vertical facies context, successions comprising facies A3, B1 and B2 and B3 are interpreted as resulting from deposition on an advancing prodelta and delta front. The scarcity of trace and body fossils suggests a high sediment supply rate, and the pronounced interbedding and presence of current-generated structures suggest that sediment was supplied intermittently by powerful unidirectional flows. The very low diversity of trace fossils present (Table 1) implies deposition in brackish water. The lateral facies changes within these successions implies that deposition took place in lobate bodies, with dimensions as recorded in isopach maps of 30–100 km long, 10–50 km wide and up to 70 km thick (Collinson & Banks 1975). In the East Midlands Oilfield the geometry of the bodies which deposited these successions is less well constrained.

Facies association C

Association C comprises six component facies which are generally underlain by rocks of

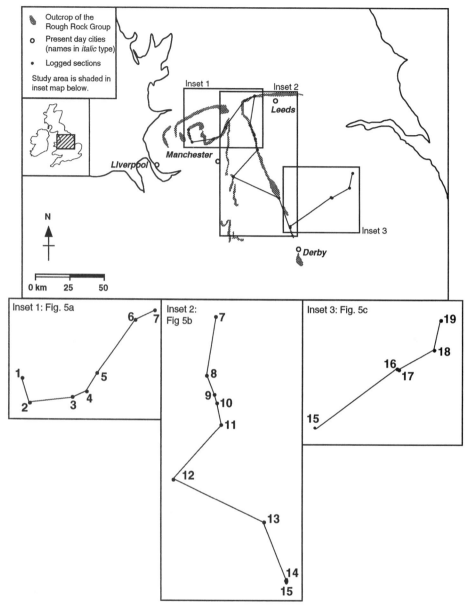

Fig. 4. Map showing the locations of logged sections and the cross-sections shown in Fig. 5. Logged section localities are listed below with national grid references: (1) Withnell Quarry [SD 642 218]; (2) Stepback Brook [SD 670 214]; (3) Scout Moor Quarry [SD 815 187]; (4) Whitworth Quarry [SD 872 203]; (5) Greens Clough [SD 890 260]; (6) Buck Park Quarry [SE 070 352]; (7) Loadpit Beck [SE 128 390]; (8) Elland Road Cutting [SE 103 214]; (9) Crossland Hill Quarry [SE 118 144]; (10) Honley Wood Quarry [SE 117 113]; (11) Harden Clough [SE 147 042]; (12) Cracken Edge [SK 037 836]; (13) Harewood Grange [SK 312 682]; (14) Ridgeway Quarry [SK 359 515]; (15) Nether Heage borehole [SK 360 512]; (16) Bothamsall No. 1 borehole [SK 659 737]; (17) Bothamsall No. 5 borehole [SK 666 734]; (18) South Leverton No. 1 borehole [SK 793 804]; and (19) Gainsborough No. 1 borehole [SK 832 902].

association B and are succeeded upward by facies A1 and A2 across a discrete bioturbated surface containing *Rhizocorallium*. Facies C1, C2 and C3 are observed to be almost exclusively confined to the uppermost Lower Haslingden Flags, Upper Haslingden Flags and Rough Rock Flags by the author and by Collinson & Banks (1975). Facies C4, C5 and C6 are generally present on a more regional scale and form the uppermost part of each cyclothem, commonly overlying facies C1, C2 and C3 in addition to facies of associations A and B.

Description. Facies C1 corresponds to facies 2 of Bristow & Myers (1989) and consists of channelized, cross-bedded, medium-grained sandstone bodies up to 7 m thick and several hundred metres wide with little or no overall internal grain size trend.

Facies C2 comprises interbeds of slightly bioturbated siltstones, fine-grained, current-rippled sandstones and medium-grained, cross-bedded sandstone sheets up to 70 cm thick.

Facies C3 has been recognized at one locality only (Figs 4 and 5a, Scout Moor Quarry, immediately underlying the lower leaf of the Rough Rock), but is considered distinctive enough to warrant definition as a facies. It comprises flat-laminated, very poorly fossiliferous, dark grey shales with bedding-parallel siderite-cemented nodules, nodular horizons and rare specimens of the non-marine bivalve *Anthracomya bellula*.

Facies C4 corresponds to facies 1 of Bristow & Myers (1989) and comprises cross-bedded medium- to very coarse-grained sandstone in bodies up to 33 m thick with either a regionally extensive sheet geometry or a strongly laterally confined geometry. Tabular and trough cross-beds approximately 1 m thick are arranged in sets, each of which is typically up to 10 m thick with a basal intraformational conglomerate lag and a fining-upwards grain size profile terminating in current ripple cross-lamination. Rare lateral accretion surfaces are also preserved (Bristow 1988). Although palaeocurrent directions vary little within individual sets, both within and between exposures the set palaeocurrent directions vary widely. Facies C4 is overlain either by facies A1 and A2 across a discrete bioturbated surface or by facies C2 and C5.

Facies C5 comprises an altered substrate approximately 1 m thick showing evidence of palaeosol development. It is rare for facies C5 to be exposed at outcrop and most data are in the form of core samples (from the East Midlands Oilfield) taken at 1 ft intervals, which prevents the study of a complete vertical profile through a facies unit. Often the uppermost horizon of the facies is bleached and silica-enriched to a variable degree. Downward-branching vertical and subvertical silica-rich tubules, some of which have carbonaceous linings and attached carbonaceous streaks, are observed. In general, the greater the degree of silica enrichment in the substrate, the less carbonaceous material is present. Some substances are silica-enriched and mottled to such a degree that they appear in hand specimen to be off-white, structureless, quartz arenitic sandstones which may be underlain by clay-enriched horizons with small slickensides. Additional structures include centimetre-scale spheroidal green and red nodules near the base of the facies and mottles, some of which are clay-coated. Facies C5 is generally overlain by facies A1, A2 and C6.

Facies C6 consists of beds of poor quality coal and coaly shale up to 1.5 m thick with very gradual thickness variations. Coals do not pass laterally into clastic facies. Two coals occur as regionally extensive single seams which thicken and sometimes split towards the basin margins. These coals are underlain by leached palaeosols (facies C5) generally developed overlying major, braided fluvial sheets (comprising facies C4) and are overlain by facies A1 and A2. One coal is locally restricted and is underlain by facies C3 and overlain by facies C4 (see earlier description of coal stratigraphy in the Rough Rock Group).

Interpretation. From the lithology, sedimentary structures and vertical facies context, facies C1 is interpreted as resulting from deposition in a deltaic distributary channel. The absence of a hierarchy of internal erosion surfaces such as those present in facies C4 suggests that the bodies comprising facies C1 are single-storey, whereas the absence of lateral accretion surfaces suggests that the channels had low sinuosities and were perhaps simple 'cut and fill' channels (Bristow 1988; Bristow & Myers 1989). These channels are interpreted to have passed down current into distributary mouth bars as represented by facies B3.

From the geometry, interbedded nature, sedimentary structures and vertical facies context, facies C2 is interpreted as mainly crevasse splay derived interdistributary bay deposits.

From the lithology, vertical and lateral facies context and scarce, monospecific fauna, facies context, facies C3 is interpreted as having been deposited in a very low energy delta plain environment with restricted oxygen and/or salinity conditions. Possibilities for this environment include a lagoon or interdistributary bay, but the apparent absence of a coastal barrier deposit favours the latter interpretation.

From the coarse grain size, cross-stratification style and widely varying palaeocurrents, facies C4 is interpreted as deposits of a powerful braided fluvial system. The presence of several sets, each with an overall fining-upwards profile, within a single body records a hierarchy of internal erosion surfaces and implies that these

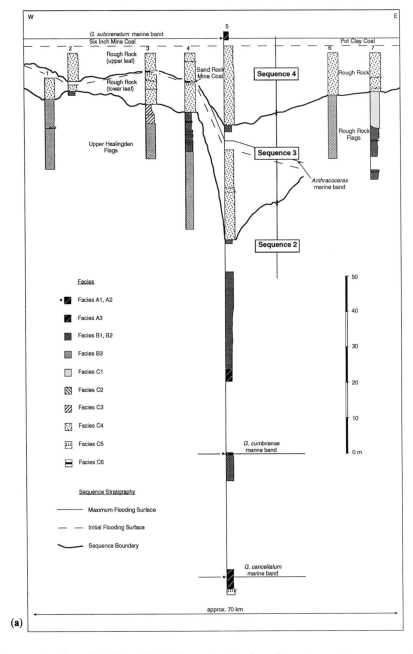

Fig. 5. Cross-sections through the Rough Rock Group showing the sedimentology and sequence stratigraphy of logged sections. See Fig. 4 for logged section names and locations. (**a**) Depositional strike section across the northern basin margin. (**b**) Oblique depositional dip section through the basin. (**c**) Depositional strike section across the southeastern basin margin.

SEQUENCE STRATIGRAPHY OF THE PENNINE BASIN

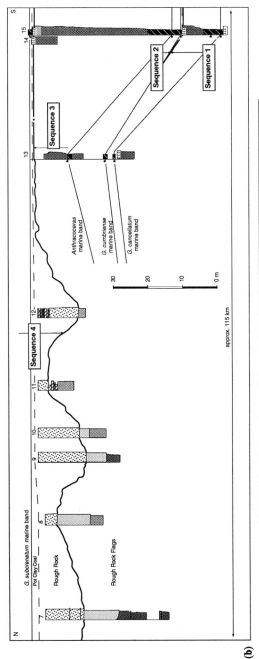

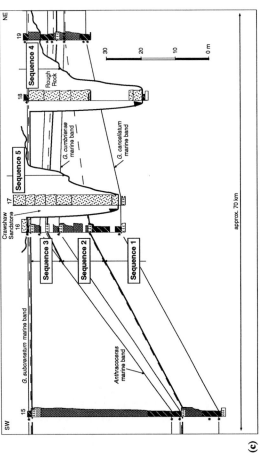

bodies are multi-storey, with each set inferred to represent a single-storey channel. The lateral margins of the channels are not preserved, but they are inferred to be of similar dimensions to those in facies C1 (Bristow & Myers 1989). The coarser overall grain size relative to facies C1 and the absence of finer grained micaceous partings between beds (Table 1) suggest that facies C4 was deposited under conditions of greater stream power and, possibly, discharge than facies C1. Two regionally extensive, multi-lateral sheets comprising facies C4 are recognized along the northern basin margin: the upper leaf of the Rough Rock (which is correlated with the Rough Rock on the eastern limb of the Pennine anticline) and the lower leaf of the Rough Rock (Fig. 2). These two sheets have areal extents of at least 6500 km^2 (Fig. 3f) and 2700 km^2 (Fig. 3b), respectively. Two laterally confined bodies comprising facies C4 are also recognized along the south-eastern basin margin at the stratigraphic levels of the upper leaf of the Rough Rock (Fig. 5c, section 18, South Leverton No. 1 borehole) and the lowermost Westphalian Crawshaw Sandstone (Fig. 5c, section 17, Bothamsall No. 5 borehole). Up to 33 m relief is observed at the bases of these laterally confined bodies over lateral distances as small as several hundred metres (Fig. 5c, between sections 16 and 17, Bothamsall Nos. 1 and 5 boreholes) and the underlying cyclothems have been eroded out beneath such bodies. These laterally confined bodies correspond to the 'major distributary channels' of Steele (1988) who observed their width to be $c.$ 10 km, which would make these bodies multilateral also, comparing their width with the single-storey channels comprising facies C1. The exaggerated lateral thickness changes and persistent directional trend (Steele 1988) of laterally restricted bodies comprising facies C4 combined with the tectonically active history of the East Midlands Oilfield, to which these bodies are confined, in the immediately preceding Marsdenian (Namurian R_2) (Edwards 1967; Smith et al. 1973) suggests that there is a synsedimentary tectonic control causing the laterally restricted geometry. It is suggested that small-scale, synsedimentary growth faults with fault scarp relief of never more than a few metres 'captured' fluvial systems. Similar processes of fluvial channel capture have been documented in other Upper Carboniferous sediments (Fielding & Johnson 1987).

From the geometry and sedimentary structures, facies C5 is interpreted to represent the deposits of leached palaeosols of varying maturity developed under a range of drainage conditions on a range of substrates. The most silica-enriched, sand-grade facies (quartz arenitic sandstones) are interpreted as ganisters (Percival 1983). These represent soils developed on a sand-grade substrate in which clays and other fine particles were removed (leached) from the upper part of the profile and translocated to a lower position in the profile by gravity-driven percolation under freely drained conditions leading to characteristic silica enrichment of the surface horizon (Percival 1983; Wright 1989). Palaeosols showing weak silica enrichment of the substrate and preservation of carbonaceous rootlet material are interpreted as gley palaeosols formed under waterlogged drainage conditions (Wright 1989). The presence of a thin bleached horizon, reflecting the leaching of clays and iron minerals, in these palaeosols indicates a relatively low water-table, which would prevent the accumulation of thick peat deposits (Wright 1989). Most palaeosols represented by facies C5 show a variable development of features associated with both ganisters and gley palaeosols. These are interpreted as representing (1) palaeosols formed under freely drained conditions on a silt-grade substrate, (2) immature ganisters and gley palaeosols, (3) palaeosols formed under intermediate drainage conditions or (4) composite palaeosols. Some of these may have been altered by (1) prolonged subaerial exposure and later marine flooding (Spears & Sezgin 1985) or (2) overprinting and reworking by later 'wetter' palaeosols (Percival 1986: 106–107). The relatively wide variation in palaeosol character is interpreted to be the result of the interplay between substrate-controlled drainage. The lack of complete vertical profiles through these palaeosols precludes more accurate interpretation, but it appears unlikely that thick peat deposits would have accumulated as part of these palaeosols.

Facies C6 is interpreted as the organic residues of *in situ* peat accumulations produced by organic hydromorphic (waterlogged) soils in delta plain swamps. The following conditions must have been met by this swamp environment for peat to accumulate: (1) the clastic sediment influx was extremely low; (2) the water-table was sufficiently high to cause reducing groundwater conditions which retarded or prevented oxidation of the organic residue; and (3) these conditions prevailed for a sufficiently long time that a substantial thickness of organic residues accumulated. For this last condition to be met the water-table must have steadily risen (in a relative sense) to provide space in which the organic residues could accumulate.

The thin, uniform, regionally extensive character of most coal seams suggests that the peat swamps were established in a regionally extensive, relatively uniform environment, rather than a locally restricted environment such as a raised mire, floating mire, lake bottom or beach ridge (Haszeldine 1989; McCabe 1991). Hence it appears that peat accumulation occurred in blanket bogs (Haszeldine 1989) or low-lying mires (McCabe 1991) overlying abandoned major, braided fluvial sheets (comprising facies C4), which may already have been colonized by leached palaeosols (facies C5). This is analogous to many previously documented ancient coals and modern peat deposits (Arditto 1987; Haszeldine 1989). Where the coals thicken locally near the basin margins this may represent deposition in a locally restricted environment. The relatively up-dip, topographically high settings of these environments implies that such locally restricted environments were probably raised mires.

The regionally extensive sharp transition from leached palaeosols (facies C5) to organic hydromorphic palaeosols (facies C6) suggests an abrupt basin-wide rise in the water-table. Where facies C6 is directly overlain by facies A1 and A2 this records peat accumulation abruptly terminated by regionally extensive flooding, implying that peat accumulation could not keep pace with the rising base level. Spears & Sezgin (1985) also report anomalous clay mineral assemblages in the Pot Clay Coal and underlying palaeosol, which they interpret as representing alteration by later marine flooding.

Sequence stratigraphic framework

The three cyclothems comprising the Rough Rock Group are reinterpreted within a high-resolution sequence stratigraphic framework (Figs 4 and 5) using the concepts and terminology of Posamentier & Vail (1988), Van Wagoner et al. (1990) and Mitchum & Van Wagoner (1991). Each high frequency sequence comprises key surfaces bounding systems tracts, which contain distinctive facies architectures identified from detailed facies analysis and interpretation. These key surfaces and systems tracts are described in the following. Figure 3 shows a series of palaeogeographical reconstructions of the Pennine Basin at specific time intervals in the sequence stratigraphic framework, illustrating the spatial organization of individual systems tracts and key surfaces. A model for the depositional history of the Rough Rock Group within the sequence stratigraphic framework is illustrated in Fig. 6. The sequence geometry is comparable with the ramp model of Van Wagoner et al. (1990, fig. 20A).

Previous sequence stratigraphic interpretations of the Rough Rock Group have been constructed using smaller databases and as a result have either taken individual components of the group out of context or have not identified some of the more subtly expressed key surfaces. Key surfaces and systems tracts have not been accurately characterized. In Maynard's reconstruction (Maynard 1992), for example, sequences 3 and 4 (Fig. 5) have been interpreted as a single sequence with the result that no sequence stratigraphic significance has been attached to the Sand Rock Mine Coal. It appears highly unlikely that a stable coal-forming environment could exist within a major, braided fluvial complex such as the Rough Rock for sufficient time to produce a regionally extensive coal, as Maynard's interpretation infers. It seems far more likely that the Sand Rock Mine Coal represents a key surface in a sequence stratigraphic context.

Sequence boundaries and lowstand systems tracts

Major braided fluvial sand bodies (facies C4) are multi-storey and, by definition, incised as the erosional relief on the basal surface of each sand body is greater than the thickness of an individual channel storey within that sand body. Incision is either localized, under laterally confined sand bodies, or regional, under regionally extensive sheet sand bodies. Up to four entire cyclothems are incised out beneath laterally confined sand bodies (e.g. Fig. 5c, incision at the base of the Rough Rock in section 18, South Leverton borehole No. 1). Regionally extensive sheet sand bodies overlie a range of facies representing delta front (association B) and delta plain (association C) sediments deposited in lobate bodies. The lateral correlatives (interfluves) to major, braided fluvial systems are leached palaeosols (facies C5) (Fig. 3b and 3c). Leached palaeosols overlie facies representing prodelta (association A), delta front (association B) and delta plain (association C) sediments. Mature leached palaeosols are unlikely to have formed in poorly drained delta plain environments, or on relatively short-lived depositional highs such as beach ridges and levees (Wright 1989). Hence the basal surfaces of leached palaeosols are interpreted as basinward facies shifts, even in delta plain settings.

Sand bodies deposited by major, braided fluvial systems are not genetically linked to the

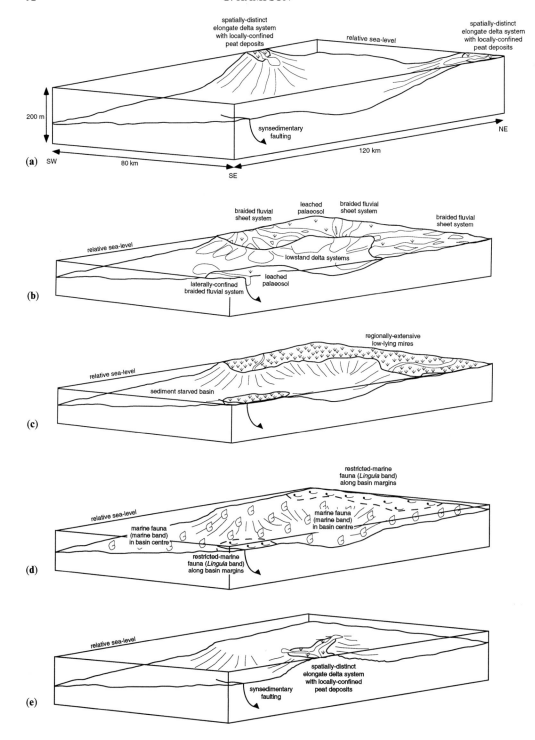

Fig. 6. Sequence stratigraphic model of deposition in the Rough Rock Group: (a) preceding highstand systems tract; (b) lowstand systems tract; (c) initial flooding surface; (d) maximum flooding surface; and (e) highstand systems tract.

underlying sediments due to (1) their incised character, (2) their regionally extensive and unconformable character relative to the laterally restricted underlying sediment bodies and (3) the basinward facies shift underlying their interfluves. The basal surfaces of these sand bodies and their interfluves are interpreted as sequence boundaries and the sand bodies themselves are reinterpreted as incised valley-fills. These valley-fills are characterized by coarse-grained sandstones in mutually erosive channel storeys which suggest deposition under conditions of low accommodation space and hence in a lowstand systems tract.

Initial flooding surfaces, transgressive systems tracts and maximum flooding surfaces

Lowstand incised valley-fills and their interfluves are overlain by either a regionally extensive coal (facies C6) with a leached palaeosol (facies C5) underlay or a bioturbation surface containing *Rhizocorallium*. In turn, these are overlain by either a marine band (facies A1) or a *Lingula* band (facies A2). These intervals are very thin (1 or 2 m thick) and represent (1) an abrupt basin-wide suppression of clastic sediment supply and (2) a synchronous basin-wide rise in relative base level/sea level. Coarse-grained clastic sediment is apparently trapped in the hinterland. Conditions (1) and (2) are characteristic of transgressive systems tracts (Van Wagoner et al. 1990). Bioturbated surfaces and regionally extensive coals at the base of such intervals are interpreted as initial flooding surfaces and their up-dip equivalents on the delta plain, respectively. This is directly comparable with other fluviatile coal-bearing successions in which regionally extensive coals are directly correlated down-dip with initial flooding surfaces (Shanley & McCabe 1991; Aitken & Flint 1994). Faunal-concentrate condensed horizons (facies A1 and A2) at the tops of such intervals represent the maximum suppression of clastic sediment supply across the basin and are interpreted as maximum flooding surfaces.

Highstand systems tracts

Maximum flooding surfaces are overlain by laterally restricted, lobate sediment bodies deposited by advancing delta systems. These delta systems were fed by widely spaced point sources and are hence spatially distinct, implying deposition under conditions of relatively high accommodation space. Such sediment bodies are interpreted as highstand systems tracts due to their vertical sequence stratigraphic context and inferred depositional conditions in terms of accommodation space.

Sequence stratigraphic evolution of the Rough Rock Group

Four high frequency sequences (*sensu* Posamentier & Vail 1988; Posamentier et al. 1988; Van Wagoner et al. 1990; Mitchum and Van Wagoner 1991) are recognized within the Rough Rock Group (Figs 2b and 5). Each sequence comprises a lowstand incised valley-fill (lowstand systems tract) overlain by a regionally extensive coal (initial flooding surface) and a marine band or equivalent (maximum flooding surface), which define the lower and upper limits of a transgressive systems tract, respectively, overlain by a laterally restricted, lobate deltaic sediment body (highstand systems tract). For example, sequence 3 comprises the lower leaf of the Rough Rock (lowstand systems tract; Fig. 3b), the Sand Rock Mine Coal (initial flooding surface; Fig. 3c), an *Anthracoceras* marine band (maximum flooding surface; Fig. 3d) and the Rough Rock Flags (highstand systems tract; Fig. 3e). Sequence 4 comprises the upper leaf of the Rough Rock (lowstand systems tract; Fig. 3f), the Six Inch Mine Coal/Pot Clay Coal (initial flooding surface; Fig. 3g), the *G. subcrenatum* marine band (maximum flooding surface; Fig. 3h) and an overlying, unidentified highstand systems tract.

Along the northern basin margin, sequences 3 and 4 are arranged in a progradational stacking pattern and in both sequences lowstand systems tract development is enhanced relative to transgressive and highstand systems tracts (Fig. 5a). Thus the Rough Rock itself is interpreted as a candidate lowstand sequence set (Mitchum & Van Wagoner 1991) along the northern basin margin. Incorporating data from the appropriate British Geological Survey memoirs, palaeogeographic reorganizations of sediment dispersal paths within the candidate lowstand sequence set are not observed across sequence boundaries. For example, the Upper Haslingden Flags highstand systems tract sequence 2; Fig. 3a) and the lower leaf of the Rough Rock (lowstand systems tract sequence 3; Fig. 3b) were both fed from the northwest, and the Rough Rock Flags (highstand systems tract sequence 3; Fig. 3e) and the upper leaf of the Rough Rock (lowstand systems tract sequence 4; Fig. 3f) were both sourced from the northeast. Instead palaeogeographic reorganizations occurred across maximum flooding surfaces; following a maximum flooding event, clastic sedimentation recommenced, not above the previously abandoned depocentre, which formed a topographic high, but in adjacent topographic lows where there was available accommodation space. For example, the lower leaf of the Rough Rock (lowstand systems tract sequence 3; Fig. 3b) is sourced from the northwest, whereas the Rough Rock Flags (highstand systems tract sequence 3: Fig. 3e) are sourced from the northeast; the palaeogeographic reorganization occurs across the *Anthracoceras* marine band (maximum flooding surface sequence 3; Fig. 3d).

In contrast with sequences 3 and 4, sequences 1 and 2 contain very poorly developed lowstand and transgressive systems tracts along the northern basin margin (Fig. 5a). Owing to this relatively suppressed lowstand and transgressive systems tract development, combined with their stratigraphic position directly underlying a candidate lowstand sequence set, sequences 1 and 2 are interpreted as a candidate highstand sequence set (Mitchum & Van Wagoner 1991).

There is a marked difference in lowstand incised valley-fill geometry between the northern and southeastern basin margins (Fig. 5a and 5c, respectively). The northern basin margin is a tectonically stable ramp, where there was no site for preferential incision, thus producing incised valleys with sheet geometries. The southeastern basin margin is tectonically active (Edwards 1967; Smith *et al.* 1973), with major fluvial channels apparently 'captured' by minor synsedimentary faults (see interpretation of facies C4). These faults acted as sites of preferential incision during falling sea level, resulting in locally confined incised valleys (the 'major distributary channels' of Steele (1988) are c. 10 km wide). There are no documented lowstand deposits down-dip of the Rough Rock Group incised valleys, probably due to the almost total lack of exposure in the extreme south of the basin.

Along the northern basin margin, a regionally extensive, mature leached palaeosol (facies C5) directly overlies the upper leaf of the Rough Rock (lowstand systems tract sequence 4) and directly underlies the Pot Clay Coal (initial flooding surface sequence 4) (Spears & Sezgin 1985; Percival 1986). This palaeosol records conditions of free drainage (and low water-table) after the deposition of a lowstand incised valley-fill and immediately preceding an initial flooding event. Although the very sandy substrate of this palaeosol (i.e. the upper leaf of the Rough Rock) is partly responsible for the freely drained nature of the palaeosol, the apparently abrupt transition from freely drained to poorly drained conditions suggests that there was not a continuous, gentle rise in relative sea level (and associated water-table) from the lowstand into the transgressive systems tract. Instead, flooding may have been abrupt.

Coals in this sequence stratigraphic framework

Depositional environments in which peat accumulations form are found in a number of sequence stratigraphic settings, particularly within transgressive and highstand systems tracts (Arditto 1987; McCabe 1993). No reassessment of coal depositional environments is required, but an additional controlling factor in the accumulation and preservation of peat deposits is the rate of accommodation space creation. For coals to form, accommodation space must be created at a rate sufficiently high for peat deposits to steadily accumulate in the delta plain, but not so high that the delta plain is flooded. In the Rough Rock Group this critical rate of accommodation space creation exists locally on the abandoned tops of highstand delta systems (e.g. the unnamed coal at the top of the Rough Rock Flags in the Harden Clough area) and regionally on the abandoned tops of lowstand incised valley-fills and their interfluves as up-dip equivalents to initial flooding surfaces in high-frequency sequences (e.g. the Sand Rock Mine Coal and the Six Inch Mine Coal/Pot Clay Coal). Arditto (1987: 11) made similar observations in the late Permian strata of the southern Sydney Basin, where 'regionally-extensive coals are related to [the early part of] a transgressive phase of the eustasy cycle.' In an intracratonic basin such as the Pennine Basin the rates of differential subsidence and, hence, accommodation space creation are lower at the basin margins than in the basin centre. This retards flooding of the basin margins and allows peat to accumulate for longer here. This may partly explain why regional extensive coals thicken towards the basin margins. In addition, locally restricted, coal-forming environments such as raised mires are likely to be more common in topographically high, up-dip settings along the basin margins.

Potential sequence stratigraphic elements, including regionally extensive coals, have been recognized in the Coal Measures of the Pennine Basin (Flint *et al.* this volume). Many of these regionally extensive coals are documented to overlie clastic delta plain sediments directly, including ganisters and major fluvial sandbodies, and directly underlie *Lingula* bands and marine bands in a manner directly analogous to the Sand Rock Mine Coal and Six Inch Mine Coal/Pot Clay Coal (e.g. Wright *et al.* 1927; Wray *et al.* 1930; Bromehead *et al.* 1933; Mitchell *et al.* 1947; Edwards *et al.* 1950; Stephens *et al.* 1953; Eden *et al.* 1957; Earp *et al.* 1961; Price *et al.* 1963; Edwards 1967; Smith *et al.* 1967; Evans *et al.* 1968; Stevenson & Gaunt 1971; Smith *et al.* 1973; Frost & Smart 1979; Aitkenhead *et al.* 1985, Chisholm *et al.* 1988). These coals are regionally extensive over several hundreds of square kilometres (Guion & Fielding 1988) and are tentatively interpreted as the deposits of blanket bogs overlying abandoned fluvial systems (Haszeldine 1989). Such coals represent deposition during periods of synchronous basin-wide suppression of clastic sediment supply and basin-wide rising base level. By implication these coals may also represent up-dip equivalents of initial flooding surfaces on the delta plain within high frequency sequences. Lower frequency sequence stratigraphic elements may also be recognized using criteria such as the vertical stacking relationships and lateral extent of regionally extensive coals (Flint *et al.* this volume).

Conclusions

The Yeadonian Rough Rock Group of the Pennine Basin was deposited by the repeated advance and abandonment of fluvially dominated delta systems into a shallow water, wave-influenced basin as controlled by relative sea-level fluctuations. Through the successful application of sequence stratigraphic concepts, key surfaces and systems tracts are recognized and clearly characterized within a high resolution sequence stratigraphic framework. This framework is similar to those of Church (1992) and Maynard (1992), but incorporates several other sedimentological elements, in particular regionally extensive coals.

Regionally extensive coals are interpreted to represent peat deposits in blanket bogs or low-lying mires overlying abandoned lowstand fluvial complexes and their interfluves. Such coals are interpreted as the up-dip equivalents of initial flooding surfaces, representing deposition on the delta plain during periods of (1) basin-vide suppression of clastic sediment supply and (2) synchronous, basin wide rising relative base level.

Regionally extensive coals in the Coal Measures are documented within similar stratigraphic settings and are tentatively interpreted to represent up-dip equivalents of initial flooding surfaces in high frequency sequences.

I thank Johnsons Wellfield, Marshall Mono, Ridgeway Quarry and Evered Quarries for access to quarry sections and the British Geological Survey for access to core data. I also thank T. Elliott, S. Flint and J. Aitken for their useful discussions regarding this work. This study was carried out under the tenure of a NERC studentship. I gratefully acknowledge the constructive reviews by W. Read and an anonymous reviewer of an earlier draft of the paper.

References

AITKEN, J. F. & FLINT, S. 1994. High frequency sequences and the nature of incised valley fills in fluvial systems of the Breathitt Group (Pennsylvanian), Appalachian basin, eastern Kentucky. *In:* BOYD, R., DALRYMPLE, R. W. & ZAITLN, B. (eds) *Incised Valley Systems: Origin and Sedimentary Sequences.* Society of Economic Paleontologists and Mineralogists, Special Publication, **51**, 353–368.

AITKENHEAD, N., CHISHOLM, J. I. & STEVENSON, I. P. 1985. *Geology of the country around Buxton, Leek and Bakewell.* Memoir, British Geological Survey, Sheet 111.

ARDITTO, P. A. 1987. Eustasy, sequence stratigraphic analysis and peat formation: a model for widespread late Permian coal deposition in the Sydney Basin, N.S.W. *Advanced Studies of the Sydney Basin: 21st Newcastle Symposium Proceedings* [abstract], 11–17.

BISAT, W. S. 1924. The Carboniferous goniatites of the north of England and their zones. *Proceedings of the Yorkshire Geological Society,* **20**, 40–124.

BRISTOW, C. S. 1988. Controls on the sedimentation of the Rough Rock Group (Namurian) from the Pennine Basin of northern England. *In:* BESLY, B. M. & KELLING, G. (eds) *Sedimentation in a Synorogenic Basin Complex: the Upper Carboniferous of NW Europe,* Blackie, Glasgow, 114–131.

—— & MYERS, K. J. 1989. Detailed sedimentology and gamma-ray log characteristics of a Namurian deltaic succession I: sedimentology and facies analysis. *In:* WHATELEY, M. K. G. & PICKERING, K. T. (eds) *Deltas: Sites and Traps for Fossil Fuels.* Geological Society, London, Special Publication, **41**, 75–80.

BROMEHEAD, C. E. N., EDWARDS, W. N., WRAY, D. A. & STEPHENS, J. D. 1993. *Geology of the Country around Holmfirth and Glossop.* Memoir, British Geological Survey, Sheet 86.

CALVER, M. A. 1968. Distribution of Westphalian marine faunas in northern England and adjoining areas. *Proceedings of the Yorkshire Geological Society,* **37**, 1–72.

CHISHOLM, J. I., CHARSLEY, T. J. & AITKENHEAD, N. 1988. *Geology of the Country Around Ashbourne and Cheadle.* Memoir, British Geological Survey, Sheet 124.

CHURCH, K. 1992. Sequence stratigraphy of late Namurian (Marsdenian–Yeadonian) delta systems in the Widmerpool Gulf, East Midlands, UK. *British Sedimentological Research Group AGM Southampton* [abstract], Department of Oceanography, University of Southampton.

COLLIER, R. E. LI. 1990. Transgressions and regressions: a model for the influence of tectonic subsidence, deposition and eustasy, with application to Quaternary and Carboniferous examples. *Geological Magazine*, **127**, 117–128

COLLINSON, J. D. 1988. Controls on Namurian sedimentation in the Central Province Basins of northern England. *In:* BESLY, B. M. & KELLING, G. (eds) *Sedimentation in a Synorogenic Basin Complex: the Upper Carboniferous of NW Europe.* Blackie, Glasgow, 85–101.

—— & BANKS, N. L. 1975. The Haslingden Flags (Namurian G_1) of south-east Lancashire: bar finger sands in the Pennine Basin. *Proceedings of the Yorkshire Geological Society*, **40**, 431–458.

——, JONES, C. M. & WILSON, A. A. 1977. The Marsdenian (Namurian R_2) succession west of Blackburn: implications for the evolution of the Pennine Delta Systems. *Geological Journal* **12**, 59–76.

EARP, J. R., MAGRAW, D., POOLE, E. G., LAND, D. H. & WHITEMAN, A. J. 1961. *Geology of the Country Around Clitheroe and Nelson.* Memoir, British Geological Survey, Sheet 68.

EDEN, R. A., STEPHENSON, I. P. & EDWARDS, W. N. 1957. *Geology of the Country Around Sheffield.* Memoir, British Geological Survey, Sheet 100.

EDWARDS, W. 1967 *Geology of the Country Around Ollerton.* Memoir, British Geological Survey, Sheet 113.

——, MITCHELL, G. H. & WHITEHEAD, T. H. 1950. *Geology of the Country Around Leeds.* Memoir, British Geological Survey, Sheet 70.

EVANS, W. B., WILSON, A. A., TAYLOR, B. J. & PRICE, D. 1968. *Geology of the Country Around Macclesfield, Congleton, Crewe and Middlewich.* Memoir, British Geological Survey, Sheet 110.

FIELDING, C. R. & JOHNSON, G. A. L. 1987. Sedimentary structures associated with extensional fault movement from the Westphalian of northeast England. *In:* COWARD, M. P., DEWEY, J. F. & HANCOCK, P. L. (eds) *Continental Extensional Tectonics.* Geological Society, London, Special Publication **28**, 511–516.

FLINT, S. S., AITKEN, J. F. & HAMPSON, G. J. 1995. The application of sequence stratigraphy to coal-bearing fluvial successions: implications for the UK Coal Measures. *In:* WHATELEY, M. K. G. & SPEARS, D. A. (eds) *European Coal Geology.* Geological Society, London, Special Publication, **82**, 1–16.

FROST, D. V. & SMART, J. G. O. 1979. *Geology of the Country Around Derby.* Memoir, British Geological Survey, Sheet 124.

GUION, P. D. & FIELDING, C. R. 1988. Westphalian A and B sedimentation in the Pennine Basin, U.K. *In:* BESLY, B. M. & KELLING, G. (eds) *Sedimentation in a Synorogenic Basin Complex: the Upper Carboniferous of NW Europe.* Blackie, Glasgow, 153–177.

HASZELDINE, R. S. 1989. Coal reviewed: depositional controls, modern analogues and ancient climates. *In:* WHATELEY, M. K. G. & PICKERING, K. T. (eds) *Deltas: Sites and Traps for Fossil Fuels.* Geological Society, London, Special Publication, **41**, 289–308.

HOLDSWORTH, B. K. & COLLINSON, J. D. 1988. Millstone grit cyclicity revisited. *In:* BESLY, B. M. & KELLING, G. (eds) *Sedimentation in a Synorogenic Basin Complex: the Upper Carboniferous of NW Europe.* Blackie, Glasgow, 132–152.

JONES, C. M. 1980. Deltaic sedimentation in the Roaches Grit and associated sediments (Namurian R_{2b}) in the south west Pennines. *Proceedings of the Yorkshire Geological Society*, **43**, 39–67.

MCCABE, 1991. Geology of coal; environments of deposition. *In:* GLUSKOTER, H. J., RICE, D. D. & TAYLOR, R. B. (eds) *The Geology of North America.* Vol. P-2, *Economic Geology*, Geological Society of America, Boulder, 469–482.

—— 1993. Sequence stratigraphy of coal-bearing strata. *In:* BREYER, J. A., ARCHER, A. W. & MCCABE, P. J. (eds) *Sequence Stratigraphy of Coal-bearing Strata: Field Trip Guidebook.* Short Course Supplement, Energy Minerals Division American Association of Petroleum Geologists, Tulsa.

MARTINSEN, O. J. 1993. Upper Carboniferous sedimentary successions of the Craven–Askrigg area, northern England: implications for sequence stratigraphic models. *In:* POSAMENTIER, H. W., SUMMERHAYES, C. P., HAQ, B. U. & ALLEN, G. P. (eds) *Sequence Stratigraphy and Facies Associations.* International Association of Sedimentologists, Special Publication, **18**, 247–281.

MAYNARD, J. R. 1992. Sequence stratigraphy of the upper Yeadonian of Northern England. *Marine and Petroleum Geology*, **9**, 197–207.

—— & LEEDER, M. R. 1992. On the periodicity and magnitude of Late Carboniferous glacio-eustatic sea level changes. *Journal of the Geological Society*, **149**, 303–311.

MILLER, G. D. 1986. The sediments and trace fossils of the Rough Rock Group on Cracken Edge, Derbyshire. *Mercian Geologist*, **10**, 189–202.

MITCHELL, G. H., STEPHENS, J. V., BROMEHEAD, C. E. N. & WRAY, D. A. 1947. *Geology of the Country Around Barnsley.* Memoir, British Geological Survey, Sheet 87.

MITCHUM, R. M. & VAN WAGONER, J. C. 1991. High-frequency sequences and their stacking patterns: sequence stratigraphic evidence of high-frequency eustatic cycles. *Sedimentary Geology*, **70**, 131–160.

PERCIVAL, C. J. 1983. A definition of the term ganister. *Geological Magazine*, **120**, 187–190.

—— 1986. Paleosols containing an albic horizon: examples from the upper carboniferous of northern England. *In:* WRIGHT, V. P. (ed.) *Paleosols: their Recognition and Interpretation.* Blackwell Scientific, Oxford, 87–111.

POSAMENTIER, H. W. & VAIL, P. R. 1988. Eustatic controls on clastic deposition II—sequence and systems tract models. *In:* WILGUS, C. M., HASTINGS, B. S., KENDALL, C. G. St. C., POSAMENTIER, H. W., ROSS, C. A. & VAN WAGONER, J. C. (eds) *Sea-level Changes—an Integrated Approach.* Society of Economic Paleontologists and Mineralogists, Special Publication, **42**, 125–154.

——, JERVEY, M. T. & VAIL, P. R. 1988. Eustatic controls on clastic deposition I—conceptual framework. *In:* WILGUS, C. M., HASTINGS, B. S., KENDALL, C. G. St. C., POSAMENTIER, H. W., ROSS, C. A. & VAN WAGONER, J. C. (eds) *Sea-level Changes—an Integrated Approach.* Society of Economic Paleontologists and Mineralogists, Special Publication, **42**, 109–124.

PRICE, D., WRIGHT, W. B., JONES, R. C. B., TONKS, L. H. & WHITEHEAD, T. H. 1963. *Geology of the Country Around Preston.* Memoir, British Geological Survey, Sheet 75.

RAMSBOTTOM, W. H. C. 1977. Major cycles of transgression and regression (mesothems) in the Namurian. *Proceedings of the Yorkshire Geological Society*, **41**, 261–291.

——, CALVER, M. A., EAGAR, R. M. C., HODSON, F., HOLLIDAY, D. W., STUBBLEFIELD, C. J. & WILSON, R. B. 1978. *A Correlation of Silesian Rocks in the British Isles.* Geological Society, Special Report, **10**, 82 pp.

READ, W. A. 1991. The Millstone grit (Namurian) of the southern Pennines viewed in the light of eustatically controlled sequence stratigraphy. *Geological Journal*, **26**, 157–166.

SHANLEY, K. W. & McCABE, P. J. 1991. Predicting facies architecture through sequence stratigraphy—an example from the Kaiparowits Plateau, Utah. *Geology*, **19**, 742–745.

SMITH, E. G. 1967. The Namurian and basal Westphalian rocks of the Beeley–Holymoorside area. *In:* NEVES, R. & DOWNIE, C. (eds) *Geological Excursions in the Sheffield Region.* University of Sheffield Press, 103–108.

——, RHYS, G. H. & EDEN, R. A. 1967. *Geology of the Country Around Chesterfield, Matlock and Mansfield.* Memoir, British Geological Survey, Sheet 112.

——, ——, & GOOSENS, R. F. 1973. *Geology of the Country Around East Retford, Worksop and Gainsborough.* Memoir, British Geological Survey, Sheet 101.

SPEARS, D. A. & SEZGIN, H. I. 1985. Mineralogy and geochemistry of the subcrenatum marine band and associated coal-bearing sediments, Langsett, South Yorkshire. *Journal of Sedimentary Petrology*, **55**, 570–578.

STEELE, R. P. 1988. The Namurian sedimentary history of the Gainsborough Trough. *In:* BESLY, B. M. & KELLING, G. (eds) *Sedimentation in a Synorogenic Basin Complex: the Upper Carboniferous of NW Europe.* Blackie, Glasgow, 102–113.

STEPHENS, J. V., MITCHELL, G. H. & EDWARDS, W. 1953. *Geology of the Country Between Bradford and Skipton.* Memoir, British Geological Survey, Sheet 69.

STEVENSON, I. P. & GAUNT, G. D. 1971. *Geology of the Country Around Chapel-en-le-Frith.* Memoir, British Geological Survey, Sheet 99.

VAN WAGONER, J. C., MITCHUM, R. M., CAMPION, K. M. & RAHMANIAN, V. D. 1990. *Siliciclastic Sequence Stratigraphy in Well Logs, Cores and Outcrops.* American Association of Petroleum Geologists, Methods in Exploration Series, **7**, 55 pp.

WIGNALL, P. B. 1987. A biofacies analysis of the *Gastroiceras cumbriense* marine band (Namurian) of the central Pennines. *Proceedings of the Yorkshire Geological Society*, **46**, 111–128.

WRAY, D. A., STEPHENS, J. V., EDWARDS, W. N. & BROMEHEAD, C. E. N. 1930. *Geology of the Country Around Huddersfield and Halifax.* Memoir, British Geological Survey, Sheet 77.

WRIGHT, V. P. 1989. Palaeosols in deltaic settings. *In:* WRIGHT, V. P. (ed.) *Palaeosols in Siliciclastic Sequences.* Postgraduate Research Institute for Sedimentology, University of Reading, 70–79.

WRIGHT, W. B., SHERLOCK, R. L., WRAY, D. A., LLOYD, W. & TONKS, L. H. 1927. *Geology of the Rossendale Anticline.* Memoir, British Geological Survey, Sheet 76.

Newly developed techniques to determine proportions of undersized (friable) coal during prospective site investigations

IAN H. HARRIS

University of Wales College Cardiff, PO Box 914, Cardiff, CF1 3YE, UK

Abstract: Site investigations to date have been unable to determine the presence of friable (undersized) coal within areas of extreme tectonic deformation. This is especially important in the northwest of the South Wales coalfield, where most of the coal extracted is for the smokeless fuel domestic market, which requires a consistent grade size. Coal in the proximity of the deformation zone is unlikely to be suitable for its intended market due to its small grain size.

The deformation process and fracture styles which contribute to the development of friable coal are discussed and a remote method utilizing the drilling parameters of speed of advance, thrust and torque is investigated to ascertain its suitability in determining the amount and position of friable coal within a prospective coal site. If areas and volumes of friable coal can be determined before the extraction stage of an opencast coal site, site boundaries can be amended so that they do not contain large volumes of the friable product.

New opencast coal sites are developed with predetermined markets. In the anthracite region of the South Wales coalfield the market is principally the smokeless fuel domestic market. The domestic market requires a strict control of the size of the coal product extracted.

In areas of intense structural deformation coal can become severely fractured by tectonic shearing during orogenesis and by mechanical abrasion on extraction; during subsequent transport it is degraded to a size unsuitable for its intended market. If the deformation associated with tectonism is aerially widespread, large proportions of the coal extracted from newly developed opencast coal sites may be unsuitable for its intended market. Finding new markets for this small grained anthracite is both difficult and expensive.

Ideally, in areas of known structural complexity, it would be beneficial to estimate accurately (quantify) the proportion of undersized coal before extraction, i.e. during the site exploration stage. There are currently no known methods which can accomplish this. In an attempt to determine the volume and position of friable coal two newly developed methods were investigated: (1) a drill rig penetrometer; and (2) structural cross-sections.

The drill penetrometer is attached to the drill rig during exploratory drilling. The penetrometer measures the speed of advance of the drill bit, the thrust applied to the drill string and the torque supplied by the staffer motor to rotate the drill string. An increase in the amount of thrust can increase the speed of advance. Higher speeds of advance are normally associated with less competent lithologies; however, in this example, the speed of advance has increased due to increases in the downforce on the drill string and not because the lithology is less competent. Large-scale increases in the torque on the drill string may also affect the speed of advance; conversely, decreases in either thrust or torque can reduce it. Previous research (Smart 1976, British Coal internal document) has indicated that thrust and torque are interrelated with the speed of advance.

Structural cross-sections can be used to determine whether thickening or thinning of coal seams occurs. The construction of cross-sections perpendicular to strike indicates areas which are under- or overlain by thrusts. They also enable the thickness of the coal to be determined accurately.

In tectonically over-thickened or over-thinned areas, thrust discontinuities shear the coal, making the average product grade much smaller than normal; thrusts occurring under or over the seam can also cause degradation of the seam because of strain transfer from the thrust to the over- or underlying strata.

Undeformed coal seams have a normal stratigraphic thickness which may vary as a result of depositional processes. Any variation not attributable to sedimentary processes may be associated with tectonic movements causing thrust-thickening/thinning of the coal seam.

South Wales coalfield

The South Wales coalfield developed along the northern margin of the Variscan orogenic belt. It

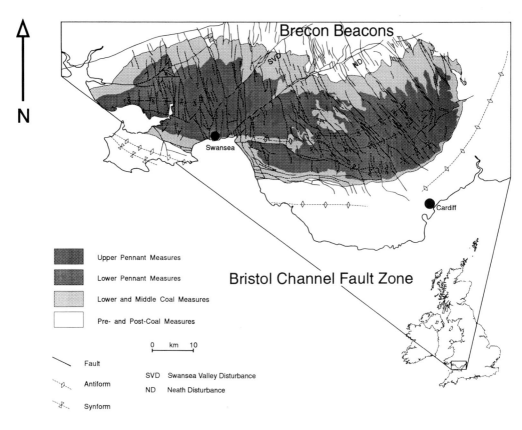

Fig. 1. Location map of South Wales coalfield.

is an elongated asymmetrical syncline measuring 96 × 30 km, lying north of the Bristol Channel Fault Zone and south of the Brecon Beacons (Fig. 1). The current outcrop pattern is an erosional remnant of the original Silesian basin. Recent basinal studies have indicated that the basin formed as a response to tectonic loading ahead of the northward propagating Variscan front (Gayer et al. 1991) caused by a late Dinantian compressive phase. From west to east the amount and style of Variscan deformation varies greatly. The structures in the eastern half of the coal basin are west–east trending folds and thrusts which rotate to SW–NE and become less intense towards the eastern margin. Tectonic shortening reduces from approximately 40 to 10%. The structure in the western half is characterized by the Ammanford and Cwmtyrch compression belts and the Llannon and Trimsaran disturbances; the latter are reactivated thrust structures. Generally, the structures trend NE–SW or NW–SE depending on whether reactivation of the underlying pre-Variscan structures occurred. Large-scale (> 50%) shortening has been calculated for structures in the northern area (Frodsham 1990). Deformation in the northwest of the coalfield appears to be far greater than any other area within the coalfield; this may be due to the presence of the Welsh Massif creating a buttressing effect on the Variscan propagation (Frodsham 1990). The Coal Measures in the northwest have shown throughout their extraction history a tendency towards 'outburst' conditions in deep mines and very granular 'friable' coal in opencast operations. Most of the friable seams within the South Wales coalfield are situated within this area of structural deformation.

The sedimentary pile is up to 3.5 km thick in the centre of the basin and dates from early Namurian to basal Stephanian times. The basin-fill sequence coarsens and shallows upwards from marine mudstones and sandstones through coastal plain coal-bearing sandstones and mudstones to coarse-grained sandstones and conglomerates attributable to an alluvial braidplane

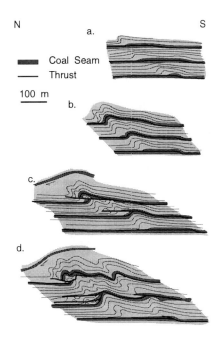

Fig. 2. Stages in progressive easy slip thrusting. After Frodsham (1990).

environment (Hartley 1993). The basin-fill facies are dominated by fluvial and eustatic processes.

Coal rank varies across the coalfield regardless of the stratigraphic level of the coal seam. Contoured isovol maps have shown a higher stage of thermal evolution in the northwest of the coal basin (White 1991); vitrinite reflectance values for the coals across the basin show concordant relations with the isovol data, reiterating the higher thermal maturity of the coals in the northwest. White (1992) argued that the increased coal rank in the northwest was developed as a response to burial before the onset of Variscan deformation. Other models which try to explain the higher rank in the northwest include percolating hot fluids through the coal measure strata and the presence of an intrusive body beneath the northwest of the coal basin (there is no geophysical evidence for the presence of this intrusive body).

Processes and fracture styles which contribute to coal friability

An unusual deformation style is seen within the northwest of the South Wales coalfield; it is unlike any deformation within the rest of the coalfield. The presence of several unusual fracture styles has also been described. The relationship between the deformation style and these fractures and their significance to coal friability are discussed in the following sections.

Processes

Variscan deformation in the coalfield is intense, especially in the northwest. Until recently the complex deformation has been described as an 'incompetent deformation', suggesting relatively low stresses acting on the predominantly mudrock sequence. More recent evidence has shown that the structures have been systematically developed in an unusual and unique way (Frodsham 1990).

The style of thrusting is very different from normal thrust propagation, with long thrust flats and short, low angle ramps. The flats are generally parallel to subparallel (within 5°) to the coal bedding and unlike normal thrust systems it appears that movement has not occurred on a single discontinuity but on several discontinuities simultaneously. This deformation system is referred to as easy slip thrusting (EST) and in extreme areas of deformation is termed progressive easy slip thrusting (PEST) (Frodsham 1990). Easy slip thrusting within the South Wales coalfield has been described as being similar in operation to that outlined by Hubbert & Rubey (1959). For EST to occur, overpressuring of the strata is necessary otherwise the overburden pressure would be too great to allow movement on more than one discontinuity. Hubbert & Rubey (1959) recognized that the frictional stress is not a function of the normal stress acting across the fault surface, but rather of the effective normal stress, i.e. the normal stress minus the pore fluid pressure. Therefore if the pore fluid pressure approaches the solid overburden pressure the frictional resistance to movement is negligible.

Coal seams within the South Wales coalfield are generally overlain by very fine-grained mudstone and siltstone and are underlain by impervious seat earths. The coal seams are recognized aquifers. Cleat mineralization studies conducted on the South Wales coalfield have proved the presence of fluids within the coal seams during orogenesis (Gayer et al. 1991). During orogenesis the fluids trapped within the coal seams could easily become overpressured due to the inability of the fluid to escape.

A comparison of the EST deformation outlined above with the piggy-back model of Boyer & Elliott (1982) shows differences in three important aspects (Fig. 2) (Gayer et al. 1991):

1. Thrust sequences normally consist of flat detachments from which ramps cut upwards in an imbricate fashion. In these deformed coal regions there are no major ramps and there are multiple thrust detachments.
2. In piggy-back thrusting the thrusts develop from the hinterland to the foreland, with the highest thrusts forming first. This causes the early thrusts to be deformed by later thrusts; this is not seen in South Wales, where the higher thrusts often lock up and affect the lower thrusts.
3. In normal thrust sequences, a thrust stops moving when thrusting propagates forwards to a lower thrust. In EST there is evidence for simultaneous movement along several thrusts.

Fracture styles

There are several styles of coal fracture noted within the South Wales coalfield; these include cleat and conchoidal fractures, which are common throughout the basin. Within the northwest less common fracture styles, termed feather and slickenside fractures, are seen. These latter fracture styles have been closely related with coal friability (Barker-Read 1980; Dumpleton 1989) and are absent in areas of only slight deformation.

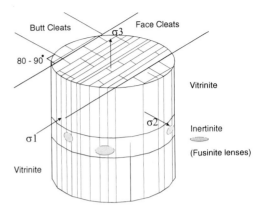

Fig. 3. Geometric relationship of face and butt cleat. Adapted from Law (1993).

Cleat fracture. In most coalfields the predominant fracture in coal seams is the joint system referred to as cleat. These are planar fractures which occur on a variety of scales. In the South Wales coalfield they rarely intersect fusinite-rich bands; this limits their duration. Two sets of cleat fracture develop during burial and compaction, termed face and butt cleat. The face cleat is more continuous and highly developed. Face cleat develops during the late stages of burial and compaction and is thought to form as a mode 1 failure, striking parallel to the maximum principal stress. During the development of cleat there is little difference in the magnitude of the principal, intermediate and minimum stresses. Butt cleat is rarely as well developed as face cleat and is sometimes curved as opposed to planar. Butt cleat often terminates against two face cleats (McCulloch et al. 1974; Law 1993). Face and butt cleat normally show an orthogonal relationship (Fig. 3). Variations from this orthogonal nature occur where the cleat has been developed within a fold, rarely seen within the South Wales coalfield, or where differential compaction has occurred beneath the seam during coalification.

Variations in cleat spacing have been attributed to the rank and thickness of the coal. Thin coals have a closer spacing. Law (1993) stated that the cleat spacing decreases with rank and showed that the average face cleat spacing for lignite was 22.0 cm compared with 2.0 cm for anthracite; similar results were reported for butt cleat. This contradicts the work of Ammonsov & Eremin (1963) and Ting (1977), who stated that the cleat spacing decreased from lignite to

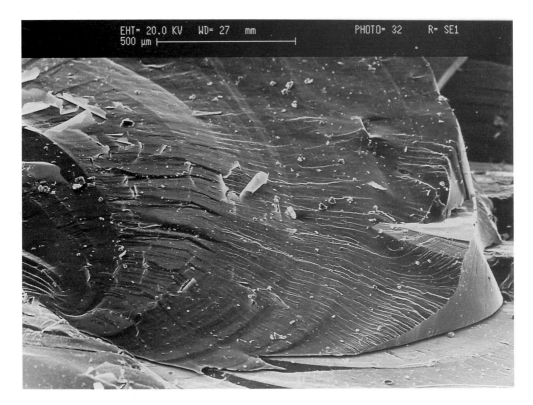

Fig. 4. Scanning electron photomicrograph of a conchoidal fracture in anthracite.

medium volatile coal, but then increased from medium volatile coals to higher rank coals. Law (1993) has attributed the development of more cleat in higher rank coals to a relationship with the mechanical strength of the material, i.e. the rheological properties of the material during coalification.

The mineral composition of coals also affects cleat development. The ash content within two seams of the Mesaverde Group showed a change with cleat spacing between two seams. The high ash coal (12.3%) had an average cleat spacing of 6.43 cm, whereas the lower ash coal (6.0%) had an average cleat spacing of 3.99 cm. This is associated with the rheological properties of the low ash coal (Law 1993).

Little research has been undertaken on the cleat relationships within the South Wales coalfield. Field evidence to date has shown that the cleat spacing decreases at maceral boundaries (at this scale, cleat may be referred to as micro-cleat, i.e. fractures which are planar and parallel with cleat), e.g. between vitrinite and inertinite. Cleat is also more highly developed within the glassy vitrinite areas than in the duller inertinite areas. Cleat has not been seen to continue through fusinite bands.

Conchoidal fractures. Conchoidal fractures, also known as rib markings, are common in the South Wales coalfield (Fig. 4). There is no evidence of movement on these fractures. This type of fracture develops in homogeneous substances, e.g. the vitrinite bands within the anthracite coals in the northwest of the coalfield. They are less well developed within the inertinite bands. The fracture has a similar morphology to those seen in semi-brittle homogeneous substances, e.g. obsidian.

Cleat and conchoidal fractures rarely contribute to coal friability, except where the cleat is very closely spaced and open. Cleat and conchoidal fractures are rarely detrimental to coal quality; it is later tectonically overprinted fractures which contribute to the friability of the coal within the northwest of the South Wales coalfield. These fractures have been termed feather, slickenside feather and slickenside fractures.

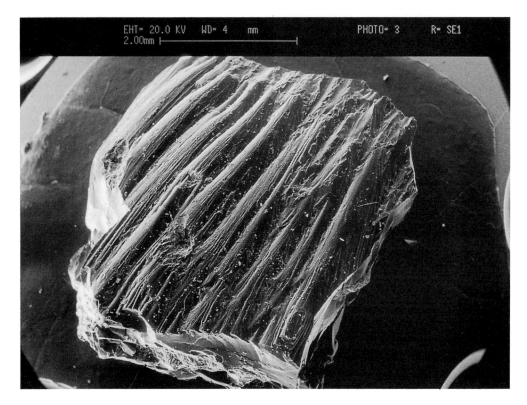

Fig. 5. Scanning electron photomicrograph of a feather fracture in anthracite.

Fractures associated with friability

Feather fractures. The development of feather fractures is not fully understood. Field evidence suggest that they overprint cleat and conchoidal fractures. Within the South Wales coalfield there are few reported instances of feather fractures east of Methyr Tydfil within the bituminous coal region. Feather fractures are more commonly seen in the higher grade anthracite areas. Their development may be dependent on the rheological properties associated with the higher thermal maturity of the coals within the anthracite belt.

Feather fractures are conical surfaces striated radially from an apex. They range in size from millimetres to centimetres at the base of the cone. There is no evidence of displacement along the cones, although micro-movement cannot be ruled out. Individual striations originate at the apex of the cone and can be seen to bifurcate in some examples; they consist of alternate grooves and ridges of similar proportions which intersect. To date cones have not been seen to overlap each other; generally, one abuts another and is terminated by the dominant cone. The overall effect of their morphology is a horsetail pattern on the cones (Fig. 5), which is especially apparent in highly developed examples. This horsetail pattern also develops on elongate prismatic surfaces, its surface morphology identical to the conical examples. The symmetry of the conical examples may suggest that they have developed with the principal compressive stress parallel to the long axis of the cone. The formation of a conical shape suggests that the intermediate and minimum compressive stresses are identical, i.e. a uniaxial stress field. The elongated prismatic variety may have developed in a local stress field which was not uniaxial. Cone orientations can become more complex during thrust events due to the reorientation of the coal seam during EST. To date the process which creates the feather fracture is unknown.

In relatively undeformed areas of anthracite it has been noted that the long axis of the feather fracture is perpendicular to the coal bedding; this indicates that the maximum principal compressive stress is vertical and substantially greater than the intermediate and minimum

Fig. 6. Photograph of slickenside fracture. Note the highly developed face cleat to the right of the fracture and the curved nature of the fracture.

stresses. From the evidence above it appears that the feather fracture is developed in a different stress field to the cleat fracture. One explanation for this difference may be the increased subsidence of the basin, increasing the overburden weight and therefore rotating the principal compressive stress from horizontal to vertical.

As with cleat fractures, mineralization has been noted on these feather fractures, indicating their open nature during orogenesis. It is highly likely that they developed under high pore fluid pressures.

Slickenside fractures. Slickenside fractures are characterized by a polished surface, indicative of in-seam movement (Fig. 6). Striations are present on most of the polished surfaces; dip-slip and oblique-slip examples are observed. Slickenside fractures are easily delimited from feather fractures by the absence of features on the failure surface. In some seams there is evidence of slickenside fractures following a previous line of weakness, i.e. feather fractures. Small-scale movements polish the surface of these fractures but do not totally destroy the characteristic horsetail pattern of the feather fracture. These are known as slickensided feather fractures. It appears that slickenside and slickensided feather fractures were developed during EST and it is these fractures which make the coal seam friable. The tectonic shearing associated with EST severely affects the strength of the coal seam, creating large amounts of granular product. The process which creates this granular product is termed nodular encapsulation (Barker-Read 1980).

It is unlikely that the stress field which created feather fractures contributed to the development of slickenside fractures. Field evidence has suggested that these slickenside fractures are associated with the thrust deformation termed EST. During thrusting the principal compressive stress was horizontal, necessitating a rotation of the stress field from a vertical to a horizontal principal compressive stress.

Relationship between friable coal, petrography and geological structure

The amount of friability within a seam is rarely homogeneous and is ultimately dependent on several interrelated parameters. In moderately deformed coal seams, the duller maceral inertinite has shown a greater inclination to become friable. The rheological properties of a seam have also been identified as being influential in creating friable coal. Comparison of a cannel and humic coal within similar tectonically affected areas has shown that the humic coal is more susceptible to friability. The largest influence on the development of friable coal is the extent of deformation which has occurred in proximity to the coal seam. Several factors affect the amount of strain a seam undergoes; these include the presence of rigid units above or below a seam which concentrate strain into the coal seam, the scale of displacement on the thrust discontinuity and the proximity of the coal seam to the thrust discontinuity. These influences are discussed in detail below.

Field evidence has shown that in moderately deformed coal seams the dull inertinite macerals are usually more friable than the shiny vitrinite macerals. Generally in these slightly deformed areas, the discontinuities which create friable coal occur at the junction of the vitrinite and inertinite macerals. In areas of heavy deformation all macerals show the same degree of friability, therefore the occurrence of friable coal cannot solely be attributed to the petrographic properties of the coal. It appears that the physical properties of vitrinite are less sensitive to smaller scale strains than inertinite.

There also appears to be rheological controls on friability, noted when comparing cannel and humic coal from similar tectonic settings. It was observed that cannel coal was less deformed than humic coal (Frodsham 1990) and was less likely to become friable. An area within a coal seam will become friable only when the stress applied during deformation is greater than the intact strength of the coal either side of the discontinuity. In areas of low strain, e.g. <10% shortening, EST may only occur on several discrete discontinuities, with the resultant nodular encapsulation creating distinct areas of friability either side of the discontinuity. Conversely, in areas of high strain (>40% shortening), the whole seam may be affected by EST; the nodular encapsulation related to the in-seam movement creates a completely friable seam.

A good correlation between the friability of a seam and the displacement of the thrust has been documented (Frodsham 1990). In areas which underwent large displacements the scale of strain was great enough to create total friability in the coal seam. This compares markedly with small-scale thrusts which show limited friability patterns, where the friable coal is related to the dicontinuities. Therefore friability can be attributed to strain magnitude. The recorded magnitude of strain increases with the proximity of a coal seam to a thrust and the amount of displacement that occurred on a thrust.

Friability has also been recognized in the limbs and hinges of folds. In small-scale folds (<2 m amplitude) friable coal has been noted in the forelimb, and appears to be created by multiple slickenside surfaces indicative of movement. However, on large-scale folds (>7 m amplitude) it is rare to recognize friable coal within the limb region; friability is formed by the flexural slip in these folds and occurs only in the hinge region. Normally the deformed coal in the hinge is connected with later thrust accommodation structures which have made the hinge region thicker than the limbs.

The effect of overlying competent lithologies cannot be underestimated. In thrust discontinuities overlain by competent sandstone units, small thrust displacements on large-scale structures can cause high levels of friability. The overlying competent lithologies act as a rigid unit to strain which channels the strain into the relatively weak coal seam. Similarly, some seams have dirt (mudrock) partings between two leaves, which guide the strain into the underlying and overlying coal seams, creating sizeable areas of friable coal.

In regions of intense deformation levels of friability can change markedly from one area to the next due to the localized development of fold and thrust structures.

The effects of tectonic overprinting must also be accounted for. For example, in an early deformation stage, a fold may create friability within a certain area, e.g. the limb of a small fold. If this structure is later overprinted by a thrust, then the friability pattern and the extent of the friability may not be comparable with the scale/type of overprinted structure. Thus in areas of high progressive deformation the structure originally associated with the friability may no longer exist. This can affect the predictive accuracy of determining friability by cross-section construction.

Drill rig penetrometer

The previous sections have discussed the processes and fracture styles which contribute to coal friability. These have been determined when the coal is in outcrop. Any technique developed to estimate the amount of friability present within a prospective coal site will have to rely on other methods due to the inaccessible nature of the coal. Coal can be investigated from core, but this is impractical on a large scale due to poor recovery in tectonically disturbed areas and its prohibitive cost.

One technique which has been developed to quantify friability by a remote method is a penetrometer device which differentiates the ease with which coal can be drilled. From *in situ* evidence it has been recognized that friable coal gives a higher penetration rate than coal which is competent, thus coal friability can be differentiated via a speed of advance measurement.

This section describes the initial testing of the penetrometer device and describes the para-

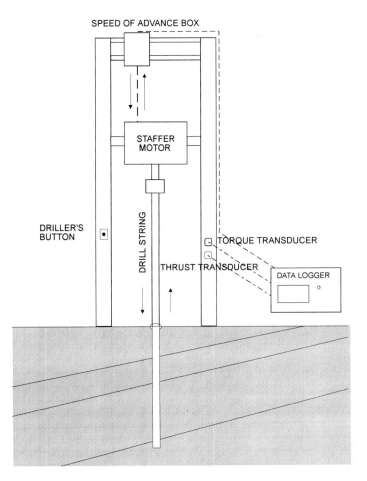

Fig. 7. Schematic diagram of the logging instrumentation measure penetrometer.

meters which are used to delimit coal seams from the mudrock sequence surrounding them. A case study is presented in which the ease with which coal can be drilled is compared with a particle size distribution analysis. This was undertaken on the coal flushed out of the drill top during excavation to ascertain if there is any correlation between the speed of advance and particle size distribution of the seam sites tested.

Present geophysical logging techniques have proved to be inadequate in delimiting areas of friable coal within normal measures. This is partly due to the unsuitability of using unlined holes in these heavily deformed areas because of concerns over the loss of the radioactive source and also due to the high cost of such analyses. Previous work has attempted, with little success, to adapt the techniques of combined gamma, long spacing density, bed resolution density, neutron and calliper logs to delimit friable coal (Frodsham 1990).

Previous research to investigate friable coals by techniques other than wireline logs has also been attempted. One technique was the enhancement of the drillers log (Smart 1976, British Coal internal document). This recorded changes in the ease of drilling related to differences in the competence of the lithological units. One of the parameters tested was the accurate monitoring of penetration rate. The initial results were compared with point load index records taken from an adjacent core and showed a correlation between the speed of advance (normalized for torque and thrust differences) and the competence of the lithological unit.

A similar technique was attempted by Frodsham (1990). Using a standard drill rig, a borehole was drilled. On reaching a coal seam, the drillstring was paused and the blast hole rig elevated off the ground. This created 230 kg loading on the drill bit. The non-rotating bit pierced the coal at a rate proportional to the strength of the coal. It was intended to compare this penetration rate to cored material taken from either side of the test hole in an attempt to correlate drill penetration rates with Schmidt hammer rebound values. The results were not conclusive because of experimental problems, which included the non-recovery of core and coal being extracted before tests around the drillhole could be completed. However, this does not infer that the test is unsuitable for the delimitation of friable coal.

Robertsons Geophysical Logging Ltd, in association with British Coal, have been conducting tests in the South Wales region using 'logging instrumentation mesure' apparatus, incorporating a penetrometer device. The 'logging instrumentation mesure' attaches directly to the drilling rig, with changes in the drilling parameters sensed by transducers and recorded electronically.

In operation the drill rig develops hydraulic power, which facilitates normal drilling functions. Rotary power for the drill string is provided by the staffer motor. Travel of the staffing box is controlled by two vertical rams connected to the staffer motor by chains. Hole flushing is provided by pressurized water/air.

The logging instrumentation measure (Fig. 7) measures three parameters which are related to the ease with which a lithological unit can be drilled.

1. Speed of advance. This is the speed at which the drill string progresses through the lithological units. The speed of advance is monitored by a potentiometer attached to the top of the drilling rig, which measures the displacement of the staffer motor. It is suitably geared to allow movement through a full 6 m (or 3 m depending on the rig used), allowing simple rod extensions. The resolution of the device is satisfactory for determining coal beds.
2. Thrust. This is the downward pressure exerted by the hydraulic ram onto the drill string. The penetration rates during drilling are directly related to the amount of downthrust on the hydraulic ram. Any differences in the thrust pressures must be noted. If the speed of advance increases but the thrust remains constant, then the increase can be attributed to a less competent lithology. The thrust value recorded by the logging instrumentation measure is the amount of downward pressure exerted by the drill rig. In deep holes the weight of the drillstring is not taken into account, therefore in any hole > 10 m the thrust reading must be normalized for the extra downward pressure exerted by the weight of the string.
3. Torque. The equipment measures changes in the torque output from the staffer motor. This can be affected by different types of lithology, e.g. very soft lithologies clog the bit, resulting in slower initial penetration rates, and a gradual increase in torque may be noted when the clogged material is flushed. However, in friable coal the bit does not clog, but because of the ease of drilling the torque values decrease quickly.

A geophysical logging suite accompanies the logging instrumentation measure penetrometer data. The gamma log is a measure of the amount of radiation within a lithology. The radiation is primarily limited to potassium, which is normally concentrated in shale and shale-rich horizons. For this reason the gamma tool is good for delimiting lithologies. The density electron log operates as follows. The rocks in the borehole are subjected to gamma ray bombardment. Any attenuation (Compton scattering) between the source radiation and subsequent arrival at the detection point is a function of the electron density. British Coal use both long spacing density and bed resolution density methods. Coal has very low densities and can be delimited by very low density peaks.

The combination of gamma ray and density (electron) logs is successful in outlining coal seams within a mudrock sequence. Calliper logs are also used, but are not discussed here.

Case study

Previous experimental data have proved the ability of the logging instrumentation measure penetrometer in delimiting coal seams. This is shown in Fig. 8, which outlines a coal seam with two leaves: an upper leaf at approximately 45.5 m, an interseam area at 47–47.1 m and a lower leaf at 47.2 m. The speed of advance shows a gradual increase in the upper leaf, a rapid decrease in the interseam area and is consistently high in the lower leaf. Thrust on the drillstring decreases markedly in these coal areas and the torque increases gradually in the interseam area. The gamma ray and density

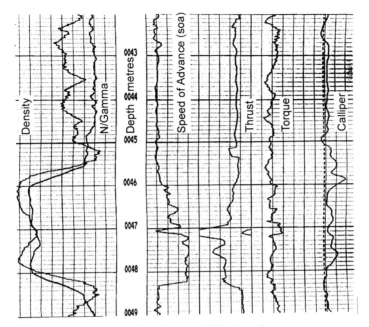

Fig. 8. Penetrometer and geolog traces of a coal seam within the coal measures.

Table 1. *Individual and total sieve proportions (<0.425 mm) for sites 1, 2 and 3*

Site	Hole/sieve size (mm)					Total proportion	Mean
	0.425	0.250	0.125	0.045	<0.045		
1a	0.05	0.15	0.32	0.01	0.02	0.55	
1b	0.04	0.11	0.10	0.15	0.01	0.40	0.45
1c	0.05	0.12	0.21	0.003	0.02	0.40	
2a	0.06	0.14	0.11	0.12	0.004	0.43	
2b	0.07	0.19	0.15	0.16	0.01	0.58	0.53
2c	0.06	0.18	0.13	0.15	0.004	0.53	
2d	0.06	0.18	0.15	0.19	0.01	0.60	
3a	0.06	0.15	0.13	0.06	0.006	0.40	
3b	0.05	0.13	0.12	0.05	0.005	0.36	0.39
3c	0.06	0.13	0.12	0.08	0.007	0.39	
3d	0.05	0.12	0.13	0.09	0.04	0.40	

(electron) logs also show the presence of two leaves in the coal seam.

At this point of the research it was unknown whether the penetrometer could delimit friable coal, even though it could outline coal seams in mudrock sequence. To determine this, several short holes were drilled on a working site, where *in situ* coal will eventually be seen. The results can thus be compared with reproducible *in situ* tests to determine positively any correlation.

The areas chosen for the short holes were in an area of an anticline which was plunging 18° towards the northeast. The northerly limb dipped between 60° and 75°; the southerly limb was more shallow at 20°. In this case study the penetrometer results were compared with drill flushings collected at known depths within the coal seam during the tests. Friable coal has a mean grain size of $<250\,\mu$m.

It was assumed that although the drill's advancement degraded coal to an artificially small grain size, areas of friable coal have greater proportions of smaller grained product due to the nodular encapsulation associated with PEST deformation. The more friable areas have a greater proportion of smaller grained product than areas which are less friable.

For each borehole the drill operator halted

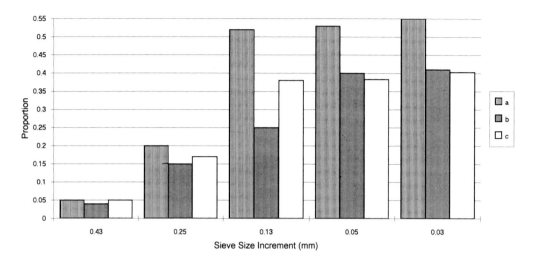

Fig. 9. Site 1: cumulative sieve proportions.

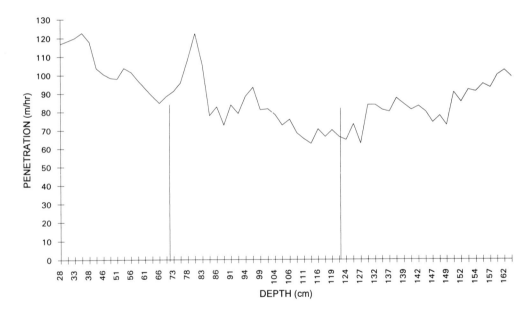

Fig. 10. Site 1: speed of advance; a, b, c, left to right.

penetration when a coal seam was entered and drilling recommenced for a set distance (generally 50 cm). The coal flushings from this area were collected during drilling. After a further 0.5 m drilling was once again halted and any remaining coal in the drillstring air-flushed so that any lower collections would not be contaminated by coal from the higher level. The procedure was repeated every 0.5 m until the seat earth was encountered. In areas of interspersed coal seam and rashings (deformed mudrock), the collected samples were rejected.

In the laboratory the flushed coal was dried at room temperature and sieved for 30 minutes in a

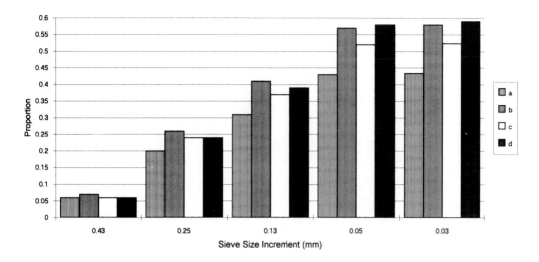

Fig. 11. Site 2: cumulative sieve proportions.

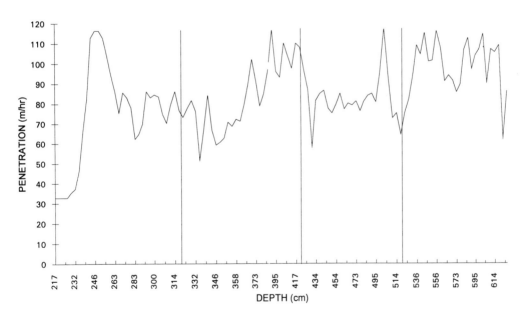

Fig. 12. Site 2: speed of advance; a, b, c, d, left to right.

series of sieves ranging from 2 mm to <0.045 mm by a laboratory sieve shaker. The results are shown in Table 1 for three of the boreholes.

Site 1

Data from Table 1 indicate that the upper area (a) of the seam has a much larger proportion of small grained (friable) coal than areas (b) and (c). This is seen graphically in Fig. 9. Area (a) has 55% product equal to or less than sieve size 0.425 mm, compared with 40% for areas (b) and (c).

The three areas delimited in Fig. 10 show two similar traces for the speed of advance, i.e. areas (b) and (c) and one notably higher (a). These

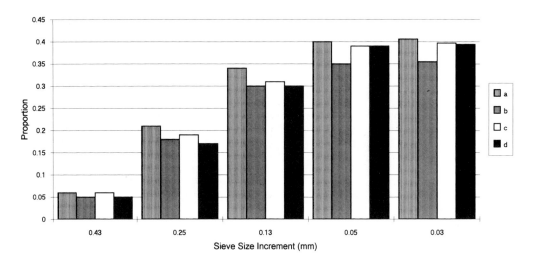

Fig. 13. Site 3: cumulative sieve proportions.

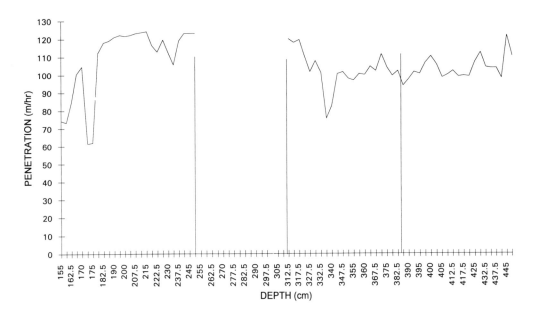

Fig. 14. Site 3: speed of advance; a, b, c, d, left to right.

observations were made without normalizing the speed of advance curve with respect to torque and thrust. Comparison of the speed of advance with drill flushings shows a good correlation for areas (a), (b) and (c). Area (a) was easily picked out by both techniques as being less competent. The spike noted above area (c) is the seat earth.

The seam in this area has an actual thickness of 1.2 m, much less than the average of 2.7 m. This may suggest that the seam has been heavily thinned, with the upper area of the seam proximal to the plane of discontinuity.

Site 2

Drill flushing data indicate that areas (a) and (c) have a lower proportion of fine-grained material than areas (b) and (d). Areas (a) and (c) have 43 and 54% and areas (b) and (d) 58 and 60%, respectively (Fig. 11).

Figure 12 shows the penetrometer trace for the four areas outlined above; although areas (a), (b) and (c) are similar for the speed of advance, subtle differences can be noted. Area (a) has the lowest speed of advance followed by (c) and (b); this correlates well with the sieve flushing data. Area (d) has the highest speed of advance of the four sample sites, correlated with the highest proportion of small sieve grades obtained in the drill flushing test.

The actual thickness of the seam in this area is 2.74 m, very close to the average stratigraphic thickness at the site. The evidence from the two tests indicates that friable coal may be present in the bottom 0.5 m, which may have been affected by EST.

Site 3

The cumulative percentage for the four areas tested at site 3 shows remarkably little difference. Area (b) has a slightly lower cumulative proportion (36%) than areas (a) and (d) (40%) and area (c) (39%). The results for the drill flushing test can be seen in Table 1 and Fig. 13. The overall sieve proportions are well below those for sites 1 and 2.

In the speed of advance analyses (Fig. 14) one area of data is missing from the on-site logs. This is directly correlatable with area (b), therefore a comparison for this cannot be made. Like the drill flushings data, the three speed of advance results are remarkably similar; comparison with earlier boreholes reiterates the similarity between the results.

One anomaly is noted when comparing the three test areas. In sites 1 and 2, an increase in the small-grained proportions is generally mirrored by the speed of advance, i.e. a lower proportion of small-grained product generally means a slower penetration rate and a more competent unit. The average speed of advance for site 3 is much higher in the coal regions than for either of the two other sites, even though the proportion of small product grade is more commensurate with a reduced speed of advance. For example, site 1 (b) and (c) and site 2 (a), (b) and (c) show lower proportions of fine-grained coal and slower penetration rates. A possible explanation may be that during moving the drill settings were altered.

The actual thickness of the seam is 2.7 m. From the drill flushing investigation it appears that the seam as a whole is fairly competent; however, the anomalously high speed of advance must also be taken into account.

Conclusions

The deformation associated with EST in the northwest of the South Wales coalfield makes the planning and development of new opencast coal sites an expensive and speculative proposition. The processes and fracture styles which contribute to friability, although understood, are unsuitable for determining the volume of friable coal in opencast sites.

To date the only remote method which can qualitatively determine the amount of friable coal is the construction of accurate cross-sections perpendicular to strike; these can show where the coal seam thickens or thins tectonically (by EST). However, evidence that these deviations from mean coal thickness are tectonic and not associated with sedimentary processes is not always obtainable. This technique suffers from major disadvantages, the main one being the production of a large number of cross-sections, which are both time consuming and expensive to prepare.

Standard geophysical methods have proved either unsuitable or too expensive for large opencast developments. The use of drillability parameters, e.g. speed of advance, thrust and torque, and their comparison with competence tests have previously been adopted to determine the presence of coal seams. However, their usefulness in detecting friable coal has not been determined.

The 'logging instrumentation mesure' penetrometer has been investigated for its use in determining the presence and thickness of coal seams within a mudrock sequence and for outlining the presence of friable coal. The case study investigated the application of the penetrometer in determining friable coal on the limbs of a plunging anticline. The results were compared with the particle size distributions of the drill flushings, which were obtained at different depths with each test site. A good correlation was achieved between the drill flushings and the speed of advance of the drillstring. Generally, the highest speed of advance was correlatable with the coal seams which had the highest proportion of smaller grained drill flushings.

In situ competence testing of the coal sites is yet to be completed. Newly developed *in situ* quantification techniques have been introduced and tested for their reproducibility; these will be compared with the drill data derived from the penetrometer tests to ascertain if the penetrometer can accurately determine areas of more friable coal. If a positive correlation is observed this will allow future larger scale penetrometer tests to be undertaken, allowing volume estimates on the amount of friable coal present within a prospective opencast coal site. With this knowledge management decisions on site progression will be scientifically based on the amount of friable product within the site and the derivation of a market for this difficult to sell undersized product.

I thank the Science and Engineering Research Council and British Coal for financial assistance in this research and R. Lisle and R. Gayer for their invaluable comments during the preparation of this paper.

References

AMMONSOV, I. I. & EREMIN, L. V. 1963. *Fracturing in Coal*. Israel Program for Scientific Translations, 11pp [in Russian].

BARKER-READ, G. R. 1980. *The geology and related aspects of coal and gas outbursts in the Gwendraeth Valley*. MSc Thesis, University of Cardiff.

BOYER, S. E. & ELLIOTT, D. 1982. Thrust systems. *American Association of Petroleum Geologists Bulletin*, **66**, 1196–1230.

DUMPLETON, S. 1990. Outbursts in the South Wales Coalfield: their occurrence in three dimensions and a method for identifying potential outburst zones. *Mining Engineer*, March, 322-329.

FRODSHAM, K. 1990 *An investigation of geologic structure within opencast cast coal sites in South Wales*. PhD Thesis, University of Wales, College Cardiff.

GAYER, R. A., COLE, J., FRODSHAM, K., HARTLEY, A. J., HILLIER, B., MILIORIZOS, M. & WHITE, S. C. 1991. The role of fluids in the evolution of the South Wales Coalfield foreland basin. *Proceedings of the Ussher Society*, **7**, 380–384.

HARTLEY, A. J. 1993. A depositional model for the mid-Westphalian A to late-Westphalian B Coal Measures from South Wales. *Journal of the Geological Society, London*, **150**, 1121–1136.

HUBBERT, M. K. & RUBEY, W. W. 1959. The role of fluid pressure in mechanics of overthrust faulting. *Geological Society of America Bulletin*, **70**, 115–166.

LAW, B. E. 1993. The relationship between coal rank and cleat spacing: implications for the prediction of permeability in coal. *Proceedings of the 1993 International Symposium, University of Alabama, May 1993*, 435–442.

McCULLOCH, C. M., DUEL, M. & JERAN, P. W. 1974. *Cleat in bituminous coal beds*. US Bureau of Mines Report of Investigations, **7910**, 25pp.

TING, F. T. C. 1977. Origin and spacing of cleats in coal beds. *Journal of Pressure Vessel Technology*, **99**, 624–626.

WHITE, S. C. 1991. Palaeo-geothermal profiling across the South Wales Coalfield. *Proceedings of the Ussher Society*, **7**, 368–374.

—— 1992. *The tectono-thermal evolution of the South Wales Coalfield*. PhD Thesis, University of Wales.

Sedimentology and its applications within the UK opencast coal mining industry

NEIL S. JONES,[1,2] PAUL D. GUION[1] & IAIN M. FULTON[3,4]

[1] *Geology and Cartography Division, Oxford Brookes University, Gipsy Lane, Oxford OX3 0BP, UK*
[2] *Present address: British Geological Survey, Kingsley Dunham Centre, Keyworth, Nottingham NG12 5GG, UK*
[3] *British Coal, Technical Services and Research Executive, Stanhope Bretby, Burton-on-Trent DE15 0QD, UK*
[4] *Present address: Golder Associates (UK) Ltd, Landmere Lane, Edwalton, Nottingham NG12 4DG, UK*

Abstract: The opencast coal industry seeks to identify and extract near-surface coal reserves. Within the UK Pennine Basin, most coal deposits suitable for opencast exploitation range from middle Westphalian A to lower Westphalian C in age. These deposits dominantly accumulated in an upper delta plain setting in which subsidence rates and depositional setting formed the primary controls on the initial distribution, thickness and quality of coal reserves. Hence an appreciation and analysis of the sedimentary setting and controls on sedimentation are important in the delineation of potential areas of thick, persistent coals.

The calculations of *in situ* coal reserves and coal to overburden ratios are important objectives during exploration for prospective opencast coal sites. Coal reserves are detrimentally affected by early, depositionally related factors such as channel wash-outs, seam splits, channel bank collapse and compaction faults. The thickness and type of interseam sedimentary rock (overburden) is directly related to the original despositional environment and the nature of the overburden can change substantially across the working area of an opencast site, reflecting sedimentary facies variations. The various sedimentary facies have unique characteristics and adversely influence mining in a variety of ways, including highwall stability problems, seam gradient changes, groundwater problems, interseam interval variations and facies-specific discontinuities.

Many of these problems are related to the presence of palaeochannels. Palaeochannels occur at different scales and contain a variety of fills, and their identification is an important aspect of site exploration. Sedimentary facies mapping, incorporating facies analysis, the construction of isopach maps and palaeocurrent studies can play an important part in an opencast exploration programme. Facies maps have many applications which may be invaluable in site planning, including the delineation of areas of sedimentary disturbance, improved understanding of seam thickness and quality variations, and changes in the interseam strata.

Within the United Kingdom, most economic coal resources available for opencast exploitation are Carboniferous in age, from middle Westphalian A (Langsettian) to lower Westphalian C (Bolsovian). During the Westphalian, many of these coal resources accumulated within the Pennine Basin, a slowly subsiding sedimentary basin located between the Southern Uplands in the north and the Wales–Brabant Barrier in the south (Fig. 1). Other areas with coal deposits shallow enough for opencast extraction include those of the South Wales coalfield and the Midland Valley of Scotland. Although the general sedimentological applications can be applied to these other areas, this paper is based on studies from the Pennine Basin.

This paper discusses UK Coal Measures sedimentology, with particular emphasis on the effects of sedimentary facies and facies variations on an opencast site scale. The overall depositional setting from the Westphalian A through to the lower Westphalian C is described, together with the sedimentary facies recognizable and the controls on sedimentation. The paper then focuses on the influence that different sedimentary facies have on coal seam character and on mining operations. In particular, the effects of channel and near-channel facies are important and are discussed in detail. Techniques are proposed for the identification and delineation of sedimentary facies to assist in the planning and extraction phases of opencast sites.

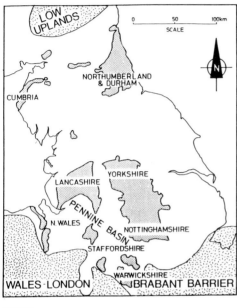

Fig. 1. Coalfields of the UK Carboniferous Pennine Basin (modified from Guion & Fulton 1993).

Opencast coal exploration

This section describes the general principles of opencast exploration to illustrate the main types of geological information available for sedimentological analysis. Opencast coal mining is the method by which coal and overburden are excavated from the surface downwards, and is restricted to the extraction of coal at, or close to, outcrop (Ward 1984). Opencast mining has a number of advantages over deep mining operations, including a greater recovery of *in situ* coal, safer working conditions and considerably lower production costs (Ward 1984). Modern opencast sites in the UK may work to depths of up to 200 m, extracting multiple seams at an average overburden to coal ratio of 22 to 1 (K. Pickup, pers. comm. 1993).

Following the initial selection of an area suitable for opencast mining, and before mining operations begin, geological exploration of the prospective site is carried out. This exploration is required to evaluate the prospective site area in terms of its economic viability, particularly with respect to parameters such as coal quality, depth and overburden ratio. In general terms, the more profitable coals are those that are continuous across the site, thick and low in ash and sulphur. During a site exploration programme, data may be available from a number of different sources, including those from previous deep and opencast working and British Geological Survey records. Each phase of exploration generates a vast amount of data. Opencast exploration currently involves drilling a network of closely spaced boreholes. Borehole spacings are determined by the complexity of the geology, and spacings averaging 30–60 m are common. Boreholes can be fully cored, partially cored or open holed. Most exploration boreholes are open holed, with information on depths and thickness of coal seams being provided by downhole geophysical logs and records of rock cuttings (MacCallum 1992). Consistent and accurate recording and presentation of geological data are essential, together with standardization and reappraisal of previous information.

Downhole, slimline geophysical logging of nearly all boreholes is now standard practice (Grimshaw 1992; MacCallum 1992). The standard geophysical logs run are gamma ray, long spaced density and high resolution density, run as a combination tool, or trisonde, within flush-jointed, metal casing (MacCallum 1992). In addition, caliper logs may be run after the casing has been pulled. Dipmeter and borehole televiewer tools have now become available, with the potential to provide much sedimentological information if analysed correctly. Information is stored on tape, allowing the geophysical logs to be viewed on a computer screen, or downloaded to give paper prints (MacCallum 1992). Geophysical logs have many uses, including seam correlation, accurate seam thickness measurements, identification of variations in seam profiles, identification of old workings in seams, water levels, identification and correlation of interseam lithologies (particularly sandstone), providing structural information (through correlation) and finally as an independent confirmation of drilling results (MacCallum 1992).

The information gathered during the exploration phase has many uses, including determining the geology of the site, proving the coal reserves in the ground and estimating the *in situ* volume of coal and overburden. Other information recorded includes variations in coal seam quality, areas of seam discontinuity, the hydro-

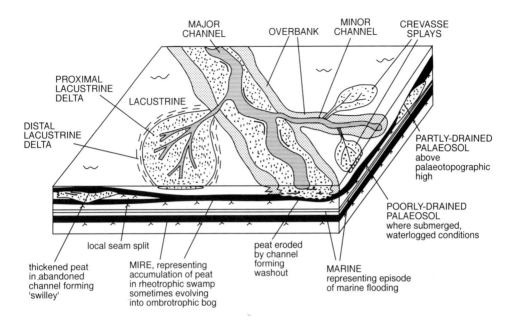

Fig. 2. Schematic facies model of the middle Westphalian A to upper Westphalian B coal-bearing upper delta plain environment, showing typical sedimentary facies recognizable. After Guion et al. (this volume, fig. 10).

geological and geotechnical properties of the strata and any variations in the overburden. The data gathered during exploration provide the basis for the schedule of amounts, the future submission of a planning application and, if favourable, a contract for competitive tendering.

Depositional setting and sedimentary facies of the Pennine Basin

Recent investigations within the Pennine Basin indicate that a lower delta plain environment occurred during the lower part of the Westphalian A, characterized by thin, laterally impersistent seams, common marine flooding surfaces and thick, coarse-grained sandstones deposited in low sinuosity channels (Fielding 1982, 1984; Guion & Fielding 1988; Guion et al., this volume). From middle Westphalian A until lower Westphalian C times, upper delta plain conditions were established, marked by finer grained, more sinuous channels, thicker and laterally persistent coals and fewer marine horizons (Fielding 1982, 1984; Guion & Fielding 1988).

The upper delta plain was a low-lying area covered by extensive, shallow freshwater lakes (Fig. 2). These lakes were infilled by sediment from overbank, crevasse and deltaic processes, supplied by a distributive network of major and minor channels. Thick, laterally continuous coals developed, indicating extensive mire development on large, infilled and abandoned sediment surfaces. Areas of reduced subsidence, typically at basin margins, gave rise to the development of thick coal seams, e.g. the Warwickshire Thick coal (Fulton 1987a, b; Fulton & Williams 1988). When regional tectonic and compactional subsidence overtook peat production, phases of mire development ended and lakes were formed. Rare glacio-eustatically driven sea-level rises gave rise to marine conditions across the basin, leading to the deposition of thin marine sequences (Leeder 1988; Collier et al. 1990).

Five facies associations, divisible into ten sedimentary facies, can be recognized from the Pennine Basin (Table 1). The depositional setting and the sedimentary facies form the primary control on the initial distribution, thickness and quality of coal reserves. Hence the lower and upper delta plain models can be used in a generalized manner to identify areas and stratigraphic horizons that are more likely to contain thick coals. However, the models are of limited use on an opencast mine scale. No

Table 1. *Characteristics of the major facies of the Westphalian (mid-Langsettian to upper Duckmantian) coal-bearing upper delta plain deposits of the Pennine Basin. After Guion et al. (this volume, table 1).*

Facies associations and facies	Lithology	Sedimentary structures	Geometry	Fossils
Pedogenic				
Mire (mainly coal)	Coal, impure coal	Lamination, banding	Extensive sheets greater than 0.1 m thick; may split or die out laterally	Plant remains
Palaeosol (mainly seat earth)	Grey to white, brown/cream or red claystone, siltstone or sandstone depending on substrate and drainage conditions. Thin impure coals	Irregular laminations disturbed by rootlets, or lamination totally destroyed. Common mottling. Abundant polished ('listric') surfaces, siderite nodules, sphaerosiderite or iron oxides, depending on drainage conditions	Generally extensive sheets	*In situ* rootlets, *Stigmaria*, *Calamites* roots in some palaeosols
Marine				
	Dark grey to black fissile carbonaceous claystone	Well-developed thin Lamination, sometimes massive	Extensive sheets	Marine goniatites, bivalves, brachiopods, gastropods, crinoids, bryozoa, forams, *Lingula*, *Orbiculoidea*, plant fragments, trace fossils
Lake fill				
Lacustrine	Medium grey to black claystone or siltstone, carbonaceous mudstone, cannel, boghead, coal	Flat lamination, rare sandy laminae and scours	Sheet-like	Non-marine bivalves, fish, ostracods, *Spirorbis*, plant fragments, trace fossils
Lacustrine delta (a) distal	Siltstones, interlaminated siltstone/sandstone forming upward-coarsening sequences	Flat lamination, current- and wave-ripple cross-lamination, climbing ripples, backflow ripples, ripple form sets, soft sediment deformation	Sheet-like to lobate deposits occur distally and beneath proximal lacustrine delta deposits, generally shows upward coarsening above lacustrine deposits	Trace fossils, plant debris, rare non-marine bivalves
(b) Proximal	Sandstone, interlaminated sandstone/siltstone, forming upward-coarsening sequences	Current-ripple cross-lamination, cross-bedding at tops of sequences, climbing ripples, wave-ripple cross-lamination, occasional trough-like scour surfaces, flat lamination, ripple form sets, backflow ripples, soft sediment deformation	Lobate to sheet-like deposits up to about 10 km across and 8m thick, with gradational or sharp bases, forming upward-coarsening sequences above distal lacustrine delta	Plant debris, trace fossils
Channel				
Major channel	Thick erosively based sandstone bodies, often with horizons of breccia of conglomerate. Subordinate siltstone, claystone	Erosion surfaces, trough and planar cross-bedding, inclined heterolithic surfaces, ripple and cross-lamination, soft sediment deformation	Elongate belts, typically 1–20 km wide, 10s of km long, greater than 8 m thick	Plant fragments and debris often abundant, rare trace fossils
Minor channel	Sandstone, siltstone, claystone, typically heterolithic, breccia, conglomerate; very variable	Erosion surfaces, ripple cross-lamination, trough and planar cross-bedding inclined heterolithic surfaces, flat lamination, soft sediment deformation, bank collapse deposits	Elongate belts typically up to 1 km wide, several km long, up to 8 m thick	Plant fragments and debris often abundant, rare trace fossils
Near-channel				
Overbank	Siltstone, claystone; sometimes interbedded sandstone	Massive or weakly laminated siltstone, passing distally into laminated claystone. Occasionally interbedded siltstone/sandstone with low dips away from adjacent channel	Elongate belts adjacent and parallel to channel margins	Abundant plant leaves and stems, *in situ* trees
Crevasse splay	Sharp based sandstones interbedded with siltstones or claystones	Current-ripple cross-lamination, flat lamination, cross-bedding, wave rippled tops	Lobate deposits adjacent to channels. Individual splays generally < 1 m thick, often vertically stacked. Thin distally away from channel	Plant debris. *Pelecypodichnus* escape shafts

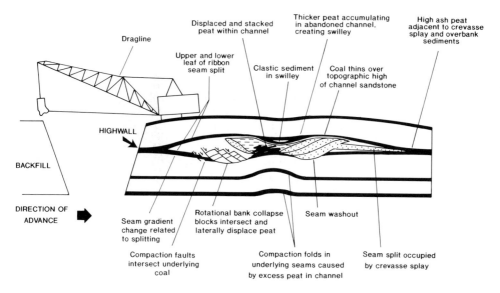

Fig. 3. Schematic diagram of an opencast highwall showing the effects of channel and near-channel facies on surrounding coal seams.

lateral change from lower to upper delta plain can be established across the Pennine Basin and, on an opencast mine scale, variations that occur are more related to local changes in sedimentary facies and local controls on sedimentation.

Controls on sedimentation within the Pennine Basin

Controls on sedimentation occur on a variety of different scales and involve a variety of different mechanisms (c.f. Fielding 1984; Fulton & Williams 1988; Guion & Fielding 1988). Of major importance on an opencast site scale are controls that could affect initial peat thickness or change the thickness and type of interseam strata, particularly the location of channel sandstones. Fielding (1987) suggested that, within a mire, the amount of peat accumulated is more closely related to subsidence patterns rather than the overall depositional environment, with optimum subsidence rates creating suitable conditions for the generation of thick coal seams. Within the West Cumbrian Coalfield, thick multi-leaf seams and reduced interseam interval thicknesses occur in the southern part of the coalfield, linked to a reduction in subsidence rates towards the Lake District block area (Taylor 1961, figs. 2 and 3; Jones 1992). Hence the optimum location for an opencast site would be in the south of this coalfield, where low overburden to coal ratios occur. A similar feature has been documented from the Warwickshire Coalfield in which various leaves of the Thick Coal unite into one seam southwards towards the slowly subsiding Wales–Brabant Barrier (Fulton 1987a, b; Fulton & Williams 1988). Hence areas characterized by reduced subsidence rates can be associated with the development of thick coal seams and reduced interseam intervals. However, subsidence rates must not be too low or the elevation too high otherwise oxidizing conditions will predominate.

Identification of syn-sedimentary tectonic controls can also have important implications for opencast exploitation. For example, areas adjacent to syn-sedimentary faults can be associated with abrupt changes in the thickness and type of interseam strata, particularly the occurrence of thick, multistorey channel sandstones, variations in seam thickness (including seam thinning and splitting) and common washouts (Bridge & Leeder 1979; Weisenfluh & Ferm 1984; Broadhurst & France 1986; Guion & Fielding 1988; Haszeldine 1989).

Contemporaneous folds can also have implications for mining, with thinning of strata across areas of reduced subsidence, typically on the crests of growth anticlines and thickening into synclines. Thicker coals can be anticipated in these areas of reduced subsidence. An example occurs across the Ashover Anticline in Derbyshire where reduced siliciclastic sedimentation is associated with a local thickening of the Alton Seam (Smith et al. 1967; Guion & Fielding 1988).

Variations in coal seam character and disruptions to seam continuity

Ideally, within the area of a prospective opencast site, seams should be thick, continuous across the site and low in ash and sulphur. However, this is rarely the case and commonly a number of variations in seam character occur, as well as disruptions to coal seam continuity. Although subsidence rates may form the primary control on coal depositional patterns, coal seams can also be affected by the presence of the surrounding sedimentary facies. The most common types of variation in coal seam character and disruptions to seam continuity are discussed in the following sections.

Wash-outs

Wash-outs occur where peat, as the precursor to coal, is partially or totally removed by the erosive action of river channels (Figs 3 and 4). Peat generally forms strongly interwoven mats of plant material that are difficult to erode. Its cohesive strength is indicated by the occurrence of palaeochannels directly overlying coal seams which commonly display little or no erosion of the underlying seam (McCabe 1984). When coal is eroded, large rip-up clasts or rafts of coal may be preserved at the base of the channel. Areas of eroded coal with resultant thin or missing coal are a major mining problem and are difficult to detect during exploration.

Seam splits

Seam slits occur where a coal divides into two or more parts (Fig. 3) and may be a consequence of: (1) increased subsidence rates related to differential compaction of sediments, tectonics or position relative to the basin depocentre (Fielding 1984, 1987; Broadhurst & France 1986). These tend to produce regional splits on scales greater than an opencast site and can be predicted from surrounding exploration; or (2) channel development close to, or within a mire (Fig. 3). Channel and associated near-channel sedimentation overlying a locally abandoned mire results in the deposition of a belt of clastic sediment on top of the mire deposits. After abandonment of the channel, peat development may be resumed above the channel and near-channel deposits, resulting in the formation of a ribbon split (Elliott 1965). These ribbon splits may occur within the area of an opencast site or extend some distance beyond it and can be linked with wash-outs and changes in seam

Fig. 4. Example of a partial wash-out of a coal seam. The coal at the base is overlain by a sandstone-filled minor channel with an irregular erosive base. The hammer for scale is 0.28 m in length. Unnamed G Seam (Westphalian C), Whitehaven, West Cumbria.

gradients (Fig. 3). Splits of this type are typically accompanied by a general thinning of the total coal and an increase in the overburden to coal ratio (Elliott 1984: p87; Ward 1984: p229). Depending on the thickness of the clastic interval in the split and the thickness of the split leaves, thin leaves of coal may have to be omitted from the extraction process, resulting in a decrease in the total amount of workable coal. In addition, Ward (1984: p153) has recorded significant bedding plane slip associated with the presence of inclined bedding within channel-related seam splits.

Channel bank collapse faults

Channel bank collapse is well documented from modern rivers and is one of the main mechanisms by which channels widen and migrate (Fisk 1947; Coleman et al. 1964; Stanley et al. 1966; Turnbull et al. 1966). Preserved channel bank collapse structures have been described from ancient sediments by numerous workers including Williams et al. (1965), Laury (1971, Guion (1978, 1987a, b), Nelson et al. (1985), Williams and Flint (1990) and Jones (1992).

Bank collapse may occur by the development of single or multiple rotational shears in cohesive near-channel sediments. This is favoured when the cut bank is unstable, typically after a flood (Laury 1971). In this situation the bank may have been oversteepened by scouring at its base and saturated and overpressured by pore waters. Rotational failure often involves the full height of the channel bank. This can occur by slope failure, where the surface of sliding intersects the slope at or above its toe position, or by base

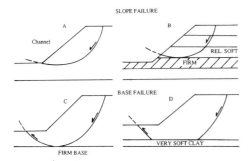

Fig. 5. The two main mechanisms of rotational channel bank collapse (from Terzaghi & Peck 1967). **(A)** Slope failure, where the fault intersects the toe of the slope. **(B)** Slope failure where the fault intersects above the toe of the slope, controlled by the position of a firm base. **(C)** Base failure, where the fault intersects below the toe of the slope, controlled by the position of a firm base. **(D)** Base failure, where the fault intersects below the toe of the slope, controlled by the position of a soft horizon.

failure where the surface of slippage passes below the toe level of the slope (Terzaghi & Peck 1967; Laury 1971) (Fig. 5). Base failure enables blocks of sediment to be emplaced below the maximum channel scour depth and these blocks will not be reworked by channel erosion.

Collapse of the bank material along the rotational failure is associated with the development of a compressive toe, resulting in the bulldozing of sediment at the base of the block (Plint 1986).

Channel bank collapse faults may have considerable economic implications because the failure surfaces accompanying collapses can intersect the underlying peat horizon (Fig. 6). Slumping of bank material along rotational failures may bulldoze the peat laterally and this peat is typically thrust horizontally over adjacent *in situ* peat (Fig. 7). Channel bank collapses can therefore be associated with irregular seam thickness variation including cut-outs, lateral seam displacement and slope stability problems. The syn-sedimentary nature of these failures is indicated by the intimate association with channels and the common 'thrusting' of peat into the actively flowing channel, resulting in its partial erosion (Fig. 3).

Differential compaction

Sand compacts to a lesser degree than mud or peat, hence abrupt changes in lithology create zones of differential compaction and can lead to the generation of compaction folds and faults (Moebs & Ferm 1982; Weisenfluh & Ferm 1991) (Fig. 3). Differential compaction around sandstone channels and adjacent mud- or coal-filled

Fig. 6. An example of a bank collapse structure at Lounge Opencast Site, Leicestershire. This comprises an inclined bank collapse block of overbank and crevasse splay sediments, pictured on the left, dipping towards a listric fault. The fault was formed by the failure of the bank sediments and truncates the Upper Lount coal (halfway up the face). This coal is absent underneath the inclined block but reappears to the right, on the exposed bench. Channel sediments occur above this bench, dipping from left to right. The coal seam at the top of the face (Smoile Seam) is undisturbed. The entire face is about 18 m in height.

Fig. 7. Vertical repetition of a coal seam as a result of channel bank collapse faulting. Both coals are identical in thickness (0.75 m) and represent the same seam. The lowermost coal is undisturbed, whereas the upper coal has been translated laterally and thrust over the *in situ* coal as a result of bulldozing during rotational failure and faulting of bank sediment. Note the undulatory base and lack of palaeosol underlying the upper coal. The associated channel forms a sandstone that overlies the coals. Cannel Seam, Top and Bottom Leaf, Nadins Opencast Site, Burton-on-Trent.

abandoned channel plugs results in the development of open folds in the surrounding strata, forming 'bow-tie' structures (Fig. 3). Differential compaction can also result in minor faulting on the margins of channels, underlying undulatory channel bases, and below channel bank collapse blocks (Fig. 3). In rare instances these faults may intersect and displace the surrounding coal seams. Compaction faults can be listric or planar and can be recognized by their limited displacement, common restriction to one inter-seam interval, association with palaeochannels and anomalous trends (typically parallel to palaeochannels) compared with tectonic fault trends.

Variation in seam thickness

Most seam thickness variations are regional and can be related to changes in subsidence rates. However, they can also be localized and facies-related. Channel sand bodies compact to a lesser degree than the surrounding strata and can form relative palaeotopographical highs, associated with the development of thinner coals (Houseknecht & Iannacchione 1982) (Fig. 3). In contrast, abandoned channels form palaeotopographical lows. These produce sites for the accumulation of thicker peat and hence the development of locally thicker coals, termed swilleys (Raistrick & Marshall 1939; Elliott 1965, 1985) (Fig. 3).

Coal seam quality

Coal seam quality can be adversely affected by the presence of moisture and impurities, including sulphur, mineral matter, chlorine and phosphorus (Elliott 1984). Sulphur in coal leads to environmental problems through the release of sulphur dioxide during combustion (Casagrande 1987). The sulphur is usually incorporated as pyrite during the peat-forming stage (Casagrande 1987). Hydrogen sulphide, formed by the microbial reduction of sulphate, reacts with organic matter to form sulphur, and reacts with ferrous iron Fe^{2+} to produce pyrite (Casagrande 1987). The sulphur content is typically higher in coals overlain by marine strata and this can be related to the availability of sulphate ions from marine waters (Williams & Keith 1963; Gluskoter & Hopkins 1970; Horne *et al.* 1978*a*; Casagrande 1987). The sulphur content may also increase where channel sandstones are in contact with coal. In this instance the sandstones form permeable conduits for the migration of sulphur-bearing solutions (Houseknecht & Iannacchione 1982).

Siliciclastic sediment can be introduced into mires from adjacent channels via overbank or crevasse splay processes (Fig. 2), or as air-fall tuffs, forming 'dirt bands'. The resulting coals close to these channels are high in inherent ash. Examples have been described by Elliott (1965, 1968, 1969), Gluskoter & Hopkins (1970) and Krausse *et al.* (1979). Coal seam quality maps, based on sulphur and ash contents, may be drawn to delineate areas of inferior coal which are unsuitable for working.

Significance and effects of sedimentary facies variations on mining operations

Before the start of mining, the working method and site layout are planned and these are influenced by local geological conditions. This section discusses how sedimentary facies and facies variations encountered during mining can influence site operations.

Highwall stability and sedimentary related discontinuities

Working of a site proceeds by the excavation of overburden strata and the exposure of successive benches of 'clean' coal. The working face is termed the highwall and the stability of this highwall is a function of many factors, including geological structure, groundwater, degree of weathering, bedding orientation, lithology,

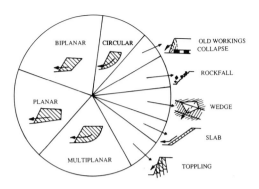

Fig. 8. Pie diagram to illustrate the main types of slope failure mechanisms within UK opencast sites (from Stead & Scoble 1983). The most common failure mechanisms (biplanar, planar and multiplanar) involve the interaction of one or more discontinuities, typically bedding and/or jointing.

mining methods, old workings, intact rock strength and the presence and strength of discontinuities (Gonano 1980; Scoble & Leigh 1982; Denby et al. 1983; Hughes & Leigh 1985).

The greatest potential for failure occurs where the highwall has numerous discontinuities or where low strength beds have a depositional dip that is greater than the dip of the interbedded coal (Gonano 1980). Hence bedding, jointing and faulting exert the major control on stability

(Mallett 1983; Stead & Scoble 1983). Failure commonly occurs by sliding along a weak horizon and at least nine types of failure mechanism are recognized (Fig. 8). Faults reduce the rock mass quality and may form the basal failure surface (Stead & Scoble 1983). In addition, they often act as a rear releasing plane for other inclined discontinuities, typically bedding. Any instability will be assisted by the presence of abundant groundwater.

Although many discontinuities are tectonic in origin and hence are not discussed further, a number are related to sedimentology. The main types include bedding produced by lithological changes, inclined bedding (bar forms), syn-sedimentary faults, compaction faults and cross-bedding (Fig. 9). Many of these are lined by plant debris, which serves to decrease the strength of the discontinuities.

Within the channel facies large-scale inclined bedding surfaces occur which form potential failure surfaces (Mallett 1983). These surfaces represent the deposits of channel bar forms and may pass through the entire thickness of a channel deposit, which could constitute virtually an entire interseam interval. Although bar forms cannot usually be detected in boreholes, the identification of channel type can generally be used as a guide to the likely orientation of the bar forms. Sand-dominated major channels typically contain downstream-accreted bar forms, dipping down-palaeocurrent and strik-

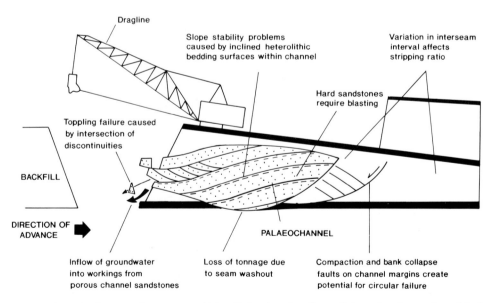

Fig. 9. Schematic diagram of an opencast highwall showing the effects of channel and near-channel facies on opencast site workings and highwall stability.

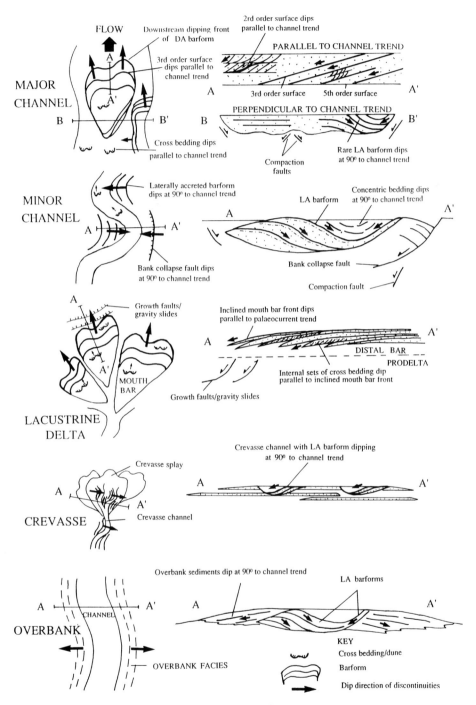

Fig. 10. Summary diagram to illustrate the main orientations of sedimentary discontinuities possible within selected facies. LA, Laterally accreted; and DA, downstream accreted. Second-, third- and fifth- order surfaces refer to the bounding surface hierarchy of Miall (1988).

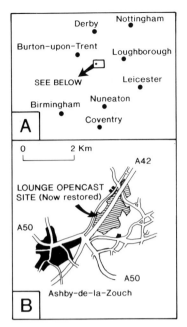
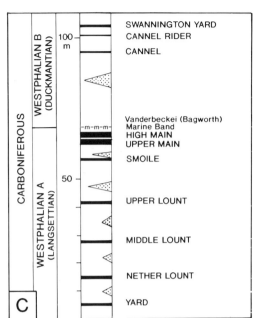

Fig. 11. (A) General map of the Midlands area. (B) Location map for the Lounge Opencast Site, Leicestershire Coalfield. (C) Stratigraphy for the site. The study interval for Figs 12–15 is between the Upper Lount and Smoile seams.

ing perpendicular to the channel trend (Fig. 10). Within minor channels heterolithic, laterally accreted bar forms are more common, producing discontinuities that dip at right angles to the channel trend (Fig. 10).

Discontinuities generated by bank collapse and compaction faults are also associated with channels and trend parallel to the channel directions. Of less importance are minor discontinuities formed by the presence of mica and organic debris on the foresets of cross-bedding. These are inclined down-palaeocurrent and failures along these small-scale discontinuities probably do not have any significant economic implications, but may be hazardous to site staff by causing minor rock falls.

The crevasse splay facies may be accompanied by minor discontinuities in the form of cross-bedding, dipping parallel to the general palaeocurrent trend, and rare heterolithic, laterally accreted bar forms within crevasse channels, dipping at right angles to the general trend (Fig. 10). The interface between individual sandy splays and interbedded mudrocks can also provide horizontal weak zones, especially if they are accompanied by organic debris. However, the discontinuities found in the crevasse splay facies are not likely to give major problems.

The overbank facies is typically massive and is relatively strong and difficult to excavate. Discontinuities are rare and would be inclined at right angles to the trend of the adjacent channel (Fig. 10). These surfaces dip gently away from the accompanying channel and represent accretion surfaces, formed by breaks in sedimentation or changing conditions.

The major discontinuities within the lacustrine delta facies consist of low angle bedding surfaces (bar forms) and sets of cross-bedding, both of which are inclined down-palaeocurrent (Fig. 10). The low angle-bedding surfaces are often mud-draped and potentially weak. In addition, small, erosively based channels may also be present proximally, with their associated sets of discontinuities. Delta deposits underlain by listric growth faults or gravity slides may also occur (c.f. Elliott & Ladipo 1981), although these are believed to be extremely rare. These would dip orthogonal to the general palaeocurrent trend of the delta (Fig. 10).

Other sedimentary facies are typically mudrock-dominated (Table 1). Mudrocks are generally weak, but lack major discontinuities and therefore tend to form stable highwalls. However, they generally deteriorate if allowed to stand for long periods of time.

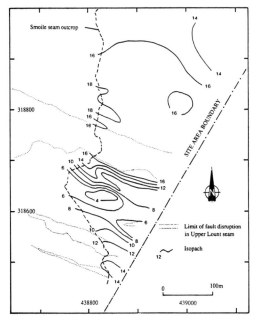

Fig. 12. Interseam interval isopach map, drawn for strata between the Upper Lount and Smoile seams, Lounge Opencast Site. Isopachs are based on surveyed data points. Note the NW–SE trending isopachs and rapid changes in interseam interval. Area of reduced interval (less than 4 m) represents an abandoned channel reach. Also note the areas of disruption to the Upper Lount seam.

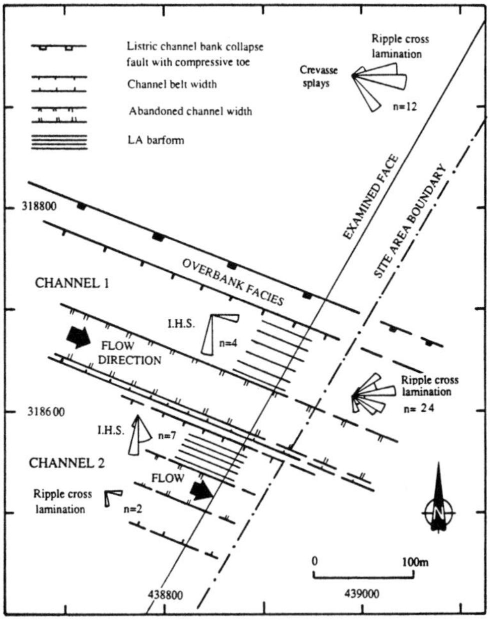

Fig. 13. Interpretation of the Upper Lount to Smoile interval based on field observations from the examined face. Channel trends are based on the extrapolation of isopachs from Fig. 12. Palaeocurrent measurements indicate that both minor channels flowed to the ESE. I.H.S., inclined heterolithic stratification; LA, laterally accreted.

Groundwater problems

The presence of groundwater can lead to the build up of pressure in discontinuities, resulting in a reduction of shear strength and the potential for instability (Houghton 1983). Old deep mine workings are one of the major sources of goundwater inflow. However, sand-dominated facies can act as aquifers if they are permeable and may provide a significant secondary source. Working these water-bearing sandstones can lead to problems, including flooding as a result of water-inflow, highwall stability problems as a result of a reduction in the effective stress along discontinuities (Walton & Atkinson 1978) and heaving of the pit floor due to groundwater pressure. Mapping of sandstone-dominated facies has value in identifying potential sources of groundwater inflow.

Interseam interval variations

Interseam interval variations are common, both in terms of thickness and lithology. Regional variations are most common, whereas abrupt variations are rare and typically associated with channel and channel-related facies (Guion 1987b, fig. 13). Sandstone-filled channels compact less than the laterally adjacent mud-dominated facies, creating a thicker interseam interval locally (Figs 11–13). The area of thickening typically forms a linear or a curvilinear feature, parallel to the channel margins (Fulton et al., this volume, fig. 14). Mud-filled abandoned channels generally form thinner interseam intervals (Figs 12 and 13), hence thicker and thinner intervals can be laterally juxtaposed. Increasing the overburden to coal ratio can have important economic consequences for the working area or the site as a whole.

Compaction around channel facies is generally associated with abrupt variations in the gradient of surrounding seams (Figs 14 and 15). These rapid changes can influence slope stability, slow down coal production and may influence coal recovery. In addition, lithological variations

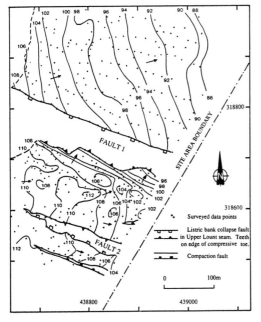
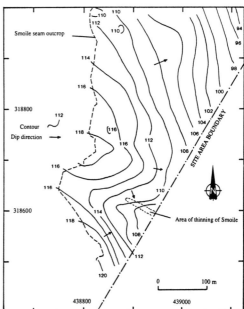

Fig. 14. Structure contour diagram, based on surveyed information (heights above OD), for the top of the Upper Lount Seam, Lounge Opencast Site. Rapid changes in seam gradient are associated with the presence of the channels and the bank collapse fault documented in Figs 12 and 13 (Fault 1).

Fig. 15. Structure contour diagram for the top of the Smoile Seam. Note the change in seam dip direction and gradient in the lower part of the diagram, above the minor channel described in Fig. 13. Localized thinning of the Smoile also occurs on one of the channel margins.

can influence plant selection and increase production costs. Unweathered siltstones and sandstones have high uniaxial compressive strengths (see Fulton et al., this volume, table 2), that may be too high for conventional plant to work. However, modern hydraulic face shovels can cope with strong strata, although costly blasting may be required to create additional discontinuities before excavation.

Sedimentology and its applications to exploration and working opencast sites

From the previous discussions of the influence of sedimentary facies on coal seams and site workings, it can be seen that the recognition and characterization of sedimentary facies, particularly channels, should form an important component of an opencast exploration programme. Sedimentological techniques can be used in the exploration phase to predict the likely depositional facies that will be encountered and their effect, if any, on adjacent coal seams and on proposed mining methods.

Sedimentology can also assist during the development of a site through continued monitoring and appraisal as working proceeds. Sedimentary features missed or poorly understood during the exploration stage can be resolved. Occasionally, further drilling may be necessary during the working of a site. The need may arise for a number of reasons, including exploration within previously poorly explored areas or in unexplored areas that may be suitable for extensions to the existing site. Strata exposed within current working faces may prove useful for the interpretation of geological problems encountered during this drilling.

Techniques used for sedimentary facies analysis are summarized in Table 2. Sedimentology can be used to provide information on the following: (1) prediction and mapping of interseam strata; (2) identification of channel zones; (3) location of resistant strata; (4) orientation and type of sedimentary discontinuities; and (5) areas of reduced seam quality.

Table 2. *Scheme for the sedimentological assessment of Coal Measures successions. Techniques used enable sedimentary facies analysis of both seam and interseam strata*

Technique	Remarks
Interseam	
Application of sedimentary facies characteristics	Application of sedimentary facies characteristics to the data set (e.g. boreholes, outcrop) to identify the sedimentary facies within individual interseam intervals
Sedimentary facies maps	Recognition of sedimentary facies can be used to create sedimentary facies maps for individual interseam intervals. Mapping palaeochannels is of particular importance. These maps can have many applications, including locating areas of potential seam variation, likely orientation of discontinuities, etc
Interseam interval isopach maps	Provides information on the trend of sediment bodies plus variations in overburden thickness. Particularly useful for identifying palaeochannels which are characterized by abrupt variations in interseam interval. These maps can be produced from borehole data using computer programs such as Geomodel
Lithology maps	Geotechnical applications. Provides information on the extent and thickness of hard and soft which affects 'diggability', plant selection and the amount of blasting required
Palaeocurrent studies	Palaeocurrents can be used to predict the likely trend of sedimentary facies within the area of a site. This generally requires study of existing outcrop, which is not common during an exploration phase. However, can be useful for extensions to existing sites
Seam	
Seam isopach maps	Can be used to identify areas with variations in seam thickness and continuity, particularly wash-out zones and swilleys
Coal seam quality maps	Can be used to identify areas with variations in seam quality
Seam structure contour maps	Gives the regional tectonic dip and identifies areas characterized by rapid seam gradient changes

Prediction and mapping of interseam strata

It is important to have an understanding of the variations in interseam strata before starting the development of an opencast site. This understanding can be greatly assisted by sedimentary facies mapping. This involves correctly identifying the different facies, establishing their dimensions and determining any lateral variations that occur. In particular, mapping of sand bodies is extremely important. The main types of data that must be considered when carrying out mapping are described in the following sections.

Outcrop and palaeocurrent data. Sedimentary facies identified from natural outcrops or adjacent sites can be extrapolated into areas of site exploration. Palaeocurrent measurements can be taken and used to determine facies trends. In general, there is a close relationship between the trend of sedimentary facies and measured palaeocurrents. Within channel facies, palaeocurrents measured from medium- to large-scale sedimentary structures are generally parallel or subparallel to palaeochannel orientations. Palaeochannel orientations can be determined from wash-out trends, mapping abandoned channels and swilleys and from interseam interval isopach maps. Palaeocurrents from proximal lacustrine delta facies are unidirectional and show little variation. Down-palaeocurrent extrapolation coincides with a thinning of the facies and a passage into more distal, less sandy deposits and lacustrine mudrocks. The crevasse splay facies generally shows more complex palaeocurrent patterns and this can be related in part to the small scale of sedimentary structure measured, i.e cross-lamination.

Palaeocurrent studies can thus be used to predict the general orientation of sedimentary facies in each interseam interval and these facies can then be extrapolated into exploration areas. One problem with this method is that some minor channels are sinuous, making extrapolation difficult. Hence other techniques such as isopach maps need to be used in conjunction with palaeocurrent analysis.

Borehole data. Cored boreholes enable the identification of sedimentary facies, using the criteria established in Guion *et al.* (this volume). However, geophysical logs are the main source of geological data from boreholes and careful interpretation of the patterns of geophysical log responses enables sedimentary facies to be inferred. It is important to identify sandstone-dominated facies and the position of the sandstone within the interseam interval. Sandstone can generally be reliably identified by its low gamma ray response. Major channels typically occupy entire interseam intervals and are sand-dominated. Minor channels may be sand-dominated at the base and typically fine upwards, with a corresponding increase in gamma ray response, but other patterns are possible if the lithological fill is complex. Lacustrine delta deposits show the reverse relationship, coarsening upwards and decreasing in gamma ray response. Crevasse splays usually form a series of thin, low gamma spikes representing the individual sandy splays.

Interseam isopachs. Sedimentary facies can be extrapolated into and interpolated across opencast sites by the analysis of interseam interval isopach patterns. The shape of the sediment body forming the facies is often reflected in the pattern of isopachs, which may be a function of both original thickness variation and differential compaction. Lacustrine deltas, which form lobate sand bodies, often have lobe-shaped isopachs. Similarly, crevasse splay facies may also have lobate isopach patterns, but on a smaller scale. This pattern may be complicated where a number of splays are stacked vertically and laterally. Channels can often be delineated by a thickening of the interval as a function of the increased sandstone component which undergoes reduced compaction relative to the surrounding lithologies. Alternatively, there may be a local reduction of the interval, representing the position of the abandoned channel that is commonly filled with compactable clays or peat (Guion 1978*b*, fig. 13; Fulton *et al.*, this volume, fig. 14). In comparison, the lacustrine facies forms sequences of mudrock with little variation in interseam thickness.

Identification of channel zones

The identification of channel facies is an important exploration requirement. The large size of major channels and their sandstone fill makes their prediction relatively straightforward. However, much more vital is the prediction of minor channels, which can considerably affect coal production if they remain undetected. Palaeochannels can be identified using the general characteristics discussed in Guion *et al.* (this volume) and summarized in Table 3. The following criteria may prove useful in identifying and delineating palaeochannels: (1) isopach maps and wash-out trends; (2) presence of overbank facies; and (3) presence of crevasse facies.

Isopach maps and wash-out trends. As discussed previously, interseam interval isopach maps can be used to detect palaeochannels, based on the recognition of anomalous variations in interseam thickness (Guion 1987 *a, b*). In addition, coal seam isopachs can be used to identify palaeochannels. These should be constructed routinely to determine variations in seam thickness. The seam underlying a palaeochannel may be affected by whole or partial seam wash-outs and the trend of the seam isopachs generally corresponds to the trend of the channel (Houseknecht & Iannacchione 1982). A seam overlying a palaeochannel may thicken into an abandoned channel reach, forming a swilley, or may thin across the top of the channel, as a function of greater palaeotopography and reduced subsidence above the relatively uncompactable sand body.

Presence of overbank facies. This facies is found in elongate belts up to about 500 m wide bordering palaeochannels and its presence indicates proximity to a palaeochannel. It is characterized by grey, unlaminated siltstones containing abundant, well preserved plant leaves and stems, and casts of *in situ* tree trunks and roots are fairly common (Fig. 16).

Presence of crevasse facies. Crevasse splays form by the breaching of a distributary channel and deposit lobe-shaped bodies of sand up to about 1 km across. Their presence can be used to infer proximity to an adjacent channel. Splays show thinning away from their source. Hence determining the direction of thickening of splays should enable the location of the feeding distributary channel (Fig. 17).

Splays may show lobate-shaped interseam isopachs adjacent to a linear belt of isopachs representing the palaeochannel. In addition, palaeocurrent readings from individual crevasse splays commonly show a radial pattern (Fig. 17) and detailed palaeocurrent analysis may enable extrapolation back to the channel source. However, sedimentary structures within crevasse splays are small scale (cross-lamination) and

Table 3. *Characteristics indicating the presence of palaeochannels (modified from Guion 1987a, table 2)*

Channel indicators	Characteristics
Erosion surfaces	Channel always have abrupt bases, generally irregular and erosive
Seam wash-outs	Wash-outs may occur in seams underlying channels. Wash-outs typically trend subparallel to the channel
Breccio-conglomerate	Erosive bases are often overlain by breccio-conglomerate lags of coal clasts, plant material, siderite and mudstone clasts. Breccio-conglomerates may also occur at various horizons within the channel fill
Fining-upward sequences	Minor channels typically fine upwards, reflecting waning flow up channel point bars or related to channel abandonment
Thick, sand-dominated successions	Major channels commonly form thick (up to 15 m) sand bodies
Bank collapse deposits	Form large tilted blocks adjacent to palaeochannels, often laterally displace coal seams, leading to seam repetition. They have trends parallel to the adjacent channel and are typically anomalous to the regional tectonic trend
Compaction faults	Small compactional faults occur above and below channels. They typically have trends anomalous to the regional tectonic trend. Faults usually have small displacements and are limited to the same interseam interval
Plant remains	Common to abundant, typically as carbonaceous debris, abundant preserved leaves and stems within abandoned channel, rare *in situ* tree trunks
Massive siltstones	Common in abandoned channels and marginal to the channel, as overbank facies
Brown coloration	Occasional feature of channels, related to the presence of siderite
Soft sediment deformation	Common, particularly in abandoned channels, related to slumping of sediment or dewatering
Anomalous dips	Related to the presence of downstream and laterally accreted bar forms, bank collapse blocks, slumps or compactional features
Isopach pattern	Channels commonly detected by construction of seam and interseam isopach maps. Interseam interval characterized by thicker and thinner isopachs, parallel to the trend of the palaeochannel. Seam isopach maps may indicate seam thinning and/or wash-out parallel to the trend of the channel
Palaeocurrent pattern	Medium- to large-scale sedimentary structures are typically unidirectional and subparallel to channel trend. Minor channels have greater variation than major channels
Thickness variation in overlying seam	Thickening of seam commonly occurs overlying an abandoned channel, forming a swilley. Thinning of seam may occur on channel margins related to the greater topography and/or to lesser compaction of sand
Overbank facies	The presence of overbank facies typically indicates proximity to a channel. The facies is characterized by the presence of massive siltstones containing abundant plant and stems and leaves
Anomalous faults trends	Related to the presence of compactional or bank collapse faults, usually subparallel to fault trend and anomalous to tectonic trend
Changes in seam gradient	Related to compaction effects around channels
Crevasse splays	The presence of crevasse splays within an interseam interval indicates proximity to a channel. Splays show increase in thickness proximal to the channel and radial palaeocurrents away from the channel

difficulties in obtaining reliable palaeocurrents occur, combined with the inherent palaeocurrent variability of this scale of structure.

Location of resistant strata

Unweathered sandstone typically has a high uniaxial compressive strength (typically from 60 to 120 MPa) compared with unweathered mudrocks which are relatively weak (30–70 MPa) (Fulton *et al.*, this volume, table 2). These values have significant implications for the ease of digging and may affect plant selection. The facies most likely to be sandstone-dominated are major and minor channels, crevasse splays and proximal lacustrine deltas. Hence facies analysis and mapping are important for the prediction of areas characterized by high uniaxial compressive strengths.

Fig. 16. Overbank facies consisting of structureless siltstones with an upright, *in situ* tree trunk. Hammer for scale is 0.39 m in length.

Orientation and type of sedimentary discontinuities

As previously discussed, discontinuities are important in influencing highwall stability. Of most significance are discontinuities that have a component of dip into the excavation, and prediction of their orientation and intensity can assist in the planning of highwall layouts. Where two sets of discontinuities intersect, e.g. bedding and faulting, stability problems are exacerbated and careful analysis is necessary.

The main types of sedimentary discontinuities and their orientation relative to their associated sedimentary facies are illustrated in Fig. 10. Most discontinuities trend parallel or orthogonal to the orientation of the sedimentary facies. Characterization and mapping of sedimentary facies trends may have important implications for the prediction of the orientation of sedimentary discontinuities. Sedimentary facies analysis can be used to create facies maps and the predicted orientation of discontinuities can be superimposed onto these maps. For instance, it should be possible to predict the likely orientations of cross-bedding, bar forms and synsedimentary faults.

Areas of reduced seam quality

Seam quality is reduced by minerals within the coal, creating high ash coals. Detrital minerals can be introduced by penecontemporaneous nearby channel and near-channel facies and incorporated into the peat. Hence palaeochannel identification would be useful for predicting potential areas of reduced seam quality, for instance where dirt bands are present on the margins of a seam split. Seam quality can also be reduced by a high sulphur content such as that derived from overlying marine horizons (Williams & Keith 1963; Gluskoter & Hopkins 1970). It is important to construct contour maps to identify any areas of high ash and sulphur.

Conclusions

There are many factors detrimental to the operation of an opencast site, including seam wash-outs and splits, slope stability problems, variations in the thickness and strength of the overburden and groundwater inflow. Some of these are directly related to the inherent characteristics and variations in sedimentary facies present in a site, particularly the channel facies. Many of these detrimental factors can be predicted during the drilling phase of exploration using a combination of sedimentological techniques.

Sedimentary facies maps are used as a basis for interpretation and may be constructed using data from cored boreholes and geophysical logs. Simple techniques such as the production of isopach maps and palaeocurrent analysis will provide information on sedimentary facies trends, thus allowing the extrapolation of facies distributions across sites. Continuous updating of isopach maps should be carried out during the exploration programme and any extreme variations detected during drilling indicate the need for more detailed exploration. The recognition of palaeochannels is a particularly important factor and may yield vital information about the distribution and trends of wash-outs, changes in rock quality and variations in overburden to coal ratios.

Once sedimentary facies have been identified and delineated, the likely orientations of discontinuities can be determined. In structurally simple areas, this could form the basis for

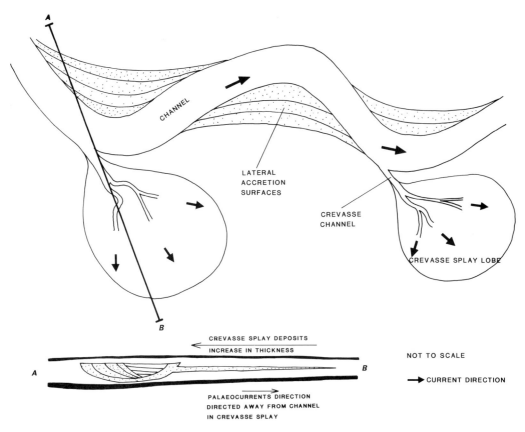

Fig. 17. Relationship between crevasse splay deposits and channels. Crevasse splays thicken towards feeding channels and show a radial palaeocurrent pattern. Careful mapping of individual crevasse splay lobes and analysis of palaeocurrents and isopachs should allow positions of channels to be identified.

planning highwall orientations. In structurally complex areas, these discontinuities may be relatively less important than tectonically induced features such as faulting and jointing.

The application of sedimentology can lead to an improved understanding of variations in the thickness and quality of an individual seam across the site, as well as variations in the interseam strata, which can be invaluable in site planning and the development of a working method. Sedimentology has much to contribute to opencast coal exploration and development, and the application of the techniques discussed is likely to be cost effective as many of them are simple and inexpensive.

British Coal Opencast are thanked for access to data and opencast sites. This paper has benefited from numerous discussions with British Coal geologists and site staff over many years. In addition, reviews by R. W. O'B. Knox, L. Knight and K. Philpott greatly improved the manuscript. Much of this work was carried out while N. S. Jones was in receipt of an NERC/CASE Research Studentship in collaboration with British Coal at Oxford Brookes University, which is gratefully acknowledged. S. Deadman is thanked for help with photographs and figures. K. M. Pickup, Head of Geology, British Coal Opencast is thanked for permission to publish, and N. S. Jones publishes with the permission of the Director of the British Geological Survey (NERC). The views expressed are those of the authors, and not necessarily those of British Coal.

References

BRIDGE, J. S. & LEEDER, M. R. 1979. A simulation model of alluvial stratigraphy. *Sedimentology*, **26**, 617–644.

BROADHURST, F. M. & FRANCE, A. A. 1986. Time represented by coal seams in the Coal Measures of

England. *International Journal of Coal Geology*, **6**, 1–12.
CASAGRANDE, D. J. 1987. Sulphur in peat and coal. *In:* SCOTT, A. C. (ed.) *Coal and Coal-bearing Strata: Recent Advances*. Geological Society, London, Special Publication, **32**, 87–105.
COLEMAN, J. M., GAGLIANO, S. M. & WEBB, J. E. 1964. Minor sedimentary structures in a prograding distributary. *Marine Geology*, **1**, 240–258.
COLLIER, R. E. Ll., LEEDER, M. R. & MAYNARD, J. R. 1990. Transgressions and regressions: a model for the influence of tectonic subsidence, deposition and eustasy, with applications to Quaternary and Carboniferous examples. *Geological Magazine*, **127**, 117–128.
DENBY, B., HASSANI, F. P. & SCOBLE, M. J. 1983. Shear strength of weak zones in Coal Measures slopes. *In: Surface Mining and Quarrying*. The Institution of Mining and Metallurgy, London, 171–181.
ELLIOTT, R. E. 1965. Swilleys in the Coal Measures of Nottinghamshire interpreted as palaeoriver courses. *Mercian Geologist*, **1**, 133–142.
—— 1968. Facies, sedimentation successions and cyclothems in productive Coal Measures in the East Midlands, Great Britain. *Mercian Geologist*, **2**, 351–372.
—— 1969. Deltaic processes and episodes: the interpretation of productive Coal Measures occurring in the East Midlands, Great Britain. *Mercian Geologist*, **3**, 111–135.
—— 1984. *Procedures in Coal Mining Geology*. National Coal Board Mining Department, London.
—— 1985. Quantification of peat to coal compaction stages, based especially on phenomena in the East Pennine Coalfield, England. *Proceedings of the Yorkshire Geological Society*, **45**, 163–172.
ELLIOTT, T. & LADIPO, K. O. 1981. Syn-sedimentary gravity slides (growth faults) in the Coal Measures of South Wales. *Nature*, **291**, 220–222.
FIELDING, C. R. 1982. *Sedimentology and stratigraphy of the Durham Coal Measures and comparison with other British Coalfields*. PhD Thesis, University of Durham.
—— 1984. A coal depositional model for the Durham Coal Measures of N.E. England. *Journal of the Geological Society, London*, **141**, 919–931.
—— 1985. Coal depositional models and the distinction between alluvial and delta plain environments. *Sedimentary Geology*, **42**, 41–48.
—— 1986. Fluvial channel and overbank deposits from the Westphalian of the Durham Coalfield, N.E. England. *Sedimentology*, **33**, 119–140.
—— 1987. Coal depositional models for deltaic and alluvial plain sequences. *Geology*, **15**, 661–664.
FISK, H. N. 1947. *Fine Grained Alluvial Deposits and their Effects on Mississippi River Activity*. Mississippi River Commission, Vicksburg, Mississippi.
FULTON, I. M. 1987a. *The Silesian sub-system in Warwickshire, some aspects of its palynology, sedimentology and stratigraphy*. PhD Thesis, University of Aston.
—— 1987b. Genesis of the Warwickshire Thick Coal: a group of long-residence histosols. *In:* SCOTT, A. C. (ed.) *Coal and Coal-bearing Strata: Recent Advances*. Geological Society, London, Special Publication, **32**, 201–218.
—— & WILLIAMS, H. 1988. Palaeogeographical change and controls on Namurian and Westphalian A/B sedimentation at the southern margin of the Pennine Basin, Central England. *In:* BESLY, B. M. & KELLING, G. (eds) *Sedimentation in a Synorogenic Basin Complex: the Upper Carboniferous of NW Europe*. Blackie, Glasgow, 178–199.
——, GUION, P. D. & JONES, N. S. 1995. The application of sedimentology to the development and extraction of deep-mined coal. *In:* WHATELEY, M. K. G. & SPEARS, D. A. *European Coal Geology*. Geological Society, London, Special Publication, **82**, 17–43.
GLUSKOTER, H. J. & HOPKINS, M. E. 1970. Distribution of sulfur in Illinois coals. *In:* SMITH, W. H., NANCE, R. B., HOPKINS, M. E., JOHNSON, R. G. & SHABICA, C. W. (eds) *Depositional Environments in Parts of the Carbondale Formation—Western and Northern Illinois*. Guidebook for the Annual Field Trip, Coal Geology Division of the Geological Society of America, Illinois State Geological Survey Guidebook Series, **8**, 89–95.
GONANO, L. P. 1980. *An integrated report on slope failure mechanisms at Goonyella—November 1976*. Commonwealth Scientific and Industrial Research Organization Technical Report 114.
GRIMSHAW, P. N. 1992. *Sunshine Miners, Opencast Coalmining in Britain 1942–1992*. British Coal Opencast Publication.
GUION, P. D. 1978. *Sedimentation of interseam strata and some relationships with coal seams in the East Midlands Coalfield*. PhD Thesis, City of London Polytechnic.
—— 1987a. Palaeochannels in mine workings in the High Hazles Coal (Westphalian B), Nottinghamshire Coalfield, England. *Journal of the Geological Society, London*, **144**, 471–488.
—— 1987b. The influence of a palaeochannel on seam thickness in the Coal Measures of Derbyshire, England. *International Journal of Coal Geology*, **7**, 269–299.
—— 1992. Westphalian. *In:* COPE, J. C. W., INGHAM, J. K. & RAWSON, P. F. (eds) *Atlas of Palaeogeography and Lithofacies*. Geological Society, London, Memoir, **13**, 80–86.
—— & FIELDING, C. R. 1988. Westphalian A and B sedimentation in the Pennine Basin, U.K. *In:* BESLY, B. M. & KELLING, G. (eds) *Sedimentation in a Synorogenic Basin Complex: the Upper Carboniferous of NW Europe*, Blackie, Glasgow, 153–177.
—— & FULTON, I. M. 1993. The importance of sedimentology in deep-mined coal extraction. *Geoscientist*, **3** (2), 25–33.
——, —— & JONES, N. S. 1995. Sedimentary facies of the coal-bearing Westphalian A and B north of the Wales–Brabant High. *In:* WHATELEY, M. K. G. & SPEARS, D. A. (eds) *European Coal Geology*. Geological Society, London, Special

Publication, **82**, 45–78.

HASZELDINE, R. S. 1989. Coal reviewed: depositional controls, modern analogues and ancient climates. *In*: WHATELEY, M. K. G. & PICKERING, K. T. (eds) *Deltas: Sites and Traps for Fossil Fuels.* Geological Society, London, Special Publication, **41**, 289–308.

HORNE, J. C., FERM, J. C., CARUCCIO, F. T. & BAGANZ, B. P. 1978a. Depositional models in coal exploration and mine planning in Appalachian region. *American Association of Petroleum Geologists Bulletin*, **62**, 2379–2411.

——, HOWELL, D. J., BAGANZ, B. & FERM, J. C. 1987b. Splay deposits as an economic factor in coal mining. *Colorado Survey Resources Series*, **4**, 89–100.

HOUSEKNECHT, D. W. & IANNACCHIONE, A. T. 1982. Anticipating facies-related coal mining problems in Hartshorne Formation, Arkoma Basin. *American Association of Petroleum Geologists Bulletin*, **66**, 923–946.

HOUGHTON, D. A. 1983. Economic applications of geotechnics to quarrying. *In: Surface Mining and Quarrying.* Institution of Mining and Metallurgy, London, 1–12.

HUGHES, D. B. & LEIGH, W. J. P. 1985. The stability of excavations and spoil mounds in relation to opencast mining. *Quarry Management*, April, 223–232.

JONES, N. S. 1992. *Sedimentology of Westphalian coal-bearing strata and applications to opencast coal mining, West Cumbrian Coalfield, U.K.* PhD Thesis, Oxford Brookes University.

KRAUSE, H. F., DAMBERGER, H. H., NELSON, W. J., HUNT, S. R., LEDVINA, C. T., TREWORGY, C. G. & WHITE, W. A. 1979. *Roof strata of the Herrin (No. 6) Coal Member in Mines of Illinois: their Geology and Stability.* Illinois State Geological Survey, Illinois Minerals Note, **72**.

LAURY, R. L. 1971. Stream bank failure and rotational slumping: preservation and significance in the geologic record. *Bulletin of the Geological Society of America*, **82**, 1251–1266.

LEEDER, M. R. 1988. Recent developments in Carboniferous Geology: a critical review with implications for the British Isles and N. W. Europe. *Proceedings of the Geologists' Association*, **99**, 73–100.

MACCALLUM, R. 1992. Geophysical logs and the search for opencast reserves. *In:* ANNELS, A. E. (ed.) *Case Histories and Methods in Mineral Resource Evaluation.* Geological Society, London, Special Publication, **63**, 77–93.

MALLETT, C. W. 1983. Sedimentological control of mining conditions in the Permian Coal Measures of the Bowen Basin, Australia. *Proceedings of the Third International Conference on Stability in Surface Mining, June 1981*, **3**, 333–346.

MCCABE, P. J. 1984. Depositional environments of coal and coal-bearing strata. *In:* RAHMANI, R. A. & FLORES, R. M. (eds) *Sedimentology of Coal and Coal-bearing Sequences.* International Association of Sedimentologists, Special Publication, 7, 13–42.

MIALL, A. D. 1988. Architectural elements and bounding surfaces in fluvial deposits: anatomy of the Kayenta Formation (lower Jurassic), southwest Colorado. *Sedimentary Geology*, **55**, 233–262.

MOEBS, N. N. & FERM, J. C. 1982. *The Relation of Geology to Mine Roof Conditions in the Pocohontas No. 3 Coal Bed.* United States Bureau of Mines Information Circular, **8864**.

NELSON, W. J., EGGERT, D. L., DIMICHELE, W. A. & STECYK, A. C. 1985. Origin of discontinuities in coal-bearing strata at Roaring Creek (basal Pennsylvanian of Indiana). *International Journal of Coal Geology*, **4**, 355–370.

PLINT, A. G. 1986. Slump blocks, intraformational conglomerates and associated erosional structures in Pennsylvanian fluvial strata of eastern Canada. *Sedimentology*, **33**, 387–399.

RAISTRICK, A. & MARSHALL, C. E. 1939. *The Nature and Origin of Coal and Coal Seams.* English University Press, London.

SCOBLE, M. J. & LEIGH, W. J. P. 1982. Factors governing the stability of rock slopes in British coal mines. *23rd Symposium on Rock Mechanics, Berkeley*, 1091–1098.

SMITH, E. G., RHYS, G. H. & EDEN, R. A. 1967. *Geology of the Country around Chesterfield, Matlock and Mansfield.* Memoir, Geological Survey, UK.

STANLEY, D. J., KRINITZSKY, E. L. & COMPTON, J. R. 1966. Mississippi river bank failure, Fort Jackson, Louisiana. *Bulletin of the Geological Society of America*, **77**, 859–866.

STEAD, D. & SCOBLE, M. 1983. Rock slope assessment in British coal mines. *In: Surface Mining and Quarrying.* Institute of Mining and Metallurgy, London, 205–215.

TAYLOR, B. J. 1961. The stratigraphy of exploratory boreholes in the West Cumberland coalfield. *Bulletin of the Geological Survey of Great Britain*, **17**, 1–74.

—— 1978. Westphalian. *In:* MOSELEY, F. M. (ed.) *The Geology of the Lake District.* Yorkshire Geological Society Occasional Publication, 3, 180–188.

TERZAGHI, K. V. & PECK, R. B. 1967. *Soil Mechanics in Engineering Practice*, 2nd Edn. Wiley, New York.

TURNBULL, W. J., KRINITZSKY, E. L. & WEAVER, F. J. 1966. Bank erosion in soils of the lower Mississippi valley. *Journal of the Soil Mechanics and Foundations Division, Proceeding of the American Society of Civil Engineers*, **92**, (SM1), 121–136.

WALTON, G. & ATKINSON, T. 1978. Some geotechncial considerations in the planning of surface coal mines. *Transactions of the Institution of Mining and Metallurgy Section A*, **87**, A147–A171.

WARD, C. R. 1984. *Coal Geology and Coal Technology.* Blackwell Scientific, Melbourne.

WEISENFLUH, G. A. & FERM, J. C. 1984. Geological controls on deposition of the Pratt seam, Black Warrior Basin, Alabama, U.S.A. *In:* RAHMANI, R. A. & FLORES, R. M. (eds) *Sedimentology of Coal and Coal-bearing Sequences.* International Association of Sedimentologists, Special Publication, **7**, 317–330.

—— & —— 1991. Roof control in the Fireclay Coal Group, southeastern Kentucky. *Journal of Coal Quality*, **10**, 67–74

WILLIAMS, E. G. & KEITH, M. L. 1963. Relationship between sulfur in coals and the occurrence of marine roof beds. *Economic Geology*, **58**, 720–729.

——, GUBER, A. L. & JOHNSON, A. M. 1965. Rotational slumping and the recognition of disconformities. *Journal of Geology*, **72**, 534–547.

WILLIAMS, H. & FLINT, S. 1990. Anatomy of a channel bank collapse structure in Tertiary fluvio-lacustrine sediments of the Lower Rhine Basin, West Germany. *Geological Magazine*, **127**, 445–451.

An update on British Tonsteins

D. A. SPEARS[1] & P. C. LYONS[2]

[1]*Department of Earth Sciences, University of Sheffield, Sheffield S1 3JD, UK*
[2]*US Geological Survey, Reston, VA 22092, USA*

Abstract: Tonsteins are thin mudstone beds composed of kaolinite that formed from the alteration of volcanic ash in coal-forming environments. In the Coal Measures of Britain, tonsteins have been recognized as volcanic in origin based on bedform, volcanic textures, field relationships with other volcanics, distinctive volcanic minerals and an igneous geochemistry. The sharp contacts, together with a contrasting composition with adjacent sediments and lateral continuity, are diagnostic of an altered air-fall volcanic ash. Passage into less altered volcanics and into a K-bentonite have both been shown in Britain. A more direct determination of volcanic origin is possible from the identification of volcanic minerals, whole-rock geochemistry and analyses of silicate melt inclusions in volcanic minerals such as β quartz. Once a volcanic origin is established the stratigraphic value of a tonstein is enhanced. Immobile trace elements in tonsteins also provide information on the tectono-magmatic setting of the parent volcano. Two tonstein groups are identified in the Coal Measures of Britain, one formed from basic (basaltic) ash of local origin and the other from acid (rhyolite or rhyodacite) ash, probably associated with distant Plinean or ultraplinean eruptions at a destructive plate margin to the south or southeast.

Tonsteins are thin beds of kaolinite mudstone which occur in coals and associated sediments. Typically kaolinite occurs to the exclusion of other clay minerals; non-clay minerals, such as quartz and zircon, are only present in trace amounts. The composition is therefore extremely unusual for sedimentary rocks and the bimodal grain size is distinctive in thin section. Microphenocrysts of euhedral β quartz, zircon and other volcanic minerals, some replaced by kaolinite, are typically set in a very fine-grained matrix. Tonsteins dominated by the fine-grained matrix have a flinty appearance with a conchoidal fracture. Although tonsteins are clay-rich, some appear granular and could possibly be mistaken for sandstones: however, the granules and the sand-sized grains are composed of kaolinite and the hardness associated with quartz-rich rocks is lacking. Tonsteins are distinctive and early workers recognized the lateral continuity of the beds and therefore the great stratigraphic value.

The early work on tonsteins, which goes back over a hundred years, has been reviewed by a number of workers including Hoehne (1953), Scheere (1956), Williamson (1961), Moore (1964), Kimpe (1966), Burger & Damburger (1985) and Bohor & Triplehorn (1993). Most of the early work was concentrated on the coalfields of continental western Europe. Tonsteins were not recognized in the coalfields of Britain until the 1960s. Interest in British tonsteins was stimulated by Williamson's review of tonsteins (1961) which coincided with a period of active coalfield exploration. Tonsteins in British coalfields were reported and described by Francis (1961), Francis & Ewing (1962), Eden et al. (1963), Salter (1964), Barnsley et al. (1966), Wilson et al. (1966), Mayland & Williamson (1970), Strauss (1971) and Worssam et al. (1971). More geochemically and mineralogically orientated studies were undertaken by Spears (1966, 1970, 1971), (Price & Duff (1969), Spears & Rice (1973) and Spears & Kanaris-Sotiriou (1979).

Tonsteins are of stratigraphic importance in the British coalfields, but they have not been used as a primary means of coal seam identification and to elucidate structures. This is in marked contrast with studies elsewhere in Europe in which, according to Kimpe (1966), correct coal seam identification would be inconceivable without the application of tonsteins. There are two main reasons why this does not apply to the British coalfields; firstly, structural complexity is generally lacking, particularly in the northern coalfields; and, secondly, there are fewer widespread tonsteins.

The origin of tonsteins has been considered by many workers and a number of proposals made, essentially reflecting the diverse environments and processes which could lead to the formation of kaolinite (Spears 1970). An influential advocate of a sedimentary origin was Hoehne (1953), whereas Bouroz (1966, 1967) was a leading proponent of the origin of tonsteins from the alteration of a volcanic ash in a peat-swamp environment. The volcanic origin has now gained general acceptance (Spears & Duff 1985).

In North America tonsteins as such were not recognized until relatively recently, although Rogers (1914), Ashley (1928) and Seiders (1965) all proposed a volcanic origin for kaolinite partings in North American coal beds. Important contributions to the recognition of tonsteins in the USA have been made by Bohor & Triplehorn (1981, 1993), Chesnut (1985) and Lyons and coworkers (1992, 1993).

In studies of North American tonsteins, the mineralogy and geochemistry of the earlier work on European tonsteins have been extended and the opportunity has therefore been created for an update on British tonsteins. In other geological disciplines significant advances have been made in the use of immobile elements to determine the original composition of altered volcanic rocks and also the tectonic setting of the magmatism (Pearce & Cann 1973; Winchester & Floyd 1977; Wood 1980; Leat et al. 1986). The use of immobile elements has been successfully applied to tonsteins in coalfields in western Canada (Spears & Duff 1984) and to K-bentonites in Lower Palaeozoic sediments (Huff et al. 1993). K-bentonites and tonsteins both form from the alteration of volcanic ash, but the clay mineralogy is different due to differences in the composition of the diagenetic pore waters. The latter is mainly a function of the depositional environment and thus tonsteins occur in dominantly non-marine sequences and K-bentonites in marine sequences (Spears 1971). The geochemical discrimination methods for tonsteins and K-bentonites are applied to some British tonsteins in this paper. The opportunity is also taken of highlighting areas of progress and uncertainty.

There are several lines of evidence which convincingly show that tonsteins result from the *in situ* alteration of air-fall ash to kaolinite in the coal-precursor swamp environment. The volcanic evidence for the origin of tonsteins and also bentonites is considered by Fisher & Schmincke (1984) under four categories: (1) bedform; (2) volcanic textures and field relationships with other volcanic rocks; (3) distinctive volcanic minerals; and (4) an igneous-related geochemical composition. These categories are considered below with specific reference to tonsteins in British coalfields.

Bedform

Tonsteins are relatively thin, generally not greater than 5 cm thick in British coalfields, and the contacts with the adjacent sediments are sharp. Thus not only does a tonstein differ in appearance to the normal sediments, but there is also a lack of gradation, suggesting different depositional processes. The widespread distribution of a tonstein bed is convincing evidence of a volcanic origin, particularly if the extent far exceeds that of the associated sedimentary facies. It could be argued that this is only proof of an aeolian origin and that the kaolinite composition represents the deflation of a mature soil from adjacent land areas. However, it is doubtful whether the aeolian transport of a mature soil would consistently generate the typical tonstein bedform. Furthermore, the kaolinite in tonsteins is well ordered, whereas most sedimentary kaolinites are not, and there is also textural evidence that the kaolinite in tonsteins is diagenetic rather than detrital.

An important tonstein occurs in the Westphalian C of the coalfields of northern and central England. This tonstein was first recognized by its high radioactivity on gamma ray borehole logs (Ponsford 1955). The tonstein occurs a short distance above an important marine horizon, the Mansfield (Aegir) Marine Band. The tonstein has been correlated between different coalfields (Earp 1961; Eden et al. 1963; Hoare 1965; Mayland & Williamson 1970; Worssam et al. 1971) and named by reference to adjacent coal seams. It is therefore variously known as the Stafford, the supra-Wyrley Yard, the sub-High Main, the sub-Worsley Four Foot, the P27 and the Sharlston Muck. The diversity in nomenclature shows a greater lateral persistence than the adjacent or enclosing coal seams. It will also be apparent that lateral continuity is established at a mature stage in coalfield exploration and, although the continuity of a tonstein bed is important evidence of a volcanic derivation, the origin needs to be established at an earlier stage in exploration to realize the correlation potential.

Field relationships and volcanic textures

Francis (1961) recorded about 40 thin kaolinized tuffs in the Carboniferous of Fife (Scotland), which thicken southwards and pass into unambiguous volcanic rocks. Graded beds are present in these kaolinized tuffs, and in thin section further evidence of the volcanic origin was provided by fragmentary volcanic glass with subordinate amounts of crystalline basalt. The kaolinized tuffs were equated with tonsteins, thus demonstrating the possibility of a volcanic origin for other tonsteins, although, as noted by Francis (1961), most workers favoured a non-volcanic origin at that time. The tonsteins in the Fife Coalfield are not laterally extensive (< 40 km), unlike tonsteins known from mainland Europe (Bouroz 1967).

The field relationships of the so-called Stafford tonstein and the equivalent horizon in other coalfields in England demonstrate a volcanic origin. In this instance the lateral transition is not from tonstein to less altered volcanic rocks, as in the Fife Coalfield, but from tonstein to K-bentonite (Spears 1971). Kaolinite does occur in the Stafford 'tonstein', but the dominant clay mineral is mixed-layer illite–smectite (Spears 1971). The presence of this mixed-layer clay and the absence of discrete 10 Å mica and other detrital minerals is characteristic of K-bentonites whose volcanic origin has long been established (Weaver 1953). The lateral variation from K-bentonite to tonstein provided further evidence of the volcanic origin of this particular tonstein.

In the Stafford 'tonstein', and stratigraphically equivalent tonsteins, bedding planes are vaguely defined and are generally non-planar. In one apparently restricted area, graded beds are preserved within the Sharlston Muck tonstein, indicating a volcanic origin. Analysis of these beds by Sezgin (1982) revealed total kaolinization. Why the beds have only been preserved in this one location is unclear. If volcanic textures are preserved in tonsteins a volcanic origin is readily demonstrated. However, the diagenetic development of kaolinite tends to obscure original primary textures. In thin section a vermicular development of kaolinite is commonly observed and this diagenetic form makes the recognition of primary textures more difficult. Nevertheless, relict tuffaceous textures and volcanic minerals were recorded by Price & Duff (1969) in their study of Scottish and English tonsteins. Textures have been used more extensively in the coalfields of mainland Europe to prove a volcanic origin and as a means of identifying specific horizons, particularly by Bouroz and coworkers (1966, 1967, 1972, 1983). In the limnic coalfields of southern France the volcanic tuffs are less altered and the original textures better preserved, which enabled Bouroz and coworkers to identify volcanic textures with more confidence than in the tonsteins from the paralic coalfields to the north.

Volcanic minerals

Price & Duff (1969) noted the occurrence of quartz, biotite, feldspars, idiomorphic apatite and zircon in their tonsteins as being consistent with a volcanic origin. These workers also noted that the common presence of acicular splinters of quartz was also consistent with the same origin.

Heavy minerals were separated (specific gravity greater than bromoform) from the supra-Wyrley Yard tonstein (Spears 1966) and from the Stafford 'tonstein' (Spears 1971). The non-opaque heavy mineral suite consists of apatite and zircon. The apatite typically occurs as euhedral, prismatic crystals with many fluid inclusions parallel to the C-axis. The zircons are more varied in form, but prismatic crystals with sharp crystal faces and pyramidal terminations are common. The heavy mineral suite is restricted in the range of minerals present and is significantly different to that normally encountered in sedimentary rocks where minerals such as tourmaline, garnet and rutile are common and apatite is rare. In tonsteins euhedral grains are the norm, whereas in normal sediments grains are rounded due to transportation. The restricted heavy mineral suite corresponds very closely with those described by Weaver (1953, 1963) from bentonites known to have formed by the alteration of volcanic air-fall ash. Heavy minerals are therefore important in establishing the volcanic origin of tonsteins.

It should be stressed that the heavy minerals are accessory minerals and are insignificant in abundance compared with kaolinite. From chemical analyses the Zr concentration may be used to calculate a maximum zircon weight percentage. In the case of the Stafford and supra-Wyrley Yard tonsteins referred to above, the maximum zircon content is about 0.04 wt.%.

A recent tonstein study by Lyons et al. (1993) includes samples of the Sharlston Muck, the sub-Worsley Four Foot, the P27 and the supra-Wyrley Yard tonsteins. Accessory minerals, concentrated by dissolving kaolinite in HF, were studied by optical, X-ray diffraction, SEM/EDAX (scanning electron microscope/energy dispersive X-ray analysis) and electron microprobe techniques, details of which are given in Lyons et al. (1992). Previous studies on British tonsteins were based on heavy liquid separations. The HF digestion procedure includes the lower density minerals, although a chemically reactive mineral such as apatite may be lost. Quartz is an important component (>70%) of the sand-sized fraction of the HF

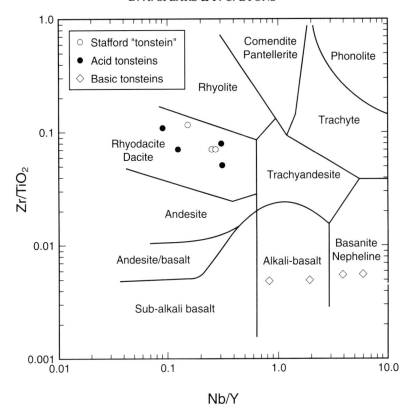

Fig. 1. Tonstein samples plotted on the magmatic discrimination diagram of Winchester & Floyd (1977). The acid tonsteins (Stafford 'tonstein' and equivalent horizons) are plotted as circles and the basic tonsteins are plotted as diamonds.

mineral residue and both detrital and volcanic quartz grains are thought to be present. The volcanic quartz is characterized by a water-clear to slightly cloudy appearance, smooth surfaces and sharply angular habit. As noted by Bohor & Triplehorn (1993), blade and flake shaped quartz splinters with sharp edges are not generally characteristic of fluvial or aeolian terrigenous detritus. Occasional grains of quartz occur in the volcanic β quartz form. Also present are β quartz forms that are partially rounded, probably due to resorption in the magma chamber (Bohor & Triplehorn 1981). A small percentage of the quartz grains contain glass inclusions and their presence is strong evidence of a volcanic origin (Sorby 1858; Tuttle 1952). Grains of detrital quartz were also noted in the four tonsteins. These were cloudy in appearance, equant to subsequent in shape and with rough surfaces. Further evidence that both volcanic and detrital quartz grains were present was provided by cathodoluminescence. The volcanic quartz grains luminesced a characteristic blue (Zinkernagel 1978), whereas the detrital quartz grains luminesced orange (Ruppert et al. 1985). More volcanic quartz was recorded than detrital quartz (Lyons et al. 1993, fig. 9), although the proportions were found to vary vertically through the tonstein with approximately equal proportions towards the top of the sub-Worsley Four Foot tonstein (Lyons et al. 1993, fig. 11). Further evidence of a detrital influence was provided by the occasional occurrence of detrital garnet (almandine) and cassiterite grains. Sanidine was not recorded by Lyons et al. (1993), although it does occur in other European tonsteins and has been used for absolute age dating (Lippolt & Hess 1985).

In the sub-Worsley Four Foot tonstein and correlatives the total quartz (free silica) content is about 1% in most locations (Spears and Kanaris-Sotiriou 1979). The near-monomineralic kaolinite composition, with only a trace of quartz, was confirmed by Lyons et al. (1993). Both volcanic and detrital quartz are therefore minor components in the bed as a whole. Trace

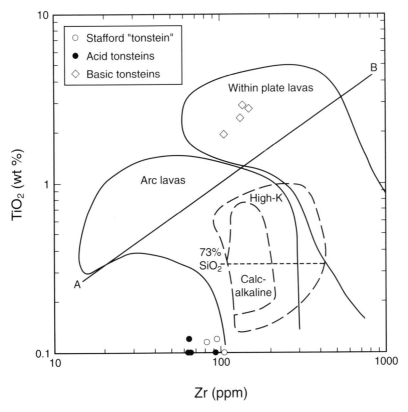

Fig. 2. Tonstein samples plotted on the Zr–TiO$_2$ tectono-magmatic discrimination diagram of Leat *et al.* (1986) after Pearce (1980). Tonstein samples have been normalized to 15% Al$_2$O$_3$ initial concentration.

amounts of detrital quartz are not unexpected in a clastic depositional environment. In earlier work on this tonstein horizon (Spears 1970) some of the samples from the East Midlands were found to contain more quartz, and detrital clay minerals, such as illite, were also recorded. In such instances there would be major dilution of the volcanic quartz and also of the heavy minerals, thus reducing the interpretative value of the latter. This problem was encountered in secondary bentonites from Iraq (Spears 1982), in which the volcanic heavy minerals were swamped by detrital heavy minerals although the total detrital input into the sediment was not more than 10%. Thus although the volcanic accessory minerals are of considerable interpretative value, this is decreased if there is a significant detrital input.

Geochemistry

Price & Duff (1969) interpreted the high concentration of TiO$_2$ in some of the English and Scottish tonsteins as indicative of a volcanic origin. The TiO$_2$/Al$_2$O$_3$ versus quartz plot was used by Spears & Kanaris-Sotiriou (1979) to distinguish tonsteins from normal detrital, non-volcanic sediments and to identify the volcanic ash as either basic or acid. Further subdivision was achieved using Ti, Zr, Cr and Ni in a discriminant function analysis. Ni would not generally be considered an immobile element and therefore is not a good indicator of the volcanic precursor. However, in the low Eh diagenetic environment of the peat swamp and in the presence of reduced S species, the retention of Ni and a number of other elements would appear to have been quantitative. Nevertheless, for general application element immobility should be independent of specific diagenetic conditions.

Based on the trace element geochemistry, Spears & Kanaris-Sotiriou (1979) identified two main groups of tonsteins in the British Coal Measures. Those formed from acid volcanic ash are comparable with other European tonsteins and the other group, apparently restricted to Britain, formed from basic volcanic ash. The

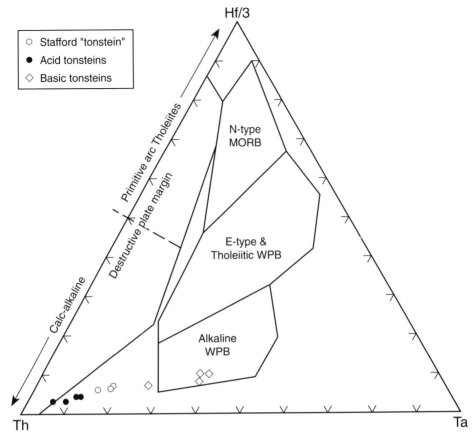

Fig. 3. Tonstein samples plotted on the Hf–Th–Ta tectono-magmatic discrimination diagram of Wood (1980). MORB, mid-ocean ridge basalt; WPB, within plate basalt.

basic tonsteins contain variable amounts of normal detrital sediment which is thought to be country rock from the vent rather than sediment incorporated in the depositional environment. The acid tonsteins, on the other hand, contain very little or no detrital sediment. The detrital sediment was originally recognized by the presence of illite, chlorite and enhanced quartz contents. More recent cathodoluminescence work, described in the previous section, enables detrital and volcanic grains to be differentiated, thus allowing the recognition of smaller detrital inputs. In the sub-Clowne tonstein (Westphalian B, northeast Derbyshire) a greater detrital input is recognized, which is unusual in that this is an acid tonstein (Spears & Kanaris-Sotiriou 1979). In this instance the ash fell into a high energy depositional environment and was mixed with normal detrital sediment. Mixing was so effective that in some locations the tonstein cannot be recognized in hand specimen. The high U and Th contents of the original, evolved acidic ash, are retained during diagenesis and the horizon can be recognized by the high radioactivity on the gamma ray borehole log (Rippon & Spears 1989).

Representative analyses of the Stafford 'tonstein' and its equivalent horizon in other coalfields (sub-Worsley Four Foot, P27 and Sharlston Muck) are shown plotted on Fig. 1, which is the immobile element discriminant diagram of Winchester & Floyd (1977). All analyses plot in the rhyodacite/dacite field, which shows the correct designation of this horizon as acid and also the lack of distinction between a K-bentonite (Stafford 'tonstein') and tonstein. The latter has undergone a greater loss of elements to achieve the kaolinite composition but, as would be predicted, immobile element ratios have not been affected. Also plotted on Fig. 1 are analyses of the Piper, Middle Lount, Deep Hard and Jersey Yellowstone tonsteins. These tonsteins have been recognized as basic on the basis of previous geochemical work and this is corroborated on Fig. 1. The Nb/Y ratios for these samples indicate an alkaline magma, and lavas with this composition are known in the Carboniferous of central England (Kirton 1984; Macdonald et al. 1984). Macdonald et al. (1984) also confirmed the intraplate setting of the basic volcanics.

In addition to defining the composition of the volcanic ashes which produced the tonsteins, immobile elements provide information on the tectonic setting of the magma generation. Figure 2 shows the Zr versus TiO_2 plot (Leat et al. 1986) on which the basic tonsteins fall in the within-plate lava field, whereas the acid tonsteins plot close to the arc lava field. The development of a tonstein from the original ash did involve the loss of mobile elements and the immobile elements are concentrated as a result. To reduce this effect the immobile elements in this plot have been normalized against a 15% Al_2O_3 concentration, on the assumption that Al is immobile and that this percentage approximates to the original concentration in the ash. Some errors are inevitable and may account for the samples plotting outside the arc lava field rather than within it. However, the samples plot close to the lower boundary of the field rather than the upper boundary, which is where the separation of arc and within-plate lavas is achieved.

The tectonic setting of the magmas is also considered in the Th–Ta–Hf diagram of Wood (1980), which has been applied to K-bentonites by Huff et al. (1993). The tonstein analyses are plotted on Fig. 3. The basic tonsteins fall in the alkaline within-plate basalt field and the acid tonsteins close to the Th apex, signifying calc-alkaline lavas in a destructive plate setting.

In earlier work (Spears & Kanaris-Sotiriou 1979) it was reasoned that the basic tonsteins were of local derivation because of restricted lateral distribution and local Carboniferous volcanics of a similar basaltic nature. The acid tonsteins, on the other hand, had to be derived from further afield based on (1) the absence of acid differentiates in the British Carboniferous, (2) the widespread extent of the horizons indicative of Plinean or ultraplinean eruptions and destructive plate margins and (3) the evidence that the thickness of French tonsteins increases towards the southeast and that tonsteins in Britain are thinner than the correlative horizons in France. A destructive plate margin to the south has therefore been proposed (Spears & Kanaris-Sotiriou 1979; Lyons et al. 1993). The proposal is substantiated by the immobile trace elements described in this paper (Figs 1–3) as well as the more local origin of the basic tonsteins.

The direct determination of the original magma composition is possible from electron microprobe analyses of glass (silicate melt) inclusions in volcanic minerals. Glass inclusions in quartz were analysed from the Fire Clay tonstein in the USA (Lyons et al. 1992) and from selected tonsteins in the UK (Lyons et al. 1993). The UK tonsteins analysed were correlative of the Stafford 'tonstein' (sub-Worsley Four Foot, P27 and supra-Wyrley Yard). The analyses all show a high silica rhyolite composition which is broadly consistent with the immobile element geochemistry (Fig. 1). In one sample (the supra-Wyrley Yard tonstein) an enrichment in halogens was observed for silicate melt inclusions in some of the quartz grains. As noted in Lyons et al. (1993) these quartz crystals could have formed near the top of the magma chamber where halogen enrichment may occur. In general, however, in large Plinean or ultraplinean eruptions the magma chamber is likely to be large (Walker 1980) and the magma more uniform in composition than in smaller magmatic systems (Smith 1979). The direct determination of magma composition from the chemical analysis of glass inclusions does offer another possibility of identifying magma composition and recognizing specific tonstein horizons (Lyons et al. 1993).

Conclusions

The volcanic origin of tonsteins in British Coal Measures has been established from several lines of evidence. The distinctive nature of the beds and their contrast with vertically adjacent sediments are suggestive of an air-fall volcanic ash origin. If coupled with a wide lateral extent then a volcanic origin is more certain, but a more direct determination of origin is required. Furthermore, some British tonsteins are known not to have a widespread distribution. The lateral transitions recorded for some tonsteins show a volcanic origin, but, as for lateral continuity, the usefulness of such evidence has been in establishing other methods of determining origin based on volcanic minerals, immobile element geochemistry and the chemistry of the silcate melt inclusions.

If the kaolinite abundance approaches 100% and, particularly, if the kaolinite is well crystallized, then a volcanic origin is likely. The origin is less certain if other minerals are present in more than trace amounts. An exception would be the presence of mixed-layer illite–smectite characteristic of K-bentonites. The early work on tonsteins in Britain pre-dated the general acceptance of a volcanic origin and in some instances tonstein recognition was based on a high kaolinite content, which would not on its own exclude all normal, non-volcanic, sedimentary mudrocks such as pedogenic fireclays.

Volcanic minerals can be concentrated from tonsteins in the laboratory by either using heavy liquids, removing the kaolinite with HF, or both. Quartz grains of volcanic origin are recognized by their form (splinters and β crystals), cathodoluminescence properties and glass inclusions. Analyses of glass inclusions by electron microprobe provide a direct determination of the original magma composition, which for the samples analysed in this work suggest an original rhyolitic composition. Other non-opaque volcanic minerals are present, notably zircon and apatite. The heavy mineral suite is restricted in that detrital minerals such as tourmaline and garnet are absent. However, the presence of detrital minerals does not preclude the volcanic origin of a tonstein. For tonsteins occurring in sediments associated with coals, detrital sediment was present in the depositional environment and some mixing of volcanic and detrital fractions was possible during and after volcanic ash deposition. The volcanic/detrital proportions may be estimated from the content of quartz, the volcanic to detrital quartz ratio and the proportion of detrital clay minerals, e.g. illite and chlorite. Estimation of volcanic–detrital proportions from the heavy minerals will be more difficult unless relative proportions in the original fractions are known. The relict volcanic minerals (e.g. sanidine, zircon and monazite) also enable absolute age determinations to be made.

Extensive alteration of the original ash and the loss of mobile elements take place in the formation of a tonstein. Nevertheless, there are elements which are immobile in the alteration process and are also diagnostic of the original volcanic ash composition and tectonic setting of the parent volcano. The concentrations of these elements are also sufficiently distinctive from normal detrital sediments that a volcanic origin can be established on the basis of the immobile elements. In the Coal Measures in Britain, the geochemistry has been used to establish two groups of tonsteins, one formed from ash of acid (rhyolitic) composition and the other from ash of basic (basaltic) composition. The latter have a restricted lateral distribution and are related to the local volcanic activity. The geochemistry of basic tonsteins and volcanic rocks is comparable and is indicative of an intraplate origin. The geographically more widespread tonsteins are those of an original acidic composition which are thought to have originated from Plinean or ultraplinean eruptions at a destructive plate margin to the south of Britain. The geochemistry of the immobile elements in these more widespread tonsteins can be used to show a rhyodacite–dacite composition and an arc-lava, destructive plate margin setting.

This paper has benefited from internal reviews within the USGS by S. Neuzil and H. Belkin and anonymous reviewers in the UK.

References

ASHLEY, G. H. 1928. *Bituminous Coalfields of Pennsylvania, Part 1.* Pennsylvania Topographic and Geologic Survey, Series 4, **Bull. M-6**, 24 pp.

BARNSLEY, G. B., CLOWES, J. M. & FOWLER, W. 1966. Kaolin tonsteins in the Westphalian of North Staffordshire. *Geological Magazine*, **103**, 509–521.

BOHOR, B. F. & TRIPLEHORN, D. M. 1981. Volcanic origin of the flint clay parting in the Hazard No. 4 (Fire Clay) coal bed of the Breathitt Formation in Eastern Kentucky. *In: Guidebook, Geological Society of America, Annual Meeting, Coal Division Field Trip, Coal and Coal-Bearing Rocks of Eastern Kentucky.* Kentucky Geological Survey, Lexington, 49–54.

——, —— 1993. *Tonsteins: Altered Volcanic Ash Layers in Coal-bearing Sequences.* Geological Society of America, Special Paper, **285**, 44 pp.

BOUROZ, A. 1966. Fréquence des manifestations volcaniques au Carbonifère supérieur en France. *Comptes Rendus Sér. D Académie des Sciences, Paris*, **263**, 1025–1028.

—— 1967. Correlations des tonsteins d'origine volcanique entre les bassins houillers de Sarre-Lorraine et du Nord-Pas-de-Calais. *Comptes Rendus Académie des Sciences, Paris*, **264**, 2729–2732.

——, ROQUES, M. & VIALETTE, Y. 1972. Etude de la cinérite au sommet de la zone 2 du bassin des Cevennes, France. *Bureau de Recherches Géologiques et Minières, Mémoires*, **77**, 503–507.

——, SPEARS, D. A. & ARBEY, F. 1983. *Essai de synthèse des données acquises sur la genèse et l'évolution des marqueurs pétrographiques dans les bassins houillers.* Société Géologique du Nord, Memoir, 16, 114 pp.

BURGER, K. & DAMBURGER, H. H. 1985. Tonsteins in the coalfields of western Europe and North America. *In: Compte Rendu, 9me. Congrès International de Stratigraphie et Géologie du Carbonifère, Champaign-Urbana, 1979*, **4**, 433–448.

CHESNUT, D. R. 1985. Source of the volcanic ash deposit (flint clay) in the Fire Clay coal of the Appalachian basin. *In: Compte Rendu, 10me. Congrès International de Stratigraphie et de Géologie du Carbonifère, Madrid, 1983*, 145–154.

EARP, J. R. 1961. Exploratory boreholes in the North Staffordshire Coalfield. *Bulletin, Geological Survey Great Britain*, **17**, 153–190.

EDEN, R. A., ELLIOTT, R. W., ELLIOTT, R. E. &

YOUNG, B. R. 1963. Tonstein bands in the coalfield of the East Midlands. *Geological Magazine*, **100**, 47–58.

FISHER, R. V. & SCHMINCKE, H. U. 1984. *Pyroclastic Rocks*. Springer Verlag, 472 pp.

FRANCIS, E. H. 1961. Thin beds of graded kaolinitized tuff and tuffaceous siltstones in the Carboniferous of Fife. *Bulletin, Geological Survey Great Britain*, **17**, 191–215.

—— & EWING, C. J. C. 1962. Skipsey's Marine Band and Red Coal Measures in Fife. *Geological Magazine*, **99**, 145–152.

HOARE, R. H. 1965. Tonsteins. *Geological Magazine*, **102**, 347–349.

HOEHNE, K. 1953. Kaolinkristalle und Quarzneubildungen im indischen Steinkohlen. *Chemie der Erde*, **16**, 211–222.

HUFF, W. D., MERRIMAN, R. J., MORGAN, D. J. & ROBERTS, B. 1993. Distribution and tectonic setting of Ordovician K-bentonites in the United Kingdom. *Geological Magazine*, **130**, 93–100.

KIMPE, W. F. M. 1966. Occurrence, development and distribution of Upper Carboniferous tonsteins in the paralic West German and Dutch coalfields, and their use as stratigraphical marker horizons. *Mededelingen van de Geologische Stichting, Nieuwe Serie*, **18**, 3–10.

KIRTON, S. R. 1984. Carboniferous volcanicity in England with special reference to the Westphalian of the E and W Midland. *Journal of the Geological Society, London*, **141**, 161–176.

LEAT, P. T., JACKSON, S. E., THORPE, R. S. & STILLMAN, C. J. 1986. Geochemistry of bimodal basalt-sub-alkaline/peralkaline provinces within the Southern British Caledonides. *Journal of The Geological Society, London*, **141**, 259–273.

LIPPOLT, H. J. & HESS, J. C. 1985. $^{40}Ar/^{39}Ar$ dating of sanidines from Upper Carboniferous tonsteins. *In: Compte Rendu, 10me. Congrès International de Stratigraphie et de Géologie du Carbonifère, Madrid, 1983*, **4**, 175–181.

LYONS, P. C., OUTERBRIDGE, W. F., TRIPLEHORN, D. M., EVANS, H. T., JR, CONGDON, R. D., CAPIRO, M., HESS, J. C. & NASH, W. P. 1992. An Appalachian isochron: A kaolinized Carboniferous air-fall volcanic-ash deposit (tonstein). *Geological Society of America, Bulletin*, **104**, 1515–1527.

——, SPEARS, D. A., OUTERBRIDGE, W. F., CONGDON, R. D. & EVANS, H. T. JR. (1993). Euramerican tonsteins: overview, magmatic origin and depositional-tectonic implications. *Palaeogeography, Palaeoclimatology, Palaeoecology*, **106**, 113–139.

MACDONALD, R., GASS, K. N., THORPE, R. S. & GASS, I. G. 1984. Geochemisry and petrogenesis of the Derbyshire Carboniferous basalts. *Journal of the Geological Society, London*, **141**, 147–159.

MAYLAND, H. & WILLIAMSON, I. A. 1970. Tonstein bands in the north-western coalfields of England and Wales. *In: Compte Rendu, 6me Congrès International de Stratigraphie et de Géologie du Carbonifère, Sheffield, 1967*, **3**, 1165–1168.

MOORE, L. R. 1964. The microbiology, mineralogy and genesis of a tonstein. *Proceedings of the Yorkshire Geological Society*, **34**, 235–292.

PEARCE, J. A. 1980. Geochemical evidence for the genesis and eruptive setting of lavas from Tethyan ophiolites. *In: Proceedings of the International Ophiolite Symposium, Nicosia, Cyprus*, 261–272.

—— & CANN, J. R. 1973. Tectonic setting of basic volcanic rocks determined using trace element analyses. *Earth and Planetary Science Letters*, **19**, 290–300.

PONSFORD, D. R. A. 1955. Radioactive studies of some British sedimentary rocks. *Bulletin, Geological Survey Great Britain*, **10**, 24–44.

PRICE, N. B. & DUFF, P. McL. D. 1969. Mineralogy and chemistry of tonsteins from Carboniferous sequences in Great Britain. *Sedimentology*, **13**, 45–69.

RIPPON, J. H. & SPEARS, D. A. 1989. The sedimentology and geochemistry of the sub-Clowne cycle (Westphalian B) of north-east Derbyshire, UK. *Proceedings of the Yorkshire Geological Society*, **47**, 181–198.

ROGERS, G. S. 1914. The occurrence of genesis of a persistent parting in a coal bed of the Lance Formation. *American Journal of Science*, **37** (4 ser.), 299–304.

RUPPERT, L. F., CECIL, C. B., STANTON, R. W. & CHRISTIAN, R. P. 1985. Authigenic quartz in the Upper Freeport coal bed, west-central Pennsylvania. *Journal of Sedimentary Petrology*, **55**, 334–339.

SALTER, D. L. 1964. New occurrences of tonsteins in England and Wales. *Geological Magazine*, **101**, 517–519.

SCHEERE, J. 1956. Nouvelle contribution a l'étude des tonstein du terrain houiller belge. *Association pour l'Etude de Paleontologie et Stratigraphie, Bruxelles*, **26**, 1–54.

SEIDERS, V. M. 1965. *Volcanic Origin of Flint Clay in the Fire Clay Coal Bed, Breathitt Formation, Eastern Kentucky*. United States Geological Survey, Professional Paper, **525-D**, D52–54.

SEZGIN, H. I. 1982. *The occurrence and formation of kaolinite and other minerals in Coal Measures rocks from the East Pennine Coalfield*. Phd Thesis, University of Sheffield.

SMITH, R. L. 1979. Ash Flow Magmatism. *Geological Society of America, Special Paper*, **180**, 5–27.

SORBY, H. C. 1858. On the microscopical structure of crystals indicating the origin of minerals and rocks. *Quarterly Journal of the Geological Society of London*, **14**, 453–500.

SPEARS, D. A. 1966. A Westphalian tonstein from south Staffordshire. *Proceedings of the Yorkshire Geological Society*, **35**, 523–548.

—— 1970. A kaolinite mudstone (tonstein) in the British Coal Measures. *Journal of Sedimentary Petrology*, **40**, 386–394.

—— 1971. The mineralogy of the Stafford Tonstein. *Proceedings of the Yorkshire Geological Society*, **38**, 497–516.

—— 1982. The recognition of volcanic clays and the significance of heavy minerals. *Clay Minerals*, **17**, 373–375.

—— & DUFF, P. McL. D. 1984. Kaolinite and mixed

layer illite-smectite in Lower Cretaceous bentonites from the Peace River coalfield, British Columbia. *Canadian Journal of Earth Sciences*, **21**, 465-476.

—— & —— 1985. Cinerites and tonsteins. Report on Symposium No. 8. *In: Compte Rendu, 10me. Congrès International de Stratigraphie et de Géologie du Carbonifère, Madrid, 1985*, **4**, 171-173.

—— & KANARIS-SOTIRIOU, R. 1979. A geochemical and mineralogical investigation of some British and other European tonsteins. *Sedimentology*, **26**, 407-425.

—— & RICE, C. M. 1973. An Upper Carboniferous tonstein of volcanic origin. *Sedimentology*, **20**, 281-294.

STRAUSS, P. G. 1971. Kaolin rich rocks in the East Midlands coalfields of England. *In: Compte Rendu, 6me. Congrès International de Stratigraphie et de Géologie du Carbonifère, Sheffield, 1967*, **4**, 1519-1532.

TUTTLE, O. F. 1952. Origin of the contrasting mineralogy of extrusive and plutonic sialic rocks. *Journal of Geology*, **60**, 107-124.

WALKER, G. P. L. 1980. The Taupo pumice, produce of the most powerful known (ultraplinean) eruption. *Journal of Volcanology and Geothermal Research*, **8**, 69-94.

WEAVER, C. E. 1953. Mineralogy and petrology of some Ordovician K-bentonites and related limestones. *Geological Society of America, Bulletin*, **64**, 921-943.

—— 1963. Interpretative value of heavy minerals from bentonites. *Journal of Sedimentary Petrology*, **33**, 343-349.

WILLIAMSON, I. A. 1961. Tonsteins: A possible additional aid to coalfield correlation. *Mining Magazine*, **104**, 9-14.

WILSON, A. A., SERGEANT, G. A., YOUNG, B. R. & HARRISON, R. K. 1966. The Rowhurst tonstein, North Staffordshire, and the occurrence of crandalite. *Proceedings of the Yorkshire Geological Society*, **35**, 421-427.

WINCHESTER, J. A. & FLOYD, P. A. 1977. Geochemical discrimination of different magma series and their differentiation products using immobile elements. *Chemical Geology*, **20**, 325-343.

WORSSAM, B. C., CALVER, M. A. & JAGO, G. 1971. Newly discovered marine bands and tonsteins in the South Derbyshire Coalfield. *Nature*, **232**, 121-122.

WOOD, D. A. 1980. The application of a Th-Hf-Ta diagram to problems of tectonomagmatic classification and to establishing the nature of crustal contamination of basaltic lavas of the British Tertiary Province. *Earth and Planetary Science Letters*, **50**, 11-30.

ZINKERNAGEL, U. 1978. Cathodoluminescence of quartz and its application to sandstone petrology. *In:* FUCHTBAUER, H., LISITZN, A. P., MILLIMAN, J. D., & SEIBOLD, E. (eds) Contributions to Sedimentology, **8**, 69 pp.

Determination of trace element affinities in coal by laser ablation microprobe–inductively coupled plasma mass spectrometry

X. QUEROL[1] & S. CHENERY[2]

[1]*Institute of Earth Science 'Jaume Almera', CSIC, C/Martí i Franqués s/n, 08028 Barcelona, Spain*

[2]*Analytical Geochemistry Group, British Geological Survey, Keyworth, Nottingham NG12 5GG, UK*

Abstract: The occurrence, association and distribution of potentially toxic trace elements (PTTEs) in coal are basic criteria for predicting the forms and amounts of trace elements that are transferred to the environment during coal combustion processes. Previously, the determination of the affinities of PTTEs in coal was carried out by laborious density separation and subsequent analysis of the density fractions. This study has evaluated a new means of directly determining PTTE affinities in coal by laser ablation microprobe–inductively coupled plasma mass spectrometry (LAMP–ICP-MS). The affinities were determined by a rapid semiquantitative methodology. Additionally, some preliminary quantitative concentrations were obtained using a novel calibration strategy. The spatial resolution of the LAMP–ICP-MS technique allows the analysis of single grains of macerals and minerals in polished blocks, leading to a wide range of applications in coal geochemistry.

Coal combustion in power stations is one of the largest sources of the anthropogenic dispersion of potentially toxic trace elements into the environment. This dispersion is carried through the emissions of combustion gases, accompanied by a portion of fly ash that escapes the particulate control mechanisms. The pathways and the amounts of the emissions depend on fuel quality and combustion technology (Smith 1987; Clark & Sloss 1992; Meij 1992). Considerable amounts of trace elements may be transferred to the surrounding environment, e.g. a 1000 MW power station typically consumes 12 000 t of subbituminous coal each day.

Since the 1960s there has been much written about the occurrence, association and distribution of trace elements in coals (Minchev & Eskenazy 1972; Gluskoter *et al.* 1977; Ward 1980; Harvey *et al.* 1983; Kojima & Furosawa 1986; Godarzi & Cameroon 1987; Miller & Given 1987*a, b*; Swaine 1990; Beaton *et al.* 1991; Querol *et al.* 1992; Huggins *et al.* 1993). These are the basic criteria for predicting the forms and amounts of trace elements that are transferred to the atmosphere during industrial coal-burning processes.

Studies of the mobility of trace elements during coal combustion have shown that their volatilities depend on their affinities and concentrations (Clark & Sloss 1992). Equally important can be physical changes and the chemical reactions of the trace elements with sulphur or other volatile components during coal combustion (Smith 1987; Meij 1992). The volatility of trace elements also depends on combustion technology parameters, such as the temperature, time of exposure and type of combustion.

Despite the numerous studies carried out on the occurrence and distribution of trace elements in coal, it is accepted that the affinities of these elements may well differ from one coal deposit to another, and that it is reasonable to expect organic associations to predominate in low ash (Finkelman 1982) and in low rank coals (Lindahl & Finkelman 1986). Common methods for the determination of trace element affinities involve the separation of different density fractions, which preconcentrates certain mineral and organic phases. This is achieved by the separation of fractions lighter and denser than the 'heavy' liquids used, and the subsequent mineralogical and chemical analysis of these subsamples. Correlation between the mineral and chemical data reveals the element affinities (Chenery *et al.* 1993). Although this conventional method has been used widely in coal geochemistry studies, there are several major disadvantages;

1. Use of toxic organic liquids and the consequent precautions and difficulties of handling, sample cleaning and liquid recovery.

2. The potential for sample contamination during successive floating/sinking density fractionations.
3. The unreliability of the organic/trace element affinity determination due to the fine-grained syngenetic minerals dispersed in the organic matrix (Querol *et al.* 1994).
4. The extreme differences in sample composition obtained from the density separation, ranging from pure organic mater to pure iron sulphides. These differences can give rise to interference or matrix effects during trace element determinations by spectrometric techniques (Chenery *et al.* 1993).
5. The technique is time consuming and requires adequate space for the fractionation of large amounts of coal, greater than 1 kg.

These disadvantages prompted consideration of an alternative methodology for the determination of trace element affinities in coal which, ideally, was rapid, free from trace element contamination and capable of discrimination between phases of the same density.

The development of the laser ablation microprobe–inductively coupled plasma mass spectrometry (LAMP–ICP–MS) technique (Darke & Tyson 1993), from the original work of Gray (1985), has provided the ideal tool for the selective analysis of individual phases in geological materials (Jackson *et al.* 1992; Perkins *et al.* 1993). Work by Lichte (1992) has demonstrated the analysis of bulk coal by LA–ICP-MS, thus overcoming problems of sample dissolution and contamination. The present study focuses on the ability of LAMP–ICP-MS to sample and analyse discrete mineral phases and coal macerals for trace elements.

Methodology

A high sulphur Spanish subbituminous coal from the Teruel mining district was selected because of the wide variety of sulphide mineral phases it contains. The trace element affinities in this coal were investigated previously using conventional density separations, digestion and

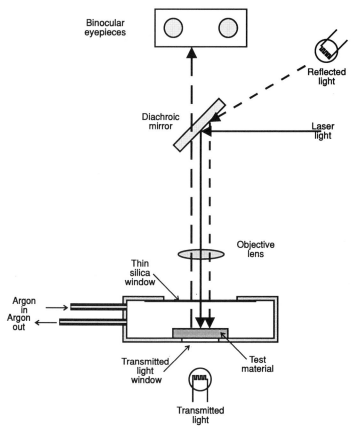

Fig. 1. Schematic diagram of the laser ablation microprobe.

analysis by ICP-MS and ICP-atomic emission spectrometry (AES) (Chenery et al. 1993).

Coal petrology polished blocks were prepared following conventional methods. Final polishing was carried out using 3 μm diamond paste and 0.05 μm pure alumina powder to avoid contamination in the polished surface of the mineral and organic particles to be analysed. A pellet of finely ground material may be prepared, as in X-ray fluorescence (XRD) spectrometry, if an analysis of a whole coal is required.

When using the LAMP instrumentation, the mineral phases and macerals of interest are identified through the microscope and then the laser is fired through the microscope column and objective lens (Fig. 1) making an ablation crater. The ablated material is swept by a stream of argon into the ICP. The plasma is an ideal atmospheric pressure ion source that totally decomposes the sampled material to predominantly singly charged ions. The quadrupole mass spectrometer provides a highly selective and sensitive filter for a large number of elements. Most ions are detected with approximately equal sensitivities, although exceptions include Se, Te and As. The detection limits are directly proportional to the amount of material sampled and are of the order of μg g^{-1} for a crater of diameter 25 μm and depth 25 μm (Chenery & Cook 1993).

The LAMP–ICP-MS system used in this study is that described in Chenery and Cook (1993). The LAMP was designed and built at Birkbeck College, University of London (UK Patent No. 91066337.0 Serial 2254444). It consists of a Nd–YAG laser, frequency quadrupled to operate in the ultraviolet (UV) light region at 266 nm. The UV light is passed into a high quality Leitz microscope and focused through a special reflecting lens. This system allows both viewing of a coal section at a quality expected by a geologist and a means of focusing the laser light on any chosen spot. The test sample is placed in a cylindrical Perspex laser ablation cell with a volume of 22 cm^3. The cell has tangential gas input and output flows and a thin quartz window at the top for both viewing and transmission of the UV laser light. The argon flow through the cell carries ablated material to the ICP-MS system. The operating conditions of the LAMP–ICP-MS are given in Table 1.

Discussion and results

Semiquantitative analysis

Thirteen trace elements (B, V, Cr, Co, Ni, Cu, Zn, As, Sr, W, Bi, U and Th) were determined semiquantitatively in vitrinite, liptinite, fusinite, calcite, clays and different types of syngenetic and epigenetic sulphides, and their affinities deduced. As shown in Fig. 2, the LAMP conditions were set to produce ablation craters with a constant diameter of 10 μm for all phases (i.e. carbonates, sulphides, clays and macerals). It was then assumed that the mass spectrometer

Table 1. *Laser ablation microprobe and inductively coupled plasma mass spectrometer operating conditions*

Laser ablation microprobe	
Laser	Spectron SL803 Nd:YAG
Wavelength	266 nm (frequency quadrupled from 1064 nm)
Maximum energy	70 mJ
Mode	Q-switched, TEM$_{00}$
Pulse length	10 ns
Laser repetition frequency	10 Hz
Microscope	Leitz Aristomet
Laser focusing objective	25 X and 36 X
Inductively coupled plasma mass spectrometer	
Spectrometer	VG PlasmaQuad 2+
Forward power	1350 W
Gas flow-rate	
Coolant	13 l min^{-1}
Auxiliary	0.8 l min^{-1}
Injector	Nominal 1 l min^{-1}
Data acquisition software	VG TRA
Data acquisition mode	Peak jumping
Points per peak	3
Dwell time per peak	10 ms
Sweeps per time unit	3
Time unit	1–5 s

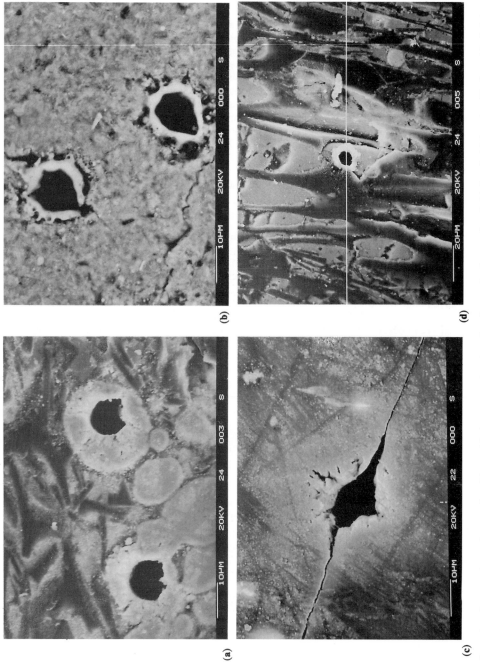

Fig. 2. Scanning electron photomicrographs of laser ablation craters in mineral and organic phases of coal: (**a**) framboidal pyrite; (**b**) clays; (**c**) vitrinite; and (**d**) fusinite. Note the constant crater diameter of 10 μm in all matrices.

Table 2. *Trace element affinities obtained by laser ablation microprobe–inductively coupled plasma mass spectrometry analysis of selected coal particles of Teruel coal. FRA, framboidal pyrite; FIB, fibrous marcasite; CIC, pyrite 'cone in cone'; CIN, massive pyrite in cellular infilling; EPI, epigenetic pyrite; and K/I, kaolinite and illite. The number of stars is a semiquantitative measure of the affinity of the trace elements for a particular phase.*

Element	Vitrinite	Liptinite	Fusinite	Sulphides FRA	FIB	CIC	CIN	EPI	Clays K/I	Calcite
B	******	***	*							
V	***	***		**					**	*
Cr				*	*				****	
Co		*		***	*				*	
Ni				***	**			*	*	
Cu				**	*				*	
Zn				**	**				*	
As			***	*	*	**	**			
Sr	**	***	***	**	*				***	**
W	*****			*	*					
Bi				***						
Th				*					****	
U	***	***	*						***	*

response was proportional to the elemental concentration because a constant volume was ablated. With the spatial resolution of this analytical tool it was possible to obtain not only the trace element affinities for different mineral phases, but also those of different diagenetic stages of the same mineral phase, e.g. framboidal, massive and cellular infilling pyrite, dendritic and massive marcasite.

The results obtained for the affinities (Table 2) agree with those acquired by conventional methods (Chenery *et al.* 1993). The elements B, V, W and U displayed major organic affinities with vitrinite and liptinite, and Sr showed some organic affinity with fusinite. Heavy metals showed greater associations with early syngenetic sulphides, such as framboidal pyrite and fibrous marcasite, than with late diagenetic sulphides. Calcite contributes a very low proportion of the trace elements to the coal. Chromium, Sr and Th have strong clay mineral affinities, whereas V, Co, Ni and U have mixed clay and organic or sulphide affinities.

Quantitative analysis

Calibration for quantitative analysis by LA–ICP-MS is normally performed using solid standards (Jackson *et al.* 1992; Perkins *et al.* 1993). This method of calibration is reasonable when either the test material is homogeneous or a large representative sample is ablated and, most importantly, the mass of material analysed after the ablation is the same in the standard and the sample. In this study, the analysis had to be performed on small particles with very different matrices, such as iron sulphides, clay minerals and macerals, for which no solid trace element standards are available. Many LA–ICP-MS users are currently using NIST glass standards to calibrate for a variety of silicate matrices, but these matrices have a restricted range of physical properties such as density. This strategy is only valid for quantitative analysis when the ablation process removes the same mass of material and transports this with a constant efficiency to the ICP-MS apparatus, for all matrices. Figure 2 shows craters of a constant diameter of 10 μm in framboidal pyrite, clays, fusinite and vitrinite, suggesting that with the ultraviolet LAMP it is possible to ablate a constant volume of material. However, it was not possible to show conclusively that the same mass of material is transported to the ICP-MS apparatus and therefore another calibration strategy was chosen.

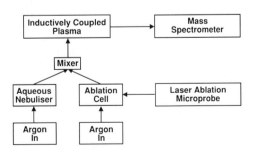

Fig. 3. Diagram of dual gas flow system for LAMP–ICP-MS quantitative calibration.

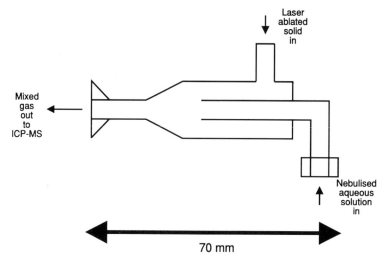

Fig. 4. Diagram of chamber used for mixing gas flows from aqueous nebulizer containing aqueous solutions and laser ablation cell carrying laser ablated solid.

An alternative calibration strategy, developed by one of the authors (Thompson et al. 1989; Chenery & Cook 1993), is to combine aqueous nebulization and laser ablation in a dual gas flow system (Fig. 3); this idea has also been developed by Moenke-Blankenburg et al. (1989). Two independent gas flows, one through an aqueous nebulizer and the other through the laser ablation cell, are combined using a mixing chamber (Fig. 4). The design is a modification of that originally produced commercially to provide a 'sheathing gas' in ICP-AES.

Essentially, this calibration methodology is based on the introduction into the ICP of a mixture of ablated material and a nebulized solution, produced in the mixing chamber. Ablated material arriving in the plasma gives rise to a response which can be compared with that produced by an aqueous standard. The dual gas flow system has the advantage that isotopic sensitivities can be obtained from homogeneous aqueous standards containing any combination of elements and concentrations. Also, the ICP-MS system can be rapidly tuned to conditions similar to normal aqueous nebulization. Polyatomic background levels are greater than with a 'dry' plasma, but still lower than normal aqueous nebulization. An essential requirement is the use of an internal standard to quantify the laser ablation response as the transport efficiency of the nebulizer is different from that for laser ablation.

The strategy followed for the calibration of LAMP–ICP-MS in the present work is based on two steps. The first is the determination of the response ratio [K_i in Equation (1)] for each trace element to be quantified and an internal standard, using a calibration solution. Iron was chosen as the internal standard as Fe is present in appreciable concentrations in all the analysed matrices, i.e. sulphides, clays, calcite and macerals. The response ratio must be determined for each analytical session as it is sensitive to many ICP-MS parameters such as plasma power and gas flows.

$$K_i = (R_i/R_{Fe})_{soln} \qquad (1)$$

where K_i = response ratio in a calibration solution; R_i = response (counts s^{-1}) of element i in a calibration solution at a concentration x; and R_{Fe} = response (counts s^{-1}) of the internal standard Fe in a calibration solution at a concentration x.

In the second step the iron content of the individual grains (C_{Fe}) is determined separately, before the ablation, either by electron probe microanalyses for clays, macerals and calcite, or by stoichiometry for the different types of pyrite and marcasite. Once the ablation has been performed, the concentrations of each element are calculated by applying the previously determined response ratio (K_i) in a conventional internal standard calculation

$$(R_i/R_{Fe})_{solid} = K_i(C_i/C_{Fe})_{solid} \qquad (2)$$

where, K_i = response ratio in a calibration solution; R_i = response (counts s^{-1}) of an element i in the ablated solid; R_{Fe} = response (counts s^{-1}) of the internal standard Fe in the ablated solid; C_i = concentration of an element i in the ablated solid; and C_{Fe} = concentration of the internal standard Fe in the ablated solid.

Table 3. Comparison of trace element content in $\mu g\,g^{-1}$ of the $>2.8\,g\,cm^{-1}$ density fraction of Teruel coal determined by inductively coupled plasma mass spectrometry (ICP-MS), instrumental neutron activation analysis (INAA) and laser ablation micropobe (LAMP)–ICP-MS for two different ablation crater diameters (10 and 25 μm). The iron content determined by INAA was used as an internal standard. In parentheses are listed the mass of the isotopes chosen for the analysis by LAMP–ICP-MS. DL, Detection limit; QL, quantification limit.

Element	LAMP–ICP-MS 10 μm crater	LAMP–ICP-MS 25 μm crater	DL 25 μm crater	QL 25 μm crater	ICP-MS	INAA
Ca (44)	1.03%	1.22%	0.02%	0.07%		0.8%
Mn (55)	296	324	2	9	270	735
Fe (57)	31.2%	31.2%	0.01%	0.04%		31.2%
Rb (85)	9.6	12	2	6	10.2	
Sr (88)	300	326	0.8	2		258
Cd (111)		3.4			1.4	
Sb (121)	1.6	2.4	0.5	2	3.7	
Tl (205)	2.1	2.0	0.3	0.9	1.9	
Pb (208)	50.4	51	0.6	2	64	
Bi (209)	<DL	0.4	0.2	1	0.3	
Li (7)	35.6	44	17	52	22	
Be (9)		1.8	0.6	2	0.7	
B (11)		27	13	30	13	
V (51)	21	33	1	4	16	19
Cr (52)	12.8	39	2	6		36
Co (59)	66.3	71	1	3		45
Cu (63)	51.7	59	2	6	48	
Zn (66)	1550	1270	1	4		1250
As (75)	288	277	8	26	168	
Al (27)	1.3%	1.3%	0.0002%	0.001%		1.5%
Y (89)	5	7.1	0.3	1.1	4.3	
Mo (95)	11	18.1	0.4	1.2	11.5	
Cs (133)	1	6.9	0.4	1.3		1.6
Ba (138)	283	521	0.2	0.8		526
La (139)	5	9.0	0.3	0.9		15
Ce (140)	13	16	0.3	1.0		19
W (182)	63	80	0.2	0.6	104	
Th (232)	3	4	0.2	0.7		2.4
U (238)	2	3	0.2	0.8		1

To evaluate the accuracy of this analytical strategy, a pressed powder pellet of a heavy density fraction of the coal studied was analysed by LAMP–ICP-MS. Table 3 compares these results with the trace element contents in the fraction as determined by neutron activation analysis (INAA) and solution ICP-MS. LAMP–ICP-MS results were obtained using Fe as an internal standard, as in the discrete phase analyses. These results show surprisingly good agreement with solution ICP-MS and INAA data considering the small mass of material sampled.

The results of analysing two sulphide phases, framboidal pyrite and fibrous marcasite, using the dual gas flow calibration strategy are shown in Table 4. The large difference in the trace element content between the two sulphides is particularly noteworthy because, in a conventional density separation, the two sulphides would be grouped together in the same denisty fraction. Figure 2a showed the ability of the LAMP to sample these sulphides selectively with the necessary resolution of 10 μm.

Conclusions

The application of LAMP–ICP-MS to the determination of elemental affinities in both mineral and organic coal phases has been demonstrated both semiquantitatively and quantitatively. The major advantages of LAMP–ICP-MS for the determination of trace elements affinities in coal are: (1) the ability selectively to ablate specific phases leading to the direct determination of the trace element contents of mineral or organic particles; and (2) rapid determination, which avoids the very time consuming sample treatment required in the density separation method. The debate as to whether a limited number of discrete point analyses of a mineral phase are representative of that phase in the bulk sample is still open.

Table 4. *Trace element content ($\mu g\,g^{-1}$) of framboidal pyrite and fibrous marcasite in the Teruel coal determined by laser ablation microprobe–inductively coupled plasma mass spectrometry. The iron content was used as an internal standard and was determined stoichiometrically (46.7%). DL, Detection limit; QL, quantification limit.*

Element	Framboidal pyrite	Fibrous marcasite
Mn	311	25
Sr	130	14
Cd	1.6 < QL	1.7
Sb	12	< DL
Tl	11	0.2 < QL
Pb	227	1.6
Co	46	4.8
Cu	32	2.2 < QL
Zn	22	12
As	36 < QL	51
Mo	27	4.6
W	5.3	1.6
Th	2.8 < QL	0.7 < QL
U	4	0.5 < qL

This study was supported by contract ERB-CHICT920063 of the Human Capital and Mobility Programme from the European Community, and by the European Coal and Steel Community contract 7220-ED/014. This paper is published with the permission of the Director, British Geological Survey (NERC). The authors thank J. M. Cook for her constructive criticism and support. D. L. Miles from the Analytical Geochemistry Group of the BGS and A. Lopez-Soler and J. L. Fernandez-Turiel from the Institute of Earth Science 'Jaume Almera' of the CSIC are also thanked for their support of this work.

References

BEATON, A. P., GOODARZI, F. & POTTER, J. 1991. The petrography, mineralogy and geochemistry of a Paleocene lignite from southern Saskatchewan, Canada. *International Journal of Coal Geology*, **17**, 117–148.

CHENERY, S. & COOK, J. M. 1993. Determination of rare earth elements in single mineral grains by laser ablation microprobe–inductively coupled plasma mass spectrometry. *Journal of Analytical Atomic Spectrometry*, **8**, 299–303.

——, ——, MILES, D. L., QUEROL, X. & FERNANDEZ-TURIEL, J. L. 1993. Potentially toxic trace elements in Spanish coal and power station by-products: the determination and mineralogical affinities. *In:* JARVIS, I. (ed.) *3rd Kingston Conference, Analytical Spectroscopy in the Earth Sciences*, 29 [abstract].

CLARK, L. B. & SLOSS, L. L. 1992. *Trace Element Emissions from Coal Combustion and Gasification*. IEA Coal Research Report, **IEACR/49**, 111 pp.

DARKE, S. A. & TYSON, J. F. 1993. Interaction of laser radiation with solid materials and its significance to analytical spectrometry. *Journal of Analytical Atomic Spectrometry*, **8**, 145–209.

FINKELMAN, R. B. 1982. Modes of occurrence of trace elements and minerals in coal: an analytical approach. *In:* FILBY, R. H., CARPENTER, B. S. & RAGAINI, R. C. (eds) *Atomic and Nuclear Methods in Fossil Energy Research*. Plenum Press, New York, 141–149.

GLUSKOTER, H. J., RUCH, R. R., MILLER, W. G., CAHILL, R. A., DREHER, G. B. & KUHN, J. K. 1977. *Trace Elements in Coal: Occurrence and Distribution*. Illinois Geological Survey Circular, **499**.

GOODARZI, F. & CAMEROON, A. R. 1987. Distribution of major, minor and trace elements in coals of the Kootenay Group, Mount Allan, Alberta. *Canadian Mineralogist*, **25**, 555–565.

GRAY, A. L. 1985. Solid sample introduction by laser ablation for inductively coupled plasma source mass spectrometry. *Analyst*, **110**, 551–556.

HARVEY, R. D., CAHILL, R. A., CHOU, C. L. & STEELE, J. D. 1983. *Mineral Matter and Trace Elements in the Herrin Springfield Coals, Illinois Basin Coal Field*. Illinois State Geological Survey Contract/Grant Re **1983–84**, 162 pp.

HUGGINS, F. E., ZHAO, M J., SHAH, N. & HUFFMAN, G. 1993. Speciation of trace elements in coal from XAFS spectroscopy. *In:* MICHDELIDIN, K. (ed.) *Proceedings of the 7th International Conference on Coal Science, Alberta, Canada*, 660–663.

JACKSON, S. E., LONGERICH, H., DUNNING, G. R. & FRYER, B. J. 1992. The application of laser ablation microprobe–inductively coupled plasma mass spectrometry (LAM–ICP-MS) to in situ trace element determinations in minerals. *Canadian Mineralogist*, **30**, 1049–1064.

KOJIMA, T. & FURUSAWA, T. 1986. Behaviour of elements in coal ash with sink-float separation of coal and organic affinity of the elements. *Nenryo Kyokai-Shi*, **65**, 143–149.

LICHTE, F. E. 1992. Analysis of coal by laser ablation inductively coupled plasma-mass spectrometry. *In:* VOURVOPOULOS. G. (ed.) *Elmental Analysis of Coal and its By-products*. World Scientific, 80–96.

LINDAHL, C. & FINKELMAN, R. B. 1986. Factors influencing trace elements variations in U.S. coals. *In:* VORRES, K. S. (ed.) *Mineral Matter and Ash in Coal*. American Chemical Society Symposium Series, **301**, 61–69.

MEIJ, R. 1992. A mass balance study of trace elements in a coal-fired power plant with a wet FGD facility. *In:* VOURVOPOULOS, G. (ed.) *Elemental Analysis of Coal and its By-products*. World

Scientific, 299–318.

MILLER, R. N. & GIVEN, H. 1987A. The association of major, minor and trace inorganic elements with lignites. II: Minerals and major and minor element profiles in four seams. *Geochimica Cosmochimica Acta*, **51**, 1843–1853.

—— & —— 1987b. The association of major, minor and trace inorganic elements with lignites. III: Trace elements in four lignites and general discussion of all data from this study. *Geochimica Cosmochimica Acta*, **51**, 1311–1322.

MINCHEV, D. & ESKENAZY, G. 1972. Trace elements in the coal basins of Bulgaria. Trace elements in the coals from Marica–Iztok basin. *God. Sofii. University Geology-Geogrphy Fak.*, **64**, 263–291; *Bulgarian Science Literature Geography*, **25** (1), 21–22 [abstract].

MOENKE-BLANKENBURG, L. & GUNTHER, D. 1992. Laser microanalysis of geological samples by atomic emission spectrometry (LM-AES) and inductively coupled plasma-atomic emission spectrometry (LM-ICP-AES). *Chemical Geology*, **95**, 85–92.

QUEROL, X., FERNANDEZ TURIEL, J. L., LOPEZ SOLER, A. & DURAN, M. E. 1992. Trace elements in high-sulphur sub-bituminous coals of the Teruel Mining District (NE Spain). *Applied Geochemistry*, **7**, 547–561.

——, ——, —— & —— 1994. Mineral transformations during combustion of Spanish sub-bituminous coals. *Mineralogical Magazine*, **58**, 113–119.

PERKINS, W. T., PEARCE, N. J. G. & JEFFRIES, T. E. 1993. Laser ablation inductively coupled plasma mass spectrometry: a new technique for the determination of trace and ultra-trace elements in silicates. *Geochimica Cosmochimica Acta*, **57**, 475–481.

SMITH, I. M. 1987. *Trace Elements from Coal Combustion: Emissions*. IEA Coal Research, London, 87 pp.

SWAINE, D. J. 1990. *Trace Elements in Coal*. Butterworth, Guildford, 278 pp.

THOMPSON, M., CHENERY, S. & BRETT, L. 1989. Calibration studies in laser ablation microprobe–inductively coupled plasma atomic emission spectrometry. *Journal of Analytical Atomic Spectrometry*, **4**, 11–16.

WARD, C. R. 1980. Mode of occurrence of trace elements in some Australian coals. *International Journal Coal Geology*, **2**, 77–98.

Geophysical exploration

Review of borehole seismic methods developed for opencast coal exploration

N. R. GOULTY

Department of Geological Sciences, University of Durham, South Road, Durham DH1 3LE, UK

Abstract: Evaluation of opencast coal reserves in the UK is accomplished by drilling boreholes on a grid spacing of 40–60 m, with even closer spacings across major fault zones and in areas of old mine workings. In spite of this intensive drilling effort, it is difficult to detect small faults with throws of less than about 2 m, which can still present a serious stability hazard at site boundaries. Borehole seismic methods could be used to detect small faults, to give structural detail across major fault zones and to locate old mine workings. Two techniques have been developed for this purpose: cross-hole seismic reflection surveying and hole to surface seismic reflection surveying. Although the cross-hole method produces sections of extremely high resolution, very close source and receiver element spacings are required to image close to the boreholes. Also the borehole separation should be less than half the length of the source and receiver arrays if complete coverage is required between the arrays. The hole to surface method produces high resolution sections and could be extended to three-dimensional surveys, but imaging is limited to a bell-shaped cross-section with its apex just below the water-table. Both methods can be used to yield continuous sections along lines of boreholes and are potentially of value in other applications.

Opencast coal mines in the UK are worked to depths of around 100 m and the largest of them produce a few million tonnes of coal. The annual production of opencast coal is about 15 million tonnes, which is about 20% of the UK's total production. The site boundaries of most mines are restricted by urban development, roads, railways or water courses. Thus they are small by world standards, but nevertheless they are very profitable.

At the exploration stage, a dense grid of boreholes is drilled to determine the site reserves and overburden ratio. The boreholes also yield useful information on the rippability of the strata and faulting, which can cause stability problems at the edge of the excavation. All the largest opencast mines are worked by independent civil engineering companies under contract to British Coal Opencast (BCO). Consequently, a detailed site specification is required as a basis for competitive tendering, as well as to justify the application for planning permission.

At many opencast mine sites there are old underground mine workings, mostly from shafts sunk in the 19th or early 20th century. In County Durham, for example, it is not unusual to encounter old mine workings in five or six seams at the same site. The mine plans dating from before nationalization in 1947 are very unreliable, so one possible application of geophysical surveying techniques would be to locate old mine workings, both to improve estimates of site reserves before going out to tender and because they are a hazard while mining is in progress. Another important role for geophysical methods would be to locate small faults, of throws less than about 2 m, which might not be identified from the borehole data. Although these small faults will not significantly affect reserves estimates, they can present a serious stability hazard at site boundaries.

The seismic reflection method naturally suggests itself as a possible technique to provide the kind of resolution required, especially in view of its success in defining structure for underground long wall coal mining (Ziolkowski & Lerwill 1979; Fairbairn *et al.* 1986). However, those surveys are designed to image horizons in the depth range 200–1000 m, whereas the depths of interest for opencast mining are generally within 100 m of the surface. Kragh *et al.* (1992) showed examples of surface seismic reflection data obtained on opencast exploration sites and found that any shallow reflected arrivals in the uppermost 100 m were swamped by reverberant refractions and ground roll so that no useful information could be obtained. The only way to improve data quality would be to drill shotholes through the overburden into the Coal Measures. This would be very expensive and would still not succeed in imaging seams close to rockhead.

From Whateley, M. K. G. & Spears, D. A. (eds), 1995, *European Coal Geology,*
Geological Society Special Publication No. 82, pp. 159–167

Recent research has concentrated instead on making use of the dense grid of boreholes already drilled on each opencast exploration site. Goulty et al. (1990) reported mixed success from investigations of three cross-hole techniques for this application. The first method tried was to propagate seam waves between boreholes, with the source and receiver at seam level in neighbouring boreholes, to check for coal seam continuity. The channel waves which propagate along coal seams in deep coal mines are horizontally polarized shear waves (e.g. Buchanan 1983; Jackson 1985); even so, they are successfully generated using explosive charges. Jackson et al. (1989) have also reported that seam waves can be generated by explosive charges in boreholes at depths of more than 500 m. However, efforts to generate and record seam waves at shallow depths were generally unsuccessful, possibly due to the relaxation of joints and fractures at shallow depths which would cause scattering and the effective attenuation of shear-wave energy. The second technique investigated by Goulty et al. (1990) was travel-time tomography, but this does not have the required resolution. Thirdly, they reported the first ever successful cross-hole seismic reflection survey, with sources and receivers deployed at 2 m spacing in neighbouring boreholes 54 m apart. Two coal seams at 44 and 58 m depth were imaged right across the section between the boreholes.

Subsequently, Findlay et al. (1991) acquired further cross-hole data sets and refined the processing to include wave equation-based prestack depth migration. Over the same period, Kragh et al. (1991) acquired and processed hole to surface seismic reflection data sets. Both these techniques produce high resolution seismic reflection sections with some capability for imaging small faults and old workings. In this review, they are illustrated with published examples, their relative merits discussed and other applications briefly mentioned.

Cross-hole acquisition and processing

Data acquisition for a cross-hole seismic reflection survey is carried out in the same way as for a cross-hole seismic tomography survey. Downhole receivers are placed in one borehole at equal spacings. Hydrophones are the simplest type of receiver, although they must be placed beneath the water-table in the borehole. Alternatively, borehole geophones can be used. They have to be clamped to the borehole wall to ensure satisfactory coupling, but this has the advantage that they can be deployed above the water-table. Some kind of explosive source must be initiated in the neighbouring borehole and repeatedly fired at depth spacings equal to the receiver spacing. The source type can be a chemical explosive, a sparker or a piezoelectric source. Normally, signals at all receiver positions would be recorded from all source positions.

Findlay et al. (1991) used small charges of chemical explosive with hydrophone receivers. Both sources and receivers were spaced 2 m apart below the water-table. They also logged the boreholes with an inclinometer to measure deviations from the vertical.

A typical cross-hole record from a single shot is shown in Fig. 1a. The first arrivals are either the direct waves or head waves, the latter generally being of small amplitude and followed by the stronger direct waves. The first step in processing is to separate upgoing and downgoing energy across the receiver array in each common shot gather (CSG). This is because some energy will be reflected upwards from interfaces below the source and receiver while other energy is reflected downwards from interfaces above the source and receiver. The polarity of the reflected wave from an interface depends on whether the incident wave approaches the interface from above or below, so upgoing and downgoing reflections must be processed separately. Wavefield separation may be achieved by filtering in the frequency–wavenumber domain. The upgoing wavefield from the CSG in Fig. 1a is shown in Fig. 1b.

The next step is to mute out residual direct wave and head wave energy at the start of the traces. This is followed by deconvolution, choosing as the desired output a zero-phase wavelet with a flat amplitude spectrum over the useful signal bandwidth. The original wavelet in the data is assumed to be minimum phase and is estimated separately for the upgoing and downgoing wavefields from each shot. Finally, the wavefields are migrated using the generalized Kirchhoff integral of Dillon (1990) or the generalized Radon transform (Miller et al. 1987). Following migration, the polarity of the downgoing wavefield is reversed and the upgoing and downgoing migrated wavefields are merged together to yield a depth-migrated reflectivity image of the section between the boreholes.

The sections obtained by processing two cross-hole data sets from the same exploration site in Yorkshire are shown in Fig. 2. The first section, between boreholes 41 m apart, shows continuous reflectors in undisturbed ground (Fig. 2a). The second section, between bore-

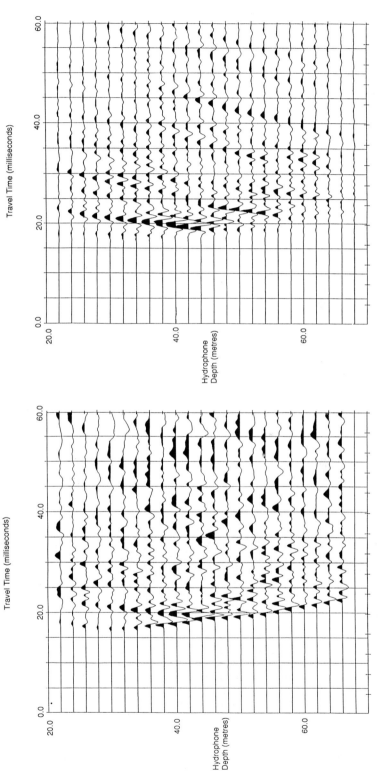

Fig. 1. Common shot gather from a cross-hole survey with the detonator at 30 m depth and the hydrophone receivers at depths of 22–66 m: (**a**) raw data with gain ramp; (**b**) upgoing wavefield only after wavefield separation in the frequency–wavenumber domain. From Findlay *et al.* (1991).

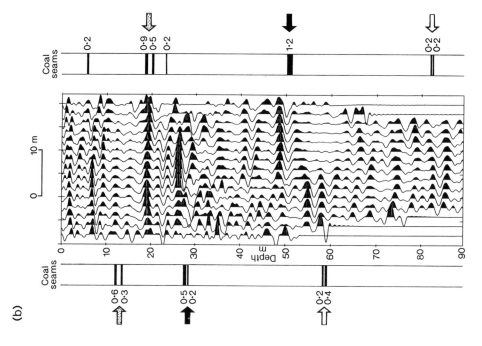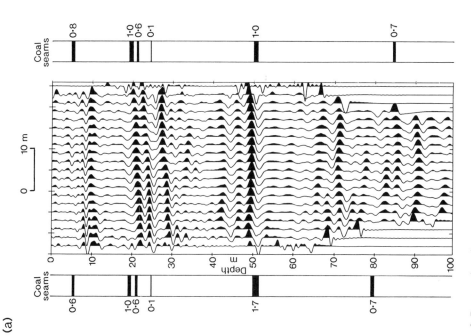

Fig. 2. Processed sections with prestack depth migration from cross-hole seismic reflection surveys: (a) across undisturbed ground; (b) across a normal fault zone with a total vertical throw of about 25 m. From Findlay et al. (1991).

holes just over 30 m apart, shows reflectors truncated at a normal fault zone with a total vertical throw of about 25 m (Fig. 2b). In both instances, single detonators were used as sources and hydrophones as receivers, spaced at 2 m intervals. In the first survey, sources and receivers were distributed over the depth range 10–60 m and in the second over 10–50 m. The polarity adopted for display is one where the reflection from the top of a coal seam should be black.

The strongest reflectors on both sections are coal seams. The slight mis-match in depth between the black peaks and the tops of the coal seams indicates that the velocity field used for migration was not estimated perfectly. In fact, mudrocks can be highly anisotropic, which means that estimating interval velocities assuming isotropy can never be perfect. At this site, the horizontal velocity through the strata was approximately 15% greater than the vertical velocity, but Findlay et al. (1991) were unable to improve the images by introducing this amount of anisotropy in each layer. From the borehole data for the second survey (Fig. 2b), it can be seen that a fault of just over 15 m throw cuts the left-hand borehole between 14 and 27 m depth and a smaller fault of about 7 m throw cuts the same borehole above 12 m depth. These faults may merge in the middle of the section between 50 and 60 m depth, where reflections from coal seams to either side have been truncated by the fault zone.

Hole to surface acquisition and processing

In a hole to surface seismic reflection survey, the source is placed in the borehole and receivers are located at the surface along a line which passes through the top of the borehole. This arrangement is known as a reverse (or inverted) multi-offset vertical seismic profile (RMOVSP or IMOVSP) in the jargon of exploration seismology. As in the cross-hole survey, an explosive source is fired below the water-table in the borehole, at several different levels, and vertical geophones are deployed at the ground surface for detecting the signals. Kragh et al. (1991) used 25 g charges of chemical explosive at depth intervals of 2 or 3 m and a spread of 24 surface geophones spaced 4 m apart.

The reason for firing shots at several different levels is to allow the separation of upgoing and downgoing wavefields, as in a regular VSP. The primary reflections which are to be imaged correspond to rays which leave the source in a downward direction, reflect at an interface and travel upwards to the geophones. However, much of the energy on the records, including the direct waves, leaves the source in an upward direction. Wavefield separation is achieved by filtering common receiver gathers (CRGs) in the frequency–wavenumber domain. The primary reflected wavefield is then deconvolved to zero-phase on a trace by trace basis, using the direct wavefield on the same trace as the input wavelet for filter design.

After wavefield separation, the data are sorted back into CSGs for calculation of static corrections and velocity analysis, which is achieved in an iterative loop. Each CSG is then migrated by wavefield extrapolation in the frequency–wavenumber domain (Berkhout 1984) and the migrated CSGs are stacked to yield the final section.

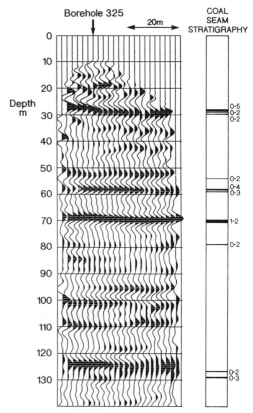

Fig. 3. Seismic depth section obtained from a survey at an exploration site in Northumberland with the coal seam stratigraphy. From Kragh et al. (1991).

The first example, shown in Fig. 3, is from a hole to surface survey acquired at an exploration site in Northumberland. The purpose of the survey was to try to locate a large fault known to exist to the right of the borehole, which was expected to form the site boundary. The geophone spread was laid out asymmetrically to bias the coverage to the right and shots were fired in the borehole at 3 m intervals over the depth range 7–70 m. The processed section shows strong, continuous reflections from coal seams between 28 and 71 m depth, proving that there is at least 30 m of undisturbed ground between the borehole and the fault.

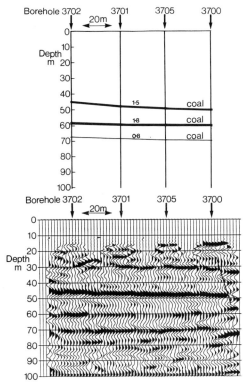

Fig. 4. Coal seam stratigraphy from four collinear boreholes at an exploration site in Yorkshire and the continuous seismic depth section obtained from hole to surface surveys in all four boreholes. From Kragh et al. (1992).

More interesting results were obtained from a line of four boreholes in Yorkshire, drilled along a causeway built part way across a shallow lake. The site geologist told the seismic team that the purpose of the investigation was to locate room and pillar mine workings in a 1.5 m thick seam at 50 m depth. The boreholes all penetrated solid coal at 50 m, although this might have been because each had encountered a pillar in the old workings. As the boreholes were only just over 30 m apart, overlapping coverage was obtained, with shots fired at 3 m spacing over the depth range 12–45 m. A single seismic section was produced by processing all four surveys and showed a strong continuous reflector between the boreholes (Fig. 4), interpreted as a continuous unworked seam. In addition, a small normal fault was revealed to the right of borehole 3700, with a downthrow of 2–3 m to the right. Further boreholes were drilled on the other side of the lake after the seismic survey had been commissioned. They showed that the old workings did not, in fact, extend beneath the lake and, although the small fault could not be identified from the borehole data, the levels of coal seams in the boreholes were consistent with the fault interpretation.

A survey to test whether the hole to surface method could detect old mine workings was conducted at a site in Cumbria. The purpose was to locate the edge of a longwall panel worked in a 0.5 m thick seam at a depth of 65 m (Fig. 5). Shots were fired at 2 m intervals over the depth range 18–60 m in borehole 4820 and the geophone spread was laid out asymmetrically to bias the coverage to the left. On the right of the processed section, where the coal seam is solid, there is only a weak reflection, but it is much stronger to the left. This contrast in reflection character, 16–18 m to the left of borehole 4820, is interpreted as the edge of the worked panel and it is interesting to note that the strong reflection comes from the worked seam. The 0.4 m thick solid seam at 55 m depth is also not associated with a clear reflection event, although the 0.55 m thick solid seam at 45 m depth does give rise to a strong reflection. A further complication here is that level information from neighbouring boreholes suggested that there might be a small fault at the edge of the worked panel. However, it is not possible to confirm, or contradict, that inference from the seismic section.

Discussion of relative merits

The reflectivity images obtained from processing cross-hole and hole to surface surveys from opencast exploration sites are remarkable compared with the virtually useless results of surface seismic reflection surveys (Kragh et al. 1992). The strongest reflections in the data arise from coal seams, including some only tens of centimetres thick and others which have been worked. However, it should be noted that thin coal seams (0.5 m or less) commonly do not give rise to coherent reflections. The resolution is high thanks to the broad bandwidth of the data: for cross-hole surveys the bandwidth extends to 500 Hz or more, and for hole to surface surveys

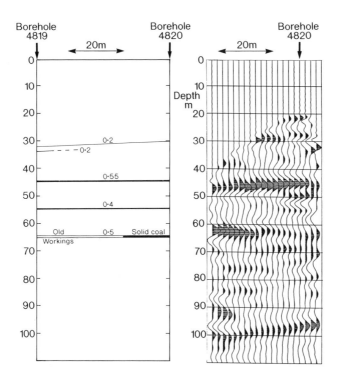

Fig. 5. Coal seam stratigraphy for two boreholes at an exploration site in Cumbria, showing the edge of a longwall panel in the seam at 65 m depth and the seismic depth section obtained from the hole to surface survey in the borehole on the right. From Kragh *et al.* (1991).

it can reach 300 Hz. Consequently, it should be possible to locate isolated small faults with throws of less than 1 m.

A very important matter regarding the applicability of such borehole seismic surveys is the actual area of cross-section which can be imaged (Fig. 6). Assuming gentle dips, a crosshole survey can image the area between the source and receiver arrays in the two boreholes and contiguous areas of tapering width above and below. A hole to surface survey provides an image over a bell-shaped section with its apex just below the water-table; at depth, the width, asymptotically approaches half the length of the geophone spread. Data quality steadily deteriorates below the bottom of the source array. Thus Fig. 3 shows the most impressive hole to surface section because shots were fired over a large depth range, whereas in Figs 4 and 5 the images are not very reliable below the target horizons.

Rowbotham & Goulty (1993) give a more comprehensive analysis of the imaging capability of cross-hole surveys. In practice, the most time-consuming task in cross-hole processing is optimization of the velocity field for migration.

If there were only vertical velocity variations (i.e. perpendicular to the bedding) it would not be so tricky, but there is certainly anisotropy and there are probably lateral variations in velocity as well. Because of this, it is desirable to mute out reflected arrivals whose raypaths lie within, say, 30 degrees of the horizontal. For these arrivals, a small error in velocity will cause a large vertical shift in the calculated position of the reflection point on depth migration. As data from several traces are stacked together at each image point, such errors will cause smearing of the image. To give complete coverage across the middle of the section, therefore, borehole separation should be no more than about half the vertical extent of the source and receiver arrays.

Another limitation is imposed by the element spacing in source and receiver arrays. Rowbotham & Goulty (1993) showed an example where a cross-hole survey failed to image a small fault cutting a reflecting horizon close to the receiver borehole, although it was evident on the hole to surface section acquired from the same hole. They deduced that this failure was because the receiver spacing of 2 m was too coarse. This

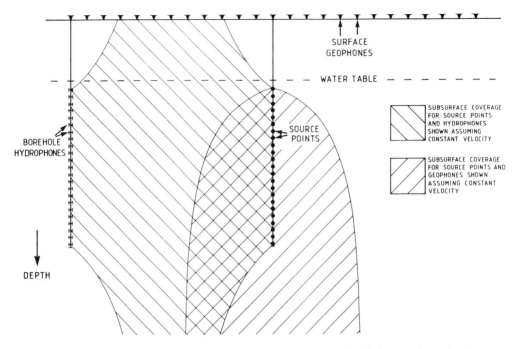

Fig. 6. Subsurface coverage obtained in a cross-hole reflection survey and a hole to surface reflection survey, assuming horizontal reflectors and constant velocity. Source positions are common for both surveys. From Goulty *et al.* (1990).

problem could be alleviated by substantially reducing both the source and receiver spacings, but only at the cost of acquiring and processing a much larger data set.

Assuming the problem of imaging close to the boreholes can be overcome, it would be possible to build up continuous coverage with cross-hole surveys along a line of boreholes, as was done with hole to surface surveys in Fig. 5. This would provide a continuous section up to the water-table, with intermittent coverage above, whereas the hole to surface section provides no coverage above the water-table and intermittent coverage immediately below it. However, the hole to surface method has other important practical advantages. When uncased boreholes are used, there is always the risk of holes collapsing so that surveys have to be aborted. Clearly the risk is greater when two boreholes have to be used together. The surface geophones may be laid out in any direction for the hole to surface survey, but the orientation of cross-hole surveys is constrained by the borehole positions. An extension of this idea is that hole to surface surveys can be acquired with an areal spread of geophones at the surface to provide a three-dimensional image over a bell-shaped volume of ground (e.g. Jackson *et al.* 1989).

At the time of writing, opencast coal exploration drilling in the UK has been savagely cut back. This is partly because the exploration programme has outstripped production, and partly because of government pressure to reduce costs in the run up to privatization. However, borehole seismic methods could still be used at the final phase of drilling: for example, to try to ensure the absence of structural disturbance at site boundaries where stability is critical.

The cross-hole method is the subject of ongoing research in the oil industry, where urgent attention is being given to enhancing recovery from existing fields. It could be used to monitor flood fronts in enhanced oil recovery (EOR) projects and also to image small-scale stratigraphic variations and sealing faults which influence fluid flow. The multi-offset VSP technique is already established in the oil industry, but the work reviewed here shows that it can be much more effective than surface seismic reflection surveys for shallow targets lying between the water-table and depths of

100–200 m.

Evidently, both methods are only applicable where a fairly intensive drilling effort is already in progress; they provide complementary information. A borehole only samples each horizon at a point, whereas seismic sections provide continuous coverage between them. The combination is a very powerful exploration tool, not only for the extractive industries, but also in site investigation where it is necessary to locate old mine workings, tunnels or structural disturbances, or to ensure that an area of ground is free from such hazards before engineering construction.

The research reviewed here was supported financially by BCO, BP, DENI, NERC, SERC and Shell UK. I thank the geological staff of BCO for professional assistance; however, the views expressed are my own and are not necessarily shared by the British Coal Corporation.

References

BERKHOUT, A. J. 1984. *Seismic Migration: Imaging of Acoustic Energy by Wavefield Extrapolation.* Elsevier, Amsterdam.

BUCHANAN, D. J. 1983. In-seam seismology: a method for detecting faults in coal seams. *In:* FITCH, A. A. (ed.) *Developments in Geophysical Exploration*, **5**, Applied Science, London, 1–34.

DILLON, P. B. 1990. A comparison between Kirchhoff and GRT migration on VSP data. *Geophysical Prospecting*, **38**, 757–777.

FAIRBAIRN, C. M., HOLT, J. M. & PADGET, N. J. 1986. Case histories of the use of surface seismics in the UK coal mining industry. *In:* BUCHANAN, D. J. & JACKSON, L. J. (eds) *Coal Geophysics*, Society of Exploration Geophysicists, Tulsa, 188–203.

FINDLAY, M. J., GOULTY, N. R. & KRAGH, J. E. 1991. The crosshole seismic reflection method in opencast coal exploration. *First Break*, **9**, 509–514.

GOULTY, N. R., THATCHER, J. S., FINDLAY, M. J., KRAGH, J. E. & JACKSON, P. D. 1990. Experimental investigation of crosshole seismic techniques for shallow coal exploration. *Quarterly Journal of Engineering Geology*, **23**, 217–228.

JACKSON, P. 1985. Horizontal seismic in coal seams: its use in the UK coal industry. *First Break*, **3**, 15–24.

JACKSON, P. J., ONIONS, K. R. & WESTERMAN, A. J. 1989. Use of inverted VSP to enhance the exploration value of boreholes. *First Break*, **7**, 233–246.

KRAGH, J. E., GOULTY, N. R. & FINDLAY, M. J. 1991. Hole-to-surface seismic reflection surveys for shallow coal exploration. *First Break*, **9**, 335–344.

——, —— & BRABHAM, P. J. 1992. Surface and hole-to-surface seismic reflection profiles in shallow Coal Measures. *Quarterly Journal of Engineering Geology*, **25**, 217–226.

MILLER, D., ORISTAGLIO, M. & BEYLKIN, G. 1987. A new slant on seismic imaging: migration and integral geometry. *Geophysics*, **52**, 943–964.

ROWBOTHAM, P. S. & GOULTY, N. R. 1993. Imaging capability of crosshole seismic reflection surveys. *Geophysical Prospecting*, **41**, 927–941.

ZIOLKOWSKI, A. & LERWILL, W. E. 1979. A simple approach to high resolution seismic profiling for coal. *Geophysical Prospecting*, **27**, 360–393.

Resources, environment and energy policies

Coals of Greece: distribution, quality and reserves

C. KOUKOUZAS[1] & N. KOUKOUZAS[2]

[1] *Department of Energy Resources, Institute of Geological and Mineral Exploration (IGME) Messogion Street, 70, Athens (115 26), Greece*

[2] *Department of Geology, University of Leicester, University Road, Leicester LE1 7RH, UK*

Abstract: Greek coals occur in a number of sedimentary basins and range in age from Eocene to Quaternary. The petrographic data indicate a wide variation in petrographic and chemical composition. The rank ranges from the transition zone peat–lignite to subbituminous. There are no deposits of hard coal in the country. Lignite constitutes the most abundant type of coal in Greece and the most important of the Greek lignite deposits formed during the Pliocene and Pleistocene in shallow lakes and marshes of closed intramontaine basins (Ptolemais, Florina, Drama in Macedonia, northern Greece and Megalopolis in Peloponnesus, southern Greece). The proved lignite reserves are currently estimated at 6750 Mt, excluding the 4300 Mm^3 of 'Phillipi' peat in Macedonia. There, 58% (about 3900 Mt) is considered to be economically recoverable. The probable and possible reserves are estimated to be of the order of 4000 Mt. The Kozane–Ptolemais–Amynteo–Florina basins in Macedonia contain most (about 64%) of the nation's coal resources. These lignites, which are all already being exploited, have a very low calorific value (at Ptolemais-Amynteo, 1400 kcal kg^{-1}; at Megalopolis, 900 kcal kg^{-1}) and high ash and low sulphur contents. The lignite production for 1992 was over 54 Mt. the greatest centres of lignite production are in Macedonia, at the opencast mines of Ptolemais and Amynteo, and in Peloponnesus, at the opencast mine of Megalopolis. The vast majority (98%) of the extracted lignite is used for electricity generation and feeds power plants which have a total capacity of 4533 MW. The lignite-based power plants account for more than 72% of the total electricity generation of the country. Today, through detailed geological exploration and evaluation, efforts are being made to locate and develop other lignite deposits throughout Greece.

Greece is a country with a deficient energy balance. The domestic energy resources, except for lignites, are limited: a few oil and natural gas fields occur on the continental shelf of Thassos island in the north Aegean Sea; a few uneconomic gas fields offshore near Katakolo in the Peloponnesus; and a few unexploited uranium deposits in the Paranesti, eastern Macedonia. There are no large-scale hard coal deposits; however, some uneconomic, lenticular, hard coal beds have been located in the coastal Permo-Carboniferous sediments in three areas of the country [Kardamila (Chios island), Central Euboea and Monemvasia Peloponnesus).

In contrast, Greece has notable lignite deposits, which are mined and contribute significantly to the energy requirements of the country. During 1992 about 30% of the country's energy consumption was supplied by domestic lignite and about 60% by, mainly imported, oil. Of the total production of electricity during 1992, the lignite contribution exceeded 72%; the remaining percentage was supplied by oil (about 21%) and hydroelectric power (about 7%).

Nuclear energy has not been introduced into the energy system of the country. Very small amounts of electricity are produced by wind parks installed mainly on the islands. Natural gas, imported from Russia, will be introduced into Greece's energy system after 1995. Electricity is not yet produced from geothermal sources because of environmental problems. There is some use of geothermal energy for heating, mainly in the agricultural sector.

The mineral rights for coal, like the rest of the energy resources, belong to the state, which issues exploration and exploitation permits to public organizations or leases to individuals. Today, coal exploration is carried out almost entirely by the Institute of Geology and Mineral Exploration (IGME) and about 95% of coal mining is undertaken by the Public Power Corporation (PPC). The rest of the lignite comes from private mines.

Geological setting

Conditions suitable for coal formation in Greece prevailed periodically from the start of the Cenozoic era to recent geological time. This has resulted in a large number of coal-bearing

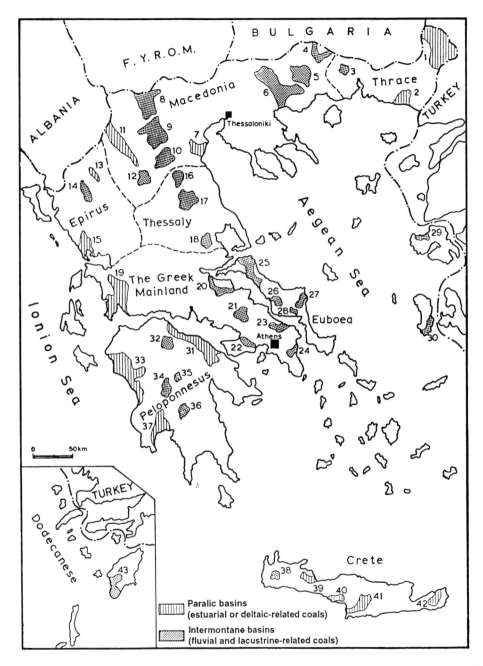

Fig. 1. Map of Greece showing location of coal-bearing basins. 1, Orestias; 2, Alexandroupolis; 3, Kotili; 4, Paranesti; 5, Drama; 6, Serres; 7, Moschopotamos; 8, Florina; 9, Ptolemais; 10, Kozane; 11, Middle Hellenic molasse; 12, Grevena; 13, Ionian flysch; 14, Ioannina; 15, Preveza; 16, Elasson; 17, Larissa; 18, Almyros; 19, Etolo-akarnania; 20, Lokris; 21, Kopais; 22, Megara; 23, Oropos; 24, Raphina; 25, Istiea; 26, Palioura; 27, Kymi; 28, Aliveri; 29, Lesvos; 30, Chios; 31, Corinth; 32, Kalavryta; 33, Pyrgos-Olympia; 34, Megalopolis; 35, Assea; 36, Pelana; 37, Messinia; 38, Kandanos; 39, Chania; 40, Plakia; 41, Erakleion; 42, Lassithi; and 43, Rhodes.

basins, both on the mainland and in the islands (Fig. 1). The numbers given in the text after the deposit names relate to the numbers given on Fig. 1.

Most of the Greek coals are classified as lignite, but the oldest are subbituminous and some of the youngest are at the division between lignite and peat. In this paper, under the term 'lignite', we also include subbituminous coal and these transitional types to peat and subbituminous coals. Pure peat deposits are not included, and where necessary these are specifically mentioned.

The oldest known lignite-bearing strata occur in Eocene flysch of the Ionian zone in western Greece (Fig. 1, No. 13 and Table 1). They are thin lignite beds, but of good quality (sapropelic, subbituminous coal). Also during the Eocene, coal-bearing rocks with the same geometry and quality were deposited in the molasse of Thrace, and beyond to the deposits of Alexandroupolis (No. 2) and Pentalofo Orestias (No. 1) (Marinos 1951; Andronopoulos 1977).

During the Oligocene, under similar geological conditions, the thin coal-bearing strata of molasse of the Middle Hellenic trough (No. 11) formed; these are of limited extent and good quality. In addition, there are very limited coal beds of lacustrine–fluvial origin in eastern Macedonia and Thrace (Paranesti, No. 4; Kotili, No. 3) (Melidonis 1980) and numerous extensive coal-forming strata formed in a molasse environment in Thrace (Dilofo Orestias, No. 1) (Fig. 1 and Table 1). The Eocene and Oligocene coals, although of good quality and high calorific value, cannot support economic mining on a large scale because of their limited reserves. However, they have supported, at various times, mainly small underground mines. These supplied local consumers and small amounts to the power stations.

The main phase of coal formation in Greece coincided with the Neogene and Quaternary, during which there was intense faulting and graben formation. The most important lignite deposits accumulated in the shallow lakes and swamps of isolated intermontane basins (Ptolemais, No. 9; Megalopolis, No. 34; Florina, No. 8; Drama, No. 5). Numerous estuarine and deltaic lignite-bearing strata, with thin lignite of limited economic interest, were formed in marine coastal zones (Pyrgos-Olympia, No. 33; Preveza, No. 15; Moschopotamos, No. 7) (Fig. 1).

The oldest Neogene lignites occur in the Miocene sediments (Table 1) of Florina (No. 8), the Vevi and Achlada deposits in the eastern margins of Ptolemais basin (Vegora, No. 9d; Komnina, No. 9c) and in those of the Moschopotamos (No. 7) basin (Koukouzas & Kouvelos 1976; Koukouzas et al. 1984; Kotis et al. 1992).

The large deposits of the central part of the Ptolemais basin (deposits of Ptolemais ss., No. 9a; Amynteo, No. 9b) and the deposits of Kozani (No. 10), Rhodes (No. 43) (Table 1) are of Pliocene age (Maratos & Koukouzas 1966; Anastopoulos & Koukouzas 1972; Anastopoulos & Broussoulis 1973; Koukouzas et al. 1979).

Finally, during the Quaternary, low rank lignites formed, such as the deposits of Drama (No. 5), Ioannina (No. 14), Kandanos (No. 38) and Megalopolis (No. 34) (Marinos et al. 1959; Broussoulis et al. 1986, 1991). In addition, there are Pleistocene to modern peat deposits, including the huge Phillipi deposit, which has a maximum thickness of 200 m and 4.3 Gm^3 of reserves (Melidonis 1969).

In general, the Eocene–Oligocene coals are developed mainly in paralic basins and the coal-forming environments were estuarine or deltaic (Andronopoulos 1977). Some lignite deposits of Miocene (i.e. Crete, Nos. 39–42) and Pliocene (i.e. Pyrgou-Olympias, No. 33) (Vagias, 1987) age are deltaic (Table 1). These deltaic coals mainly consist of non-woody plant remains. The biggest lignite deposits of Miocene (Florina, No. 8) and Pliocene (Ptolemais, No. 9) age and the large lignite deposits of the Quaternary (Drama, No. 5; Megalopolis, No. 34) are formed in intermontane, fluvial-dominated and lacustrine-dominated basins. The fluvial-dominated coals of these basins consist mainly of woody plant remains (Florina, No. 8: Komnina, No. 9c) (Koukouzas et al. 1984).

The number and thickness of the lignite beds vary significantly, not only from deposit to deposit but also within deposits. In most of the Greek basins both the number (one to five) and the thickness (0.50–5 m) of the coal beds are small, especially in the coal-bearing basins of Eocene and Oligocene age. The greatest cumulative thickness of coal beds of these ages (Orestias, No. 1a) does not exceed 15 m of strata. Of course, this does not mean that all the deposits in the younger basins are large, although those deposits with large lignite thickness, e.g. Ptolemais 60 m, Megalopolis 30 m and Amynteo 30 m, always occur in the younger coal-bearing basins.

The frequent alternation of clastic sediments (normally consisting of soft sediments such as sands, clays and marls) and lignite beds is a characteristic of the largest lignite deposits of Greece. These sediments have two main adverse effects on mining. Costs are increased because of the need to select the best coal beds and the

Table 1. Geological age, depositional environment and location of Greek coal-bearing formations (Nos. 1, 2, 9a, etc. correspond to coal basins and geographical area in Fig. 1).

Period	Depositional environments	Thrace	Macedonia – Ptolemais basin (No. 9)	Macedonia – Other basins	Epirus	Thessaly	Greek mainland	Peloponessus	Island (Euboea, Crete, etc)
Quaternary	Intermontane basins (limnotelmatic coals)		Ardassa (No. 9e)	Drama (No. 5)	Ioannina (No. 14)		Lokris (Regini) (No. 20b) Kopais (No. 21)	Pyrgos-Olympia (Vourargos) (No. 33b) Megalopolis (No. 34) Assea (No. 35)	Kandanos (No. 38)
Pliocene	Intermontane basins (limnotelmatic coals)		Ptolemais s.s. (No. 9a) Amynteo (No. 9b)	Kozane (No. 10) Grevena (No. 12)		Elasson (Pretorio) (No. 16b) Larissa (No. 17) Almyros (No. 18)	Oropos (No. 23) Raphina (No. 24) Megara (No. 22)	Pelana (No. 36) Kalavrita (No. 32)	Rhodes (No. 43)
Pliocene	Paralic basins (telmatodeltaic)				Preveza (No. 15)		Etolo-Akarnania (No. 19)	Pyrgos-Olympia (No. 33a) Corinth (No. 31) Messinia (No. 37)	
Miocene	Intermontane basins (limnotelmatic & fluviotelmatic coals)		Komnina (No. 9c) Vegora (No. 9d)	Serrai (No. 6) Florina (No. 8)		Elasson (Domeniko) (No. 16a)	Lokris (Agnadi) (No. 20a)		Kymi (No. 27), Aliveri (No. 28), Limni-Istiea (No. 25), Palioura (No. 26), Lesvos (No. 29), Chios (No. 30)
Miocene	Paralic basins (telmatodeltaic coals)			Middle Hellenic molasse (No. 11a) (Tsotylion Formation) Moschopotamos (No. 7)					Chania (No. 39), Plakia (No. 40), Eraklion (No. 42), Lassithi (No. 43)
Oligocene	Intermontane basin Paralic basins (telmatodeltaic coals)	Kotili (No. 3) Orestias (Dilofo) (No. 1a)		Paranesti (No. 4) Middle Hellenic mollasse (No. 11b) (Eptachorion Formation)					
Eocene	Paralic basins (telmatodeltaic coals)	Orestias (Pentalofo) (No. 1b) Alexandroupolis (No. 2)			Ionian flysch (No. 13)				

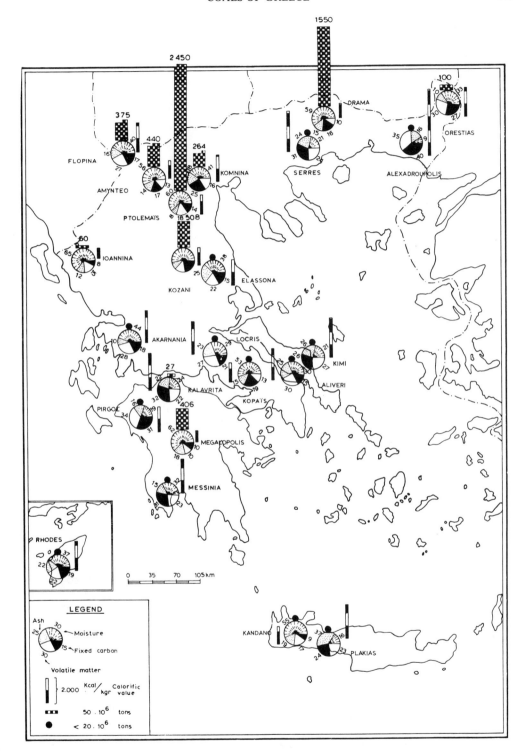

Fig. 2. Proximate analysis and calorific value of selected brown coal deposits of Greece. Modified from Koukouzas (1985).

quality is decreased due to the unavoidable mixing of lignite with the thin clastic intercalations, which cannot be separated and removed during extraction. However, their large reserves, subhorizontal bedding, continuity and proximity to the surface (depths to the first seam are often no greater than 150 m) make these younger deposits more amenable to economic mining.

Quality

The quality of Greek coals varies widely. In general, the coals of the larger deposits, which are younger in age, are lower in quality than those of the small deposits. Quality data are quoted on an 'as received' basis (Fig. 2) for most of the representative coal deposits of Greece in this paper. It must be noted that these quality data refer to mean values for the whole deposit, not for individual coal beds. There is likely to be a large variation in values at each deposit.

The net calorific value of Greek coals ranges between 830 kcal kg^{-1} (Ioannina, No. 14) to 5200 kcal kg^{-1} (Alexandroupolis, No. 2). At the mineable lignite deposits, this range may be more limited, ranging between 950 kcal kg^{-1} (Megalopolis, No. 34) and 2300 kcal kg^{-1} (Florina, No. 8), with an intermediate value of 1400 kcal kg^{-1} at Ptolemais (No. 9a) and Aminteo (No. 9b). The range of the quality parameters of the other deposits is about the same. Lignite deposits have a higher moisture content, e.g. Ioannina (No. 14) is 65%, Megalopolis (No. 34) is 62% and Drama (No. 5), Ptolemais (No. 9a) and Aminteo (No. 9b), which range between 58 and 60%. Subbituminous coals have a lower moisture content, e.g. Alexandroupolis (No. 2) with 9%.

In general, there is a decrease in the moisture content with age. Often the older coals contain a larger percentage of combustible components. The highest content of fixed carbon occurs in the Eocene subbituminous coals of Alexandroupolis (No. 2) (39.6%). The Miocene lignites of Kymi (No. 27) and Serres (No. 6) follow with about 30% fixed carbon content, whereas the lowest fixed carbon content occurs in the Pleistocene lignites of Ioannina (No. 14) (8.5%) and Megalopolis (No. 34) (9.6%). The Pliocene lignites of Ptolemais (No. 9a) have a slightly higher fixed carbon content of about 14%. The volatile matter content of these coals shows a

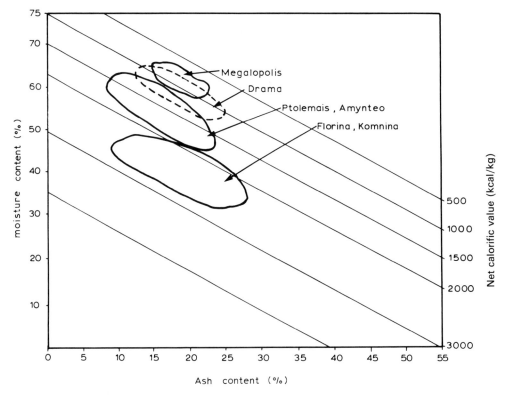

Fig. 3. Comparative quality data for the main Greek deposits (on an 'as received' basis).

similar reduction with age, e.g. Alexandroupolis (No. 2) has 35.4%, Serres (No. 6) has 38.7%, Megalopolis (No. 34) has 10.6% and Ptolemais (No. 9a) has 17.6%. There is no relationship between the ash content, which is a function of the clastic sediments interlayered and intimately mixed with the lignite, and the geographical distribution of the coal. The ash content ranges widely between individual deposits and between the lignite seams within those deposits. Hence the Alexandroupolis and Megalopolis deposits contain about the same percentage of ash. The ash content in most deposits ranges from 15 to 20%, whereas in some deposits it decreases to 8–10% and at others it may exceed 30% (Fig. 2). In Fig. 3 the coal qualities for the largest lignite deposits are plotted, based on moisture, ash content and net calorific value.

The highest total sulphur content is found in the lignites of Plakia (No. 40) with up to 4.5% sulphur, followed by the lignites of Serres (No. 6) with 3% and Orestias (No. 1), Alexandroupolis (No. 2) and Kymi (No. 27) with 2% sulphur. Most sulphur in these coals is found as pyrite and marcasite (Koukouzas & Skounakis 1990). The lignite currently mined at Ptolemais (No. 9a), Amynteo (No. 9b) and Megalopolis (No. 34) has a total sulphur content of less than 1%, whereas the combustible sulphur averages around 0.5%.

Trace element studies have been carried out on the ash from many Greek coals. These revealed high concentrations of some elements such as Pb, Zn, Ba, As, Mo, W, Sb and U in the Drama (No. 5), Serres (No. 6) and Paranesti (No. 4) deposits, whereas some lignites from southern Greece at Megalopolis (No. 34), Pyrgos (No. 33) and Kalavrita (No. 32) are enriched in fluorine (Foscolos *et al.* 1989).

The Greek coals are characterized, in general, by large amounts of macerals of the huminite group (76–95% on a mineral matter-free basis), low to intermediate amounts of macerals of the liptinite group (4–17%) and low to non-existent amounts of macerals of the inertinite group (0–4%). Notable exceptions are Ptolemais (No. 9a)

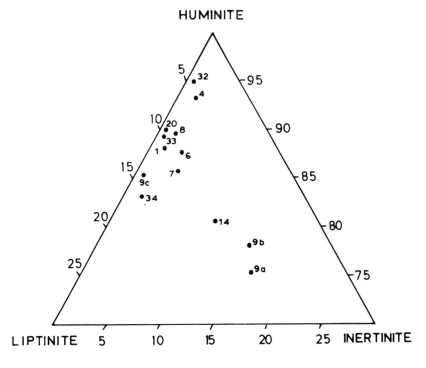

Fig. 4. Ternary diagram illustrating maceral group distribution (mineral matter-free) in Greek coals. Modified from Cameron *et al.* (1984). Nos 1, 4, 6, etc. relate to the geographical areas in Fig. 1 and Table 1.

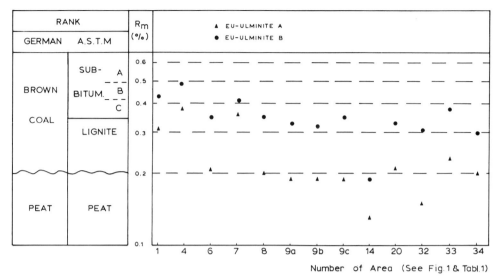

Fig. 5. Reflectances of eu-ulminite in Greek brown coals in relation to the German and ASTM rank classifications of coals. Modified from Cameron *et al.* (1984).

and Amynteo (No. 9b) lignites, where the percentage of macerals of the inertinite group averages 14% (Cameron *et al.* 1984). Figure 4 shows the distribution of the maceral groups for the most representative Greek coal deposits. Different macerals dominate the huminite group at different Greek lignite deposits. In most instances eu-ulminite and densinite dominate, e.g. Orestias (No. 1a), Paranesti (No. 4), Moschopotamos (No. 7) and older lignites with higher calorific values. Sometimes textinite dominates, e.g. Lokris (No. 20), Florina (No. 8) and Komnina (No. 9c) in the predominantly xylitic lignites. Attrinite dominates the telmatic and younger lignites, e.g. Megalopolis (No. 34), Ptolemais (No. 9a) and Ioannina (No. 14). It is these last lignites that are currently being mined.

The rank of Greek coals ranges from the transition zone peat–lignite, e.g. Ioannina (No. 14) to the subbituminous coal, e.g. Paranesti (No. 4). Rank has been determined by measuring reflectance on selected macerals (mainly eu-ulminite, Fig. 5) and by the chemical rank parameters such as volatile matter, fixed carbon and calorific value. Most Greek coals, including all of those currently mined, are classified as lignites. In addition to the lignites there are peat deposits such as the huge peat deposit of the Phillipi basin in Macedonia.

Reserves

The total proved lignite reserves of the country (Fig. 6), based on the latest exploration and most reliable data, are currently estimated at 6750 Mt (982 Mtoe), of which 58%, i.e. 3900 Mt (532 Mtoe) are considered economically recoverable (Table 2). These reserves exclude the enormous deposit of peat in the Phillipi basin in eastern Macedonia, which has reserves of 4300 Mm^3. This corresponds to about 1500 Mt of lignite with a net calorific value of 2000 kcal kg^{-1} and a moisture content of 36%.

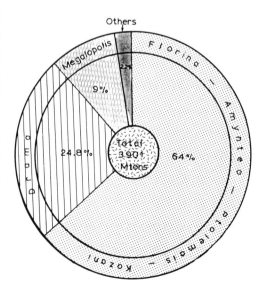

Fig. 6. Distribution of mineable lignite reserves in Greece.

Table 2. *Distribution of coal reserves of Greece*

Deposits	Area (in Fig. 1)	Net calorific value 'as received'	Reserves (10^6 tonnes) proved	(of which mineable)	No. of mines	Production in 1992 (10^6 tonnes)
Ptolemais s.s	9a	1400	2450.2	(1726.5)	3	33.7
Amynteo	9b	1400	440.1	(265.1)	1	8.4
Komnina–Anatoliko	9c	2000	469.0	(153.4)	—	—
Florina (with Vegora)	8(9d)	2200	375.6	(173.5)	3	2.0
Kozane	10	1100	508.0	(180.0)	—	—
Drama	5	1000	1550.0	(962.4)	—	—
Megalopolis	34	950	406.4	(355.6)	3	10.3
Others	—	1000–4500	554.6	(84.7)	Some small	0.5
			6753.9	(3901.2)		54.9

Probable lignite reserves are estimated at 1600 Mt (280 Mtoe), whereas possible reserves are estimated at 2300 Mt (500 Mtoe). The above reserves include those believed to exist to a depth of up to 500 m. Below this depth existing or potential lignite beds have not been added to the lignite reserves. It is known, from deep oil wells, that there are lignite beds at greater depths—for example, in the Pliocene molasse formations of the Pyrgos-Olympia basin and in the Miocene molasse formation of the Nestos basin in Thrace.

The geographical distribution of lignite-bearing basins is fairly wide (Fig. 1). However, the distribution of reserves is uneven due to the different potential of the lignite-forming processes in each basin. As already discussed, the intermontane and younger basins were the most dynamic and, hence, the largest percentage of lignite reserves accumulated in them (Fig. 2 and Table 2). For example, about 64% of the mineable reserves of Greece are found in the neighbouring basins of Florina (No. 8), Ptolemais (No. 9) and Kozani (No. 10).

Production and utilization of Greek lignite

In 1992 Greece produced 54.9 Mt of lignite, of which 95.5% was exploited by the PPC, from the opencast mines of Ptolemais and Amynteo in Macedonia and Megalopolis in Peloponnesus (Table 2). This represents 30% (6.9 Mtoe) of Greece's total primary energy needs (22.3 Mtoe). Ten years ago, the production of lignite was half this amount (27.4 Mt) and formed only 22% of the country's primary energy needs. This illustrates the intense efforts made by Greece to increase the contribution of domestic resources to the internal energy balance. Currently, most lignite production in Greece is from opencast mines. Only a very small amount (about 40×10^3 tonnes) is derived from underground workings, e.g. the lignite deposits of Kymi (No. 27).

Nearly all the domestic consumption of Greek lignite (53.8 Mt or 98%) is for electricity generation. During 1992, about 72% (24.170 GW h) of the total electricity generation of the country was produced from lignite. Twenty electricity generating units of 4533 MW total installed power (ranging from 10 MW to 310 MW) are fed with lignite. Of these, 16 (3983 MW) are installed in the area around Ptolemais and are supplied with lignite from the adjacent mines of Ptolemais, Amynteo and Florina. The remaining four (550 MW) are installed in the Megalopolis area and fed with lignite from the surrounding mines. During the next decade, five more electricity generating units of 1500 MW are to be built and supplied with lignite from the deposits of Florina (No. 8) Komnina (No. 9c) and Drama (No. 5).

Small amounts of lignite (about 1.1 Mt each year) are used to make briquette and in 'dry lignite' production (200 000 t) at the Ptolemais plant. The remainder is used for domestic heating and as a metallurgical coal in the processing of nickel/iron ores.

The non-electric uses of lignite are gradually decreasing. Although 20 years ago 13% (1.6 Mt) of the total lignite production was consumed in non-electric uses, today this consumption has decreased to 2% (1.1 Mt). For 30 years a factory in Ptolemais produced nitrogen fertilizers from lignite, but production ceased in 1992.

The large amounts of lignite are produced and consumed *in situ* in captive power stations. There is some environmental pollution, although the low combustible sulphur content of Greek lignites is low. Research and the application of new technologies are the main means of combatting pollution. Several scientific and technological research projects are currently in progress to study the trace element pollutants found in lignites. In addition, exploration for new lignite deposits is taking place in areas far from the large lignite consumption centres such as Ptolemais and Megalopolis.

References

ANASTOPOULOS, J. & BROUSSOULIS, J. 1973. Kozani–Servia lignite basin. *Mineral Deposit Research, IGME*, 1, 1–75 [in Greek, English abstract].
—— & KOUKOUZAS, C. 1972. Economic geology of the southern part Ptolemais lignite basin (Macedonia —Greece). *Geological & Geophysical Research, IGME*, 161/1, 1–189 [in Greek, English summary].
ANDRONOPOULOS, B. 1977. Geological study of Didimotichon–Pentalofos area (Orestias basin). *Geological & Geophysical Research, IGME*, 17/2, 1–58 [in Greek, English summary].
BROUSSOULIS, J., KOLOVOS, G., ECONOMOU, E. & HATZIYANNIS, G. 1986. [*Lignite deposits of Ioannina*]. Report, IGME, 1986 [in Greek].
——, YIAKKOUPIS, P., ARAPOGIANNIS, E. & ANASTASIADIS, J. 1991. *Drama Lignite Deposit. Geology, exploration, resources*. Report, IGME, Vol. 2. [in Greek, English abstract].
CAMERON, A. R., KALKREUTH, W. D. & KOUKOUZAS, C. 1984. The petrology of Greek brown coals. *International Journal of Coal Geology*, 4, 173–207.
FOSKOLOS, A. E., GOODARZI, F., KOUKOUZAS, C. & HATZIYANNIS, G. 1989. Reconnaissance study of mineral matter and trace elements in Greek lignites. *Chemical Geology*, 76, 107–130.
KOTIS, T., PLOUMIDIS, M., METAXAS, A. & VARVAROUSIS, G. 1992. *Coal Exploration of Vevi Subarea Florina District (W. Macedonia)*. Report IGME, Athens, 97 pp [in Greek, English abstract].
KOUKOUZAS, C. 1985. Greek lignite, qualities, reserves. *In: International Meeting for the Exploration of Low Calorific Value Solid Fuels*. Ptolemais, 26–28 September 1985, Public Power Corporation (PPC) [in Greek, English abstract].
—— & KOUVELOS, C. 1976. Pieria's lignite bearing district (Macedonia). *Mineral Deposit Research, IGME*, 6, 28 pp [in Greek, English abstract].
——, KOTIS, T., PLOUMIDIS, M. & METAXAS, A. 1979. Coal exploration of Anargiri Area, Aminteon (W. Macedonia). *Mineral Deposit Research, IGME*, 9, 1–69 [in Greek, English abstract].
——, ——, ——, —— & DIMITRIOU, D. 1984. Lignite deposit of Komnina area Ptolemais (W. Macedonia). *Research for Energy Resources, IGME*, 3, 103 pp [in Greek, English abstract].
KOUKOUZAS, N. & SKOUNAKIS, S. 1990. Pyrite framboids in the lignite deposit of Plakia–Lefkoja Rethymno Crete. *Bulletin of the Geological Society of Greece*, 25/2, 193–201 [in Greek, English abstract].
MARATOS, G. & KOUKOUZAS, C. 1966. Apolakkia–Rhodes lignite basin. *Geological Reconnaissance, IGME*, 37, 1–38 [in Greek, English abstract].
MARINOS, G. 1951. Alexandroupolis lignite basin. *Geological Reconnaissance, IGME*, 8, 1–15 [in Greek, English abstract].
——, ANASTOPOULOS, J. & PAPANIKOLAOU, N. 1959. Das Braunkohlebecken von Megalopolis. *Geological & Geophysical Research, IGME*, 5/3, 51 pp [in Greek, German summary].
MELIDONIS, N. 1969. The peat–lignite deposit of Philippi (Macedonia, Greece). *Geological & Geophysical Research, IGME*, 13/3, 87–250 [in Greek, German summary].
—— 1980. The geological structure of the lower Tertiary basin of Aemonio–Kotyli and its neighbouring district. (W Thrace). A study of the uranium bearing coal deposit. *Geological & Geophysical Research, IGME*, 22/2, 1–66 [in Greek, English summary].
VAYIAS, D. 1987. *Deltaic Lignite Deposit of Vassilaki-Ipso Area (West Peloponessus)*. Report, IGME, Athens [in Greek].

Environmental dust analysis in opencast mining areas

JOHN MEREFIELD,[1] IAN STONE,[1] PHILIP JARMAN,[2] GERAINT REES,[3] JO ROBERTS,[1] JEFF JONES[1] & ANDREW DEAN[1]

[1] *Earth Resources Centre, The University, North Park Road, Exeter EX4 4QE, UK*
[2] *British Coal Opencast, Heolty, Aberaman, Aberdare, Mid Glamorgan CF44 6LX, UK*
[3] *Rees Laboratories, Ponthenry, Llanelli SA15 5RE, UK*

Abstract: An atmospheric dust database has been set up for the western end of the South Wales Coalfield. This, the first of its kind in the UK, comprises dust data from British Standard 1747 four-way directional gauges, a window ledge sampler and an experimental frisbee deposit gauge. X-Ray diffraction was used to characterize the dust on the basis of mineralogy. Scanning electron microscopy was used to examine particle size and shape and an energy dispersive X-ray analysis system was used to locate coal particles.

For 18 months, a total of 15 directional and three frisbee gauges were used to collect dust on and around the Ffos Las opencast coal site in South Wales, UK. Additionally, two more directional gauges have been installed in the Brecon Beacons and at Exeter in south-west England to provide data from outside the coal mining area.

Processed data from the X-ray diffractograms, along with the weather data from an on-site weather gauge, are input to a geographical information system (ArcInfo), where they can be manipulated statistically. Rose diagram plots for the area can thus be presented, giving the mineralogical contents of dust from the four directions of each sampling station, set in a spatial context. A series of these plots introduces the temporal element.

This new dust data technology already shows considerable potential for use in nuisance complaints, planning applications, site licence renewals and for the implementation of future European environmental protection legislation.

Methods for the monitoring and assessment of dust derived from opencast coal mining operations are not well developed and no dust database has been established for the UK. In recent years mining companies and regulatory authorities have made use of the relatively inexpensive sticky pad method for dust collection, which was originally devised by WS Atkins and Partners, UK (Beaman 1981, 1984; Anon. 1990). This technique provides a planar or cylindrical adhesive surface which retains those particles that come into contact with it. It therefore provides a continual display of collected dust and readily indicates any fugitive emissions to the site management and to potential complainants (Jarman 1993). However, the sticky pad method does not necessarily capture a truly representative sample of the dust flux for use in dust characterization. Additionally, relatively little dust is held for subsequent analytical purposes. For detailed investigations, use of the British Standard directional and deposit gauges (BS 1747, Anon. 1972) has been combined with data from on-site and regional weather stations, although doubts have also been cast on their relative effectiveness (Ralph & Hall 1989, Anon. 1991a). More recently, with health effects in mind, Osiris SA respirable dust monitors, M-type samplers and the tapered element oscillating mast, for continuous monitoring of total airborne particulates, have been used (King *et al.* 1991).

With these shortcomings in mind and to set up the first dust database in the UK, British Coal Opencast commissioned the Earth Resources Centre of the University of Exeter and Rees Laboratories, Ponthenry to develop a new methodology for dust characterization. This uses an improved British Standard directional dust gauge (Merefield *et al.* 1992) and a prototype frisbee deposit gauge, based on the Warren Spring [Department of Environment (DoE)] design, for dust collection (Hall & Upton 1988). Mineralogical characterization of the dust is then carried out by X-ray diffraction analysis (XRD) and scanning electron microscopy (SEM) with energy-dispersive analysis (EDX). This paper describes the setting up of the database.

Methodology

The site chosen for this investigation was the Ffos Las Opencast coal site in South Wales, as

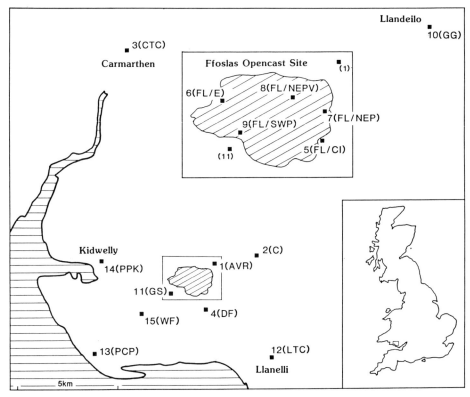

Fig. 1. Location map of the Ffos Las area showing positions of four-way directional gauges. Gauge numbers given with locality codes are: 1(AVR), 'Avril', Carway; 2(C), 'Mareal Villas', Cynheidre; 3(CTC), Carmarthen Technical College; 4(DF), Dythel Farm; 5(FL/CI), Cilferi Isaf; 6(FL/E), site entrance; 7(FL/NEP), NE New Perimeter; 8(FL/NEPV), NE Present Void; 9(FL/SWP), SW Present Void; 10(GG), 'Golden Grove', Llandeilo; 11(GS), '89 Garden Suburbs' Trimsaran; 12(LTC), Llanelli Technical College; 13(PCP), Pembrey Country Park; 14(PPK), 'Park Pendre', Kidwelly; 15(WF), Wern Farm; 16(UNI), Exeter University; and 17(BB), Brecon Beacons (off-map).

the local geology (Archer 1968; Frodsham et al. 1993) and an initial airborne dust survey there for Llanelli Borough Council (Merefield & Stone 1991) revealed a suite of minerals highly suitable for a more comprehensive provenance study. As the main aim was to compile a database of characteristic dust from Ffos Las, the immediate surroundings and the region in general, the British Standard four-way directional gauge was chosen as the principal dust collector. Early experiments showed that enough dust could be collected passively by these gauges for reliable analysis over a one calendar month period. Fifteen improved BS 1747 four-way directional gauges were set in place at monitoring stations in early October 1991 (Fig. 1), enabling 18 months of dust monitoring before project completion (Merefield et al. 1993).

The sampling locations were agreed with British Coal Opencast, who also liaised with landowners to install gauges at locations away from the opencast site. Five gauges were installed at Ffos Las around the perimeter of the workings, a further eight were sited in line with the prevailing wind direction (SW to NE) across the site and three others were located at Camarthen (off the South Wales Coalfield), at Llanelli and at Kidwelly to cover sampling from other wind directions. Additionally, two more gauges were installed after the initial trial period in the Brecon Beacons for comparison with the eastern end of the coalfield, and at Exeter University for an inter-regional comparison.

A window ledge sampler was manufactured to obtain samples from uPVC window ledges in the local community. In this way dust resulting in nuisance complaints at a variety of residential locations could be compared with dust from the directional dust database and a valuable subset of dust deposit data could be compiled from three regularly sampled locations to test temporal variations.

A prototype frisbee design manufactured for this study performed well in wind tunnel and computer modelling tests (Yacomeni 1993). Three of these deposit samplers were thus located near three of the directional gauges on, adjacent to and away from the Ffos Las site in case the dust yield from the directional gauges was too low. These gauges could be used for experimentation on dust characterization (e.g. clay mineral examination, crystallinity indices) which might otherwise prove difficult. The policy adopted for dust gauge manufacture overall has been one of constant development, but still retaining the British Standard dimensions where available.

In support of field monitoring, an XRD database on type minerals was established for reference purposes. Topsoils taken from the surrounding areas using a screw auger, which were representative of the main soil associations likely to be blown into and across the study area, were also characterized by XRD. The coarser than 125 μm fraction was sieved out, leaving only the very fine sand, silt/clay fraction for XRD analysis, as this fraction has the most potential for wind dispersal. A copper $K\alpha$ fine focus tube was used for XRD work with receiving slit at 0.1° and scatter and divergence slits set at 1°. The PA2000 APD Philips software system was used to set parameters and store the resultant patterns on hard or floppy disks. For general mineralogy (4–70°), count rates were set at 1.0 seconds with a step size of 0.02. For detailed clay investigations (4–15°), count rates were 2.0 seconds with a step size of 0.02. Identification procedures were based on Brindley & Brown (1980).

In addition to the qualitative analysis by XRD, a photographic record was made of selected samples by SEM. This, and an additional EDX analytical facility, provided high resolution and high magnification images and the geochemistry of the samples. It also assisted in any problems of mineral identification where XRD alone was inconclusive.

A second PC was added to the workstation controlling the X-ray diffractometer at Exeter and this enabled the database to be accessed while dust samples were routinely analysed by the XRD equipment. This also eased the bottleneck that was beginning to build up in the processing of monthly sample analyses. However, the software package required to map the dust in terms of provenance needed a higher disk capacity than that originally installed. Arrangements were therefore made to install a new 130 Mb capacity hard disk drive to accommodate the PC version of ArcInfo, a geographical information system (GIS) product. ArcInfo is an internationally recognized, highly sophisticated mapping package designed specifically for spatial analysis. Exploratory work was also undertaken to establish a procedure for combining the dust database, weather data (received daily in ASCII format and converted to monthly means) and mapping coordinates (using an electronic scanner) at the plotting phase.

Results

Dust database

Over the two year research period, 18 months of data were accumulated. Sampling was carried out on a monthly basis, producing the following three types of data set. These data were then ordered according to date, location and mineralogical range in varying proportions in such a way to make them readily accessible at any point.

At least 77 dust samples were prepared and analysed each month. In the course of the analysis, to allow optimum mineralogical identification, samples were often run twice. The data were then further manipulated (smoothed) by the manufacturer's diffraction software, creating up to four different data files for each sample. This necessitated the manipulation of about 308 separate data files each month, representing a considerable data management task, conceptualized in Table 1.

Table 1. *Flow chart of data management*

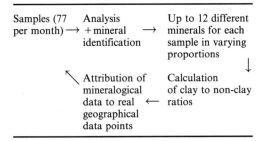

Early experimentation with a relational database (Rbase) made it clear that although an ordinary database package could handle the data and, indeed, sort them according to date, mineral or gauge, the output lacked meaning in terms of the spatial relationship of off-site, near-site and on-site dust. Therefore, the geographical details of the Ffos Las region, including main roads, coastline and soils of an approximately

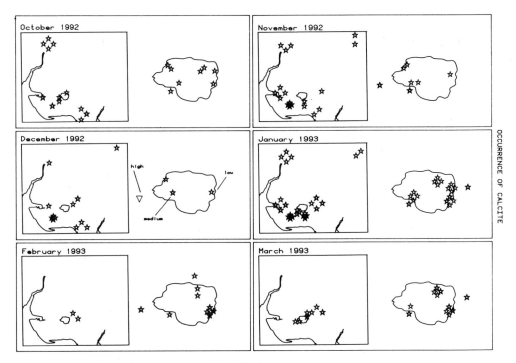

Fig. 2. Calcite distribution in airborne dust of the Ffos Las region.

200 km² area were digitized into ArcInfo, along with the gauge and window ledge coordinates.

Data were then attributed to these geographical points and stored in a tabular format which could be added to, sorted and manipulated like any other database. It is now possible to produce a series of dust distribution maps which not only show in detail the mineralogical components of all the samples collected, but also present the data in a spatially related form.

This dust database is the first of its kind and represents a major advance in the handling of empirically collected dust data; it is essential for further advances in dust data technology. It has allowed the dust data to be analysed and presented chronologically, in month by month distribution maps, representing the raw dust. It has also been possible to interrogate the database mineralogically, resulting in the production of a series of single mineral distribution maps (Fig. 2). The database exists in a number of forms: tabular, mapped and in longhand. In ArcInfo format it allows spatial analysis, but can also be exported in tabular form to other database systems.

The application of GIS software to the problem of dust characterization has proved to be a particularly successful exercise, as it is now possible to obtain pre-digitized maps of many regions of the British Isles and as part of an international network of geographical and geological software distribution it is ideally suited to environmental analysis and presentation.

Monthly mineralogical distribution

The dust maps represent the range of mineralogy identified by XRD and SEM. Each separate rose diagram represents a single collected sample, and where a cluster of four diagrams appear together they represent a single directional gauge with four samples. The crosses pin-point the actual geographical location of the gauges. The rose diagrams have no directional significance in themselves. Only when seen in a group of four do they represent a directional gauge (Fig. 3A). Ten regularly occurring minerals have been identified and their position and identifying codes are shown in (Fig. 3B).

The project brief was for a qualitative rather than a quantitative programme, and at no time were the samples weighed. However, it became clear that the relative proportion of minerals did vary and this was broadly reflected in the number of X-ray counts recorded on the

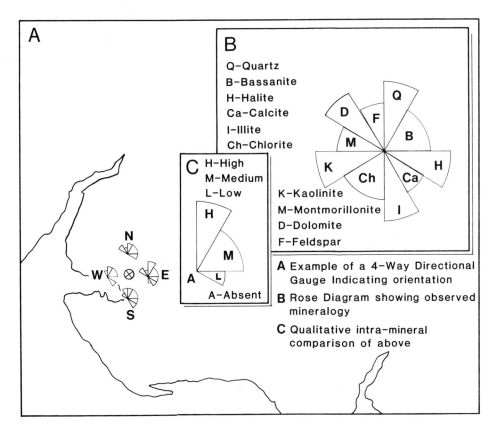

Fig. 3. Example key for rose diagram mineral plots.

diffractograms. Those minerals whose 100% peak, based on a standardized sample preparation technique, achieved greater than 500 counts were considered to be 'high', those between 50 and 500 'medium' and those under 50 'low'. This gives a qualitative indication of intra-mineral comparisons (Fig. 3C).

Mineralogical characterization

The following range of minerals is identified in airborne dust in the Ffos Las area: silicates [quartz, feldspar(s)]; evaporites [gypsum (gypsum/bassanite) halite]; carbonates [dolomite, calcite]; clays [kaolinite, illite, chlorite/dickite, montmorillonite]. Quartz (SiO_2) is ubiquitous and often comprises the largest proportion of any sample.

Feldspars [K/Na,Ca ($AlSi_3O_8$)]. Feldspar is also a ubiquitous component of continental atmospheric dust. The distribution of these minerals is uniform and although they are found in on-site gauges they are not particularly indicative of Ffos Las dust.

Gypsum ($CaSO_4 \cdot 2H_2O$). As found in the gauges, this evaporite mineral is most probably formed by the reaction between calcite and sulphate in atmospheric moisture (Pye 1987). Gypsum/bassanite is entirely an artificial product of the laboratory process of evaporation and the dehydration of gypsum at temperatures between 55° and 90°C.

Halite. Halite is artificially recrystallized by the evaporation in the laboratory of the salt (NaCl) in rain and seawater. The distribution of halite (NaCl) is particularly useful in identifying seasonal variations in mineralogy. The relative proportion decreases sharply from 'high' in the winter months of 1991–2 to minimal amounts in the summer of 1992 from June to September, and then increases sharply again in the following winter. Its general distribution is uniform and seasonal variations seem to be purely climatic

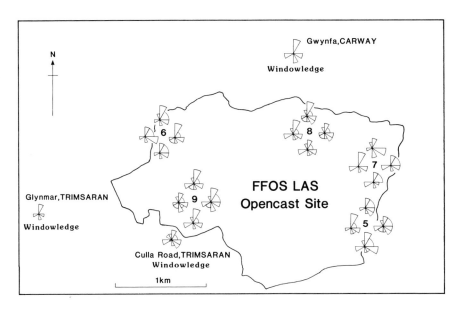

Fig. 4. Database plot of on-site and window ledge mineral distribution, May 1992, showing significant amounts of the clays kaolinite and chlorite at site 8. Key is given in Fig. 3.

and not geographical. The exception to this occurs at the Pembrey Country Park gauge site, where sea spray contributes substantial amounts of halite to the gauge collectors.

Dolomite ($MgCaCO_3$). The distribution is uniform. The very high proportion of dolomitic dust found at Abermarlais Caravan park led to an investigation of the road dust derived from the gravel driveway there, which confirmed the 'foreign' gravel as the source. It may well be that the roadways, and, in particular, imported loose gravels, represent a significant potential source of dolomitic dust contamination and is a subject which would certainly benefit from further study.

Calcite ($CaCO_3$). The distribution of calcium carbonate (Fig. 2) is similar on-site and off-site. The occurrence of calcite does not seem to be susceptible to significant seasonal variation. Calcite does occur along the coast, as was originally predicted, but not exclusively; its highest incidence was recorded about 10 km inland and other sources such as limestone quarrying should be considered.

Kaolinite [kaolinite group, including dickite— $Al_2Si_2O_5(OH)_4$)]. Kaolinite is a weathering product of other minerals, particularly feldspars, and as such is not an unusual component of atmospheric dusts. It is also to be expected and is found to be a major feature of the clay content of the shale-like material of the Ffos Las spoil heaps. As is consistently indicated by the kaolinite maps, 'medium' amounts of kaolinite occur on-site and are particularly a feature of the gauges situated at NE and SW Present Void, both of which are situated near spoil heaps. However, it is clear that the spoil heaps are not the only source of this mineral captured by the gauges. Kaolinite is also a consistent feature at Carmarthen and it is considered that road dust is the likely contaminant source in this instance. It may also be possible to tie in the occurrence of kaolinite at Kidwelly from April to October 1992 to the construction of the bypass there, which was noted by the Exeter University team as a potential dust generator.

Illite (KNa,H_2O)[($AlFe)_2(MgFe)x$] (Si_3 + xAl_{1-x}) O_{10} (OH). Illite is a clay mineral derived from the weathering of continental rocks. However, it has very similar diffraction characteristics to various micas and it takes very careful consideration of sufficient samples to differentiate them. For the purposes of characterization this is not necessary and no attempt has been made to do so. Its distribution strongly indicates that it is a characteristic of Ffos Las dust, derived from low grade metamorphosed mudstones, with 'medium' amounts only occurring on-site or within 0.5 km. However, as it is also a general feature of airborne dust, it is clear that Ffos is not the only source.

Montmorillonite. This is an infrequently occurring clay mineral, as its distribution in the Ffos Las area suggests. Its presence is likely to be detected where the sample is largest and this explains its presence on- and near-site.

Chlorite/undifferentiated dickite. It is clear that the clay component of the Ffos Las dust plays a key part in its characterization. However, clays are notoriously complex in terms of chemistry and crystallographic structure. For this reason and because the sample size has been restricted, it has not always been possible to differentiate between dickite, a fairly rare kaolinite-type clay mineral, and the occurrence of chloritic clays $[MgFeAl]_6(SiAl)_4O_{10}(OH)_8$, which are derived from low grade metamorphic rocks such as those associated with tectonic folding in the South Wales Coalfield. For the purpose of this investigation they have been treated together for presentation in distribution maps. The geographical spread shown in these maps is particularly revealing in terms of Ffos Las characterization, displaying as it does a very close geographical relationship with the site (Fig. 4).

Unidentified mineral: d-spacing 8.5–9. During the summer months of 1992, the sporadic occurrence of a 'mystery' mineral was noted. Possible explanations of this were contamination during sample collection of preparation, or the occurrence of chitin from the many insect carapaces trapped in the gauges. The high incidence of insect remains comprised by far the largest proportion of nearly all samples during this period, which would make it very difficult to weigh the dust portion accurately. This finding therefore has serious implications for quantitative dust studies under present and in future environmental legislation (Anon. 1991*b*).

Conclusions

An integrated approach to dust monitoring in a region of opencast mining and quarrying over a two-year period has generated a database of airborne dust mineralogy. This will provide a valuable reference source for similar work in the UK and elsewhere.

It is clear from the mineralogical distribution maps that some minerals occur uniformly on- and off-site, subject to seasonal variation, but do not reflect activities at Ffos Las opencast coal site. Although these are a component of the dust at Ffos Las, they are consistently found elsewhere and are characteristic of continental atmospheric dust in general. These minerals are quartz, gypsum (gypsum/bassanite), halite, calcite, dolomite and feldspar.

The distribution of clay mineralogy, however, is valuable as it can reflect events on-site, as is indicated by the relatively high incidence of chlorite/undifferentiated dickite at on-site and adjacent localities.

The dust data technology research team is particularly grateful to S. Upton of Warren Spring Laboratory for guidance on collector gauge design. R. B. Evans, K. Thomas, K. James and J. Bannister of British Coal Opencast (BCO) are thanked for their support and assistance at Ffos Las opencast site. L. Redmond (BCO) provided valuable data from the Ffos Las weather station datalogger. E. Payne is especially acknowledged for early discussions on analytical techniques. We are also indebted to M. Juster of Abermarlais Caravan Park for access to regular window ledge sampling and to R. Pun of the Storey Arms Centre for enabling the siting of a four-way directional gauge. Very grateful thanks are also extended to the residents of Trimsaran and environs who contributed much, by cheerfully making room on their properties for the dust gauges and window ledge sampling, and who waited patiently for the outcome of this investigation. J. Camp drafted the ERC illustrations. British Coal Opencast is thanked for permission to publish this paper.

References

ANON. 1972. *Directional dust gauge. BS 1747*, Part 5. HMSO, London.
—— 1990. *Assessment of the Potential Impacts from Dust Generated during Opencast Mining and Quarrying.* Wardell Armstrong Ltd, Leaflet, **24**, August.
——1991*a. Environmental Effects of Surface Mineral Workings.* DoE, HMIP Research Report, Roy Waller Associates, HMSO, London, 176pp.
—— 1991*b. Environmental Protection Act 1990, Part 1, Secretary of State's Guidance—Coal, Coke and Coal Product Processes.* DoE, PG3/5(91), HMSO, London.
ARCHER, A. A. 1968. *Geology of the South Wales Coalfield: the Upper Carboniferous and Later Formations of the Gwendreath Valley and Adjoining Areas.* Memoir of the Geological Survey of Great Britain. Special Memoir. HMSO, London, 216pp.
BEAMAN, A. L. 1981. Assessment of nuisance from deposited particulates using a simple and inexpensive measuring system. *Clean Air*, **11**, 77–81.
—— 1984. Recent developments in the method of

using sticky pads for the measurement of particulate nuisance. *Clean Air*, **14**, 74–81.

BRINDLEY, G. W. & BROWN, G. 1980. *Crystal Structures of Clay Minerals and Their X-ray Identification*. Mineralogical Society, London 495pp.

FRODSHAM, K., GAYER, R. A., JAMES, J. E. & PRYCE, R. 1993. Variscan thrust deformation in the South Wales Coalfield—a case study from Ffos Las opencast coal site. *In:* GAYER, R. A., GRIELING, R. O. & VOGEL, A. K. (eds) *The Rhenohercynian and Sub-Variscan Fold Belts*, Braunschweig-Vieweg.

HALL, D. J. & UPTON, S. L. 1988. A wind tunnel study of the particle collection efficiency of an inverted frisbee used as a dust depositional gauge. *Atmospheric Environment*, **22**, 1383–1394.

JARMAN, P. C. 1993. Dust research and development. *Mine and Quarry*, **Jan/Feb**, 31–32.

KING, A. M., FRYER, G. A. & WRIGHT, H. 1991. *Assessment of Dust Emissions from Surface Coal Mining Operations*. Final Report, Commission of the European Communities, Project No. 7263-02/077/08, 25pp.

MEREFIELD, J. R. & STONE, I. 1991. *Analysis of Particles from the Ffos Las Opencast Site and Trimsaran village*. Report ERC/91/13, University of Exeter, 24pp.

——, REES, G., STONE, I., ROBERTS, J., PARKES, C. & JONES, J. 1992. Mineralogical characterisation of atmospheric dust within and adjacent to opencast coal sites in South Wales. *In:* WILLIAMS, B. J. (ed.) *Proceedings of the Ussher Society Conference, Southampton, 1992*, **8**, 67–69.

——, STONE, I., REES, G., ROBERTS, J., DEAN, A., PARKES, C. & JONES, J. 1993. *Dust Data Technology at Ffos Las Opencast Coal Site, South Wales*. Final Report, July 1993. Report ERC/93/22, University of Exeter, 156pp.

PYE, K. 1987. *Aeolian dust and dust deposits*. Academic Press, London, 334pp.

RALPH, M. O. & HALL, D. J. 1989. Performance of the B.S. directional dust gauge. Paper No. W89001(PA), presented at the *Aerosol Society Annual Conference, March 1989*.

YACOMENI, T. P. B. 1993. The design and optimisation of frisbee-type dust collectors. *School of Engineering Project*. University of Exeter, Exeter, 107pp.

Coal production and coal reserves of the Czech Republic and former Czechoslovakia

JIŘÍ PEŠEK & JARMILA PEŠKOVÁ
Faculty of Science, Charles University, Albertov 6, 128 43 Praha 2, Czech Republic

Abstract: Czechoslovakia, following the communist takeover in 1948, became a member of the former East European COMECON group of countries. Its industrial potential, together with the heavy industry of the former USSR, represented a strategic base for the whole communist block. Concentration on production, with a high energy consumption, created a considerably higher demand for energy raw materials, namely brown coal in the case of Czechoslovakia. Production increased about five times with respect to 1937 levels, whereas the production of bituminous coal increased by only 80%. This created a considerable depletion of coal reserves. Consequently, only 1.9 Gt of mineable reserves of bituminous coal and about 3.1 Gt of brown coal reserves were registered in the former Czechoslovakia by 1 January 1992. These mineable reserves, recorded in the energy balance, can be divided into reserves in recently mined coal deposits and those occurring in the so-called reserve coal fields. The actual volume of mineable reserves in operating mines is equivalent to 822.7 Mt of bituminous coal and 1799.4 Mt of brown coal. Mineable reserves of brown and bituminous coal in the operating mines and open pits of the Czech Republic will be exhausted in the years 2027 and 2031, respectively, taking into consideration both the present and anticipated future production and recovery of coal.

Czechoslovakia, the industrial production of which, before World War II, concentrated on light and consumer industries, was the most industrially advanced country of the communist block when the East European communist community was formed. Iron and steel works located near relatively large coal deposits existed before World War II. This was probably the reason why, after the communist takeover, it was decided to reorientate production towards heavy industry which, together with the heavy industry of the former USSR, would establish a strategic base for all the East European communist countries. During the 1950s and later, the production of pig iron and steel increased dramatically, which in turn led to a higher demand for energy raw materials and initiated a substantial depletion of their reserves. Czechoslovakia, a small country, was placed 14th in world energy consumption in 1990, with a per capita consumption of 5600 kW h. This position, however, did not only express the volume of production, but also reflected the high energy demand of industry, its inefficient structure, obsolete technologies and sometimes even archaic machinery in the production plants.

The fuel basis of the former Czechoslovakia is particularly unbalanced, a problem inherited by the Czech Republic. The principal source of energy generation has been mostly brown coal. It is estimated that 40% of energy will still be generated by burning coal in the year 2000. Deposits of oil and gas are negligible. They account for less than 1% of the annual consumption of oil and approximately 2% of gas consumption. Because of a lack of large rivers, hydroelectric power generation does not exceed 2%. About 25% of the electrical energy demand was generated in nuclear power plants in 1992. This volume will increase substantially from 1996.

Production of coal and energy generation

There was an enormous increase in electrical energy consumption and coal production after 1948. The year 1937 has been selected as a basis for comparison because this was the last year of the former Czechoslovakia before World War II and before the surrender of a part of the Czechoslovakian territory to Germany.

The first substantial increase in bituminous coal production (Fig. 1), compared with 1937, was in 1950. This upward trend continued until 1970. Then production stopped increasing and more or less stagnated until 1980, when it started to decrease slowly. In 1990, production reached approximately the 1955 level and it is anticipated that output will decrease further. During the period 1955–70, the production of bituminous coal increased by 80% compared with 1937.

The output of brown coal (Fig. 2) in 1985 had

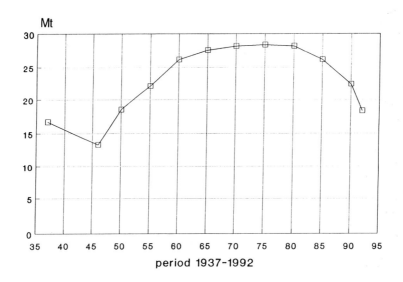

Fig. 1. Production of bituminous coal in Czechoslovakia (in metric tonnes) in the period 1937–92.

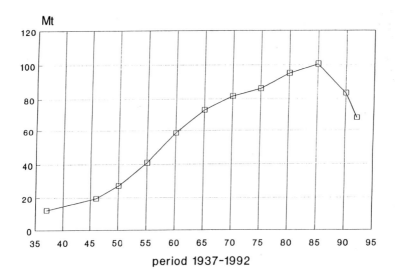

Fig. 2. Production of brown coal in Czechoslovakia (in metric tonnes) in the period 1937–92.

increased by about 5.5 times relative to 1937. The growth of production from 1950 to 1985 was 370%. The enormous increase in brown coal production was due to substantial changes in mining operations, which led to the development of numerous open pit mines at the expense of underground mining. A decrease in the production of brown coal occurred as late as 1986, and in 1992 the output was slightly lower than that in 1965. As Czechoslovakia's share of the world's mineable reserves was 0.6% (Dopita *et al.* 1985), their local intensive extraction has led to relatively fast exhaustion.

Electrical power generation (Fig. 3) in 1937 was 4.1 GkW h, whereas in 1950 it was more than double this value; in a further five-year

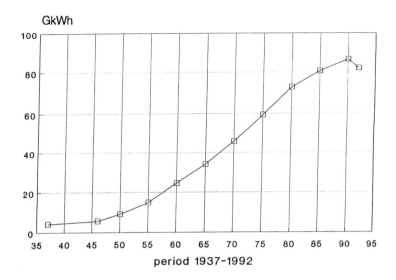

Fig. 3. Electrical power generation in Czechoslovakia (in GkW h) in the period 1937–92.

period it increased to 14 GkW h. A small decrease in electrical power generation was recorded in 1991.

Numerous power plants were built in close vicinity to the open pits because of lower transportation costs. Relatively high concentrations of sulphur in the coal, reaching 1.5–2% or higher, and huge coal production, led to an unprecedented devastation of large areas in NW Bohemia, in basins at the base of the Krušné hory Mountains (Fig. 4). A huge displacement of overburden takes place in these basins. As a result of mining operations which go down to approximately 150 m, about 250 Mm^3 of overburden were moved in 1992. Severe damage was inflicted on vegetation, particularly forests. Figure 5 shows how the slopes and peaks of the Krušné hory Mountains currently look in the vicinity of the basin. The area of the local basins is considered to be the most damaged region of the Czech Republic and perhaps in the whole of central Europe.

Coal reserves

A total of 23.6 Gt of geological coal reserves was registered by 1 Janurary 1992. The term 'economic reserves' refers to reserves which can be economically utilized. These reserves comprise 6.361 Gt of bituminous coal and 5.319 Gt of brown coal. When assessing these reserves we have to bear in mind that not all of the geological and economic reserves are useable.

Table 1. *Coal reserves of the Czech Republic registered on 1 January 1992 (in metric tonnes)*

Reserve type	Amount of coal (Gt)
Geological reserves	23.6
Economic reserves	
Bituminous coal	6.4
Brown coal	5.3
Mineable reserves	
Bituminous coal	1.9
Brown coal	3.0
Mineable reserves in operating mines and open pits	
Bituminous coal	0.8
Brown coal	2.8

If we want to establish the useable volume of coal reserves which the Czech Republic has at its disposal, then we have to consider only the mineable reserves. Mineable reserves refer to those which the miners are able to extract under present or near future economic or technological conditions. In other words, these are the reserves our industry can count on in the next few decades. These mineable reserves registered in our energy balance can be divided into reserves in recently mined coal deposits and those occurring in undeveloped coal fields. Their volume, however, does not evoke much optimism. The bituminous and brown coal reserves are 1.861 and 3.0 Gt, respectively, which are, in

Fig. 4. View into the Bílina open pit, North Bohemian Brown Coal Basin. Light grey, overburden; dark grey, coal. Photograph by M. Prokš.

Fig. 5. Characteristic view of the peaks and slopes of the Krušné hory Mountains. Spruce woods used to cover the mountains until the 1960s, but they then declined as a result of air pollution. Photograph by S. Hurník.

our opinion, very low (Table 1). Moreover, the mineable reserves of brown coal, which our energy policy is based on, cannot be substantially increased in the major coal basins.

Even the above-mentioned volume of mineable reserves cannot be relied upon completely. These values need to be further modified and adjusted. There are sound reasons for such adjustments, but all lead to a further reduction in coal reserves. Some reduced estimates of mineable reserves can be established. For example, it is possible to express more or less objectively the decrease in the mineable reserves of brown coal in the North Bohemian Brown Coal Basin because of territorial and ecological limitations issued by the Czech Government. Both the decrease in mineable reserves caused by the closing of unprofitable mines and the amount of recoverable mineable reserves affect the volume of reserves and can be estimated with some precision. On the other hand, the behaviour and attitude of the coal mining companies which will extract these reserves is difficult, if not impossible, to anticipate.

Further consideration will be focused specifically on mineable reserves in deposits which are currently being mined. The volume of these reserves is not very large. The extraction of coal reserves in undeveloped coal fields requires a huge investment that the economy of our country cannot afford within the next ten years.

The volume of mineable coal reserves which will be left in operating mines by the year 2000 can be derived from the total volume of registered mineable reserves, from data on the production of bituminous and brown coal in 1992 and from the forecast of production in 1995 and 2000 as identified by the Ministry of Economy in 1992 (Kopečný 1992; Sine 1992). The following figures are based on the assumption that no further unpredictable reduction will occur due to the aforementioned reasons. As the decrease in the production of both bituminous and brown coal in the years 1995–2000 is expected to be minimal, we anticipate that the future production will be about the same as in the year 2000. As for bituminous coal, we concluded that mineable reserves will last until 2031. The reserves of brown coal will probably last until 2027 (Tables 2 and 3).

The question is whether the next generation will face an energy crisis or even an energy disaster. Will we follow France, which now obtains nearly 80% of its energy from nuclear power? These are the questions to be answered by the Czech Administration following an assessment of all the risks and circumstances. This problem will not be solved, in any form, by special interest groups with one-sided views, regardless of the integrity of their intentions.

Table 2. *Mineable coal reserves in operating mines of the Czech Republic and their anticipated life: bituminous coal*

Mineable reserves in operating mines registered by 1 January 1993	822.7 Mt
Coal production in 1992	18.3 Mt
Anticipated production in 1995 (according to data issued by the Ministry of Economy in 1992)	16.4–16.7 Mt
Decrease in reserves in the years 1993–5 assuming a regular yearly decrease in output equal to 0.3 Mt	−51.9 Mt*
Decrease in reserves due to expected recovery equal to 60%	−20.8 Mt
Total decrease in reserves in 1993–5 at 60% recovery	−72.7 Mt
Anticipated volume of mineable reserves in operating mines by 1 January 1996 at 60% recovery	750.6 Mt
Anticipated production by the year 2000 (according to data released by the Ministry of Economy in 1992)	14.5 Mt
Decrease in reserves in 1996 to 2000 when assuming a regular yearly decrease in output equal to 0.4 Mt	−76.5 Mt †
Decrease in reserves due to expected recovery equal to 60%	−30.6 Mt
Total decrease in the years 1996–2000 at 60% recovery	−107.1 Mt
Anticipated volume of mineable reserves in operating mines by 1 January 2001 at 60% recovery	643.5 Mt
Life of reserves at the yearly output of 20.2 (i.e. 14.5 + 5.8) Mt anticipated in the year 2000, including losses due to 60% recovery, i.e. until the year 2031 at 60% recovery	31.6 years

* This figure represents an annual output in 1992, i.e. 18.3 Mt × 3 years = 54.9 Mt, reduced by 1.8 Mt due to anticipated annual decrease in production equal to 0.3 Mt in 1993, 0.6 Mt in 1994 and 0.9 Mt in 1995.
† This figure represents an anticipated annual output in 1995, i.e. 16.5 Mt × 5 years = 82.5 Mt, reduced by 6 Mt due to anticipated annual decrease in production equal to 0.4 Mt in 1996, 0.8 Mt in 1997, etc.

Table 3. *Mineable coal reserves in operating open pits and mines of the Czech Republic and their anticipated life: brown coal*

Mineable reserves in operating open pits and mines registered by 1 January 1992	2820.5 Mt
Mineable reserves in operating open pits and mines decreased by reserves blocked by regional and environmental limitations by 1 January 1992	1799.4 Mt
Output in year 1992	68.1 Mt
Anticipated production in the year 1995 (according to data issued by the Ministry of Economy in 1992)	51–54 Mt
Decrease in reserves in 1993–5 assuming a regular yearly decrease in production equal to 5.2 Mt	−173.1 Mt*
Reduction due to anticipated 90% recovery	−17.3 Mt
Total decrease in mineable reserves in 1993–5	−190.4 Mt
Anticipated volume of mineable reserves in operating mines at 90% recovery by 1 January 1996	1609.0 Mt
Anticipated production in the year 2000 (according to data issued by the Ministry of Economy in 1992)	43–48 Mt
Decrease in reserves in 1996 to 2000 when assuming a regular yearly decrease in production equal to 1.4 Mt	−241.5 Mt †
Reduction due to anticipated 90% recovery	−24.2 Mt
Total decrease in mineable reserves in 1996–2000	−265.7 Mt
Anticipated volume of mineable reserves in operating open pits and mines at 90% recovery by 1 January 2001	1343.3 Mt
Life of reserves at the yearly output of 50 Mt anticipated in the year 2000, including losses due to 90% recovery, i.e. until the year 2027	26.9 years

* This figure represents the yearly output in 1992, i.e. 68.1 Mt × 3 years = 204.3 Mt, minus 31.2 Mt due to anticipated annual decrease in production equal to 5.2 Mt in 1993, 10.4 Mt in 1994 and 15.6 Mt in 1995.
† This figure represents anticipated annual production in 1995, i.e. 52.5 Mt × 5 years = 262.5 Mt, minus 21 Mt due to anticipated annual decrease in production equal to 1.4 Mt in 1996, 2.8 Mt in 1997, etc.

The life of coal mining districts producing considerable volumes of coal in the Czech Republic could be extended by the development of new mines and open pits to extract the remaining mineable reserves, provided no rigid territorial or ecological measures are applied. Even low calorific and high ash brown coal could be extracted providing that we can make use of it for alternative energy-producing technologies (e.g. large capacity fluidized bed thermal power stations).

At the present time we should practise maximum conservation when mining and utilizing our fossil fuels. We should bear in mind that reserves in mines closed too early will no longer be accessible. Moreover, to abandon an entire coalfield would mean a loss of specific worker skills which have developed in these particular regions for several generations. We should also be aware of the devastation of large regions caused by the burning of large volumes of coal. We have been witnessing these effects for decades. Moreover, we shall be blamed by future generations for burning a raw material which could be a source of production for petrochemical products, such as various chemical, carbon, gas and synthetic fuels, in the next century.

Conclusions

Mineable reserves of brown and bituminous coal in operating mines and open pits of the Czech Republic will be exhausted by the years 2027 and 2031, respectively.

References

DOPITA, M., HAVLENA, V. & PEŠEK, J. 1985. *Ložiska fosilních paliv* [Fossil Fuel Deposits]. SNTL, Alfa, 263 pp [in Czech].
KOPEČNÝ, K. 1992. Postavení uhelného hornictví v palivo-energetické politice České republiky [Status of coal mining industry within the energy policy of the Czech Republic]. *Uhlí*, **40**, 187–188 [in Czech].
SINE 1992. *Energetická politika České republiky* [Energy Policy of the Czech Republic]. Informační Buletin MHPR ČR, 2/92 [in Czech].

European Community energy policy: import dependency and the ineffectual consensus

PHILIP WRIGHT

Division of Adult Continuing Education, University of Sheffield, 196–198 West Street, Sheffield S1 4ET, UK

Abstract: The lack of an effective European Community (EC) energy policy is discussed in the context of the different degrees and kinds of energy import dependency faced by the different member states. The way in which these contrasting energy economies have impinged on the evolution of EC energy policy, rendering it largely ineffectual, is then explored over the period 1974 to 1992.

The general theme is illustrated by focusing on the evolution of EC policy towards the coal industry, identifying how it has been impotent in the face of national government policies and perverse in terms of its recent outcomes. This argument is supported by data comparing the productivity performances of the different EC coal industries.

There are three discernible strands to European Community (EC) policy initiatives relating to the use of energy within the EC. One of them is concerned with the security of supply and was originally prompted by the first oil crisis of 1973. The second is concerned with the environmental implications of energy consumption, prompted by the prospect of global warming and the problems of acid rain, which began to be taken seriously during the 1980s. The third is also of more recent origin and is concerned with competition and deregulation in the energy sector, particularly in the context of the European Single Market.

However, although the EC has not been short of initiatives, it has been short of actual policy which has been effectively implemented. Thus the security of supply policy was driven principally by national governments and not by the EC as a whole between 1973 and 1986, and since that time security issues have faded into the background. The European Energy Charter, which purports to address the security of supply issue within the context of East–West relations, has as yet no legal instruments attached to it and its proposals are a world apart from the political realities impeding the development of energy industries in the former Soviet Union.

Environmental policy has seen spectacular proposals such as the 'carbon tax', but practice has been limited to dealing with power station emissions (the Large Combustion Plant Directive of 1988), whereas competition policy has met with strong resistance from some national governments reluctant to see their national monopolies either threatened with competition or broken up.

This failure to implement an effective energy policy by the EC, symbolized by opposition to and the exclusion of a chapter on energy in the Maastricht Treaty (EC Inform-Energy 1993: 7; Treaty on European Union 1992: 213), is the subject of this paper. It will be substantiated and explored by focusing on the different relationships between energy production and consumption faced by member states. A particular consequence of this policy failure has been the contradictory treatment of the Community's coal industry, something which will be illustrated in the final section of this paper.

Changing relationships between production and consumption

Table 1 provides details of the relationships between energy production and consumption for each of the 12 EC countries, tracing the pattern of self-sufficiency and how it has changed since 1974. The 1974 position shows a large majority of the current 12 members highly dependent on imports for meeting their energy needs—nine were importing over 80% of their energy needs. Only the Netherlands was close to energy self-sufficiency and the UK and West Germany needed to import about half of their requirements. This was reflected in an overall import dependency position of almost 62% (Table 2).

By 1988 this position had been dramatically transformed: the level of import dependency of the ten member EC had reduced to 43% and was still below 50% in 1991, despite two new members and the reunification of Germany. This transformation is also clearly reflected in Table 1—by 1990 only three of the 12 member

Table 1. *Net energy imports as percentage of energy consumption in the European Community: selected years by member state 1974–91. Source: Eurostat*

	1974	1980	1988	1989	1990	1991
Belgium	92.5	85.7	73.0	76.4	76.7	78.3
Denmark	99.5	98.6	59.6	58.9	50.2	42.7
(W)Germany	52.1	58.7	52.7	51.1	53.4	52.6
Greece	96.6	85.1	63.8	61.0	64.6	65.7
Spain	80.6	76.1	64.2	66.1	67.1	65.7
France	86.3	79.1	54.6	54.6	55.8	56.4
Ireland	88.0	79.8	66.2	66.8	70.2	67.0
Italy	84.5	86.3	81.9	85.7	85.6	82.8
Luxembourg	99.8	99.6	98.2	98.7	99.4	98.6
Netherlands	7.3	7.1	27.0	21.6	22.7	18.0
Portugal	90.2	99.1	88.5	94.6	96.5	93.9
United Kingdom	52.0	6.3	−8.9	3.7	3.4	5.1

The consumption data used in producing these percentages include stocks.
There is a discontinuity in the German series between 1990 and 1991 caused by the unification of Germany.

Table 2. *Energy production, consumption and imports in the European Community: selected years 1974–91. Millions of tonnes of oil equivalent. Source: Eurostat*

Year	Primary production	Gross consumption	Net imports (implied)	Net imports/ consumption(%)
Ten member EC				
1974	346.7	904.3	557.6	61.7
1980	460.2	953.6	493.4	51.7
1988	559.8	984.5	424.7	43.1
1989	544.3	1000.2	455.9	45.6
1990	541.9	1014.3	472.4	46.6
1991	596.8	1106.4	509.6	46.1
Twelve member EC				
1988	591.1	1077.1	486.0	45.1
1989	575.4	1098.4	523.0	47.6
1990	572.6	1114.9	542.3	48.6
1991	627.6	1212.4	584.8	48.2

Net imports are calculated as the difference between production and consumption. These data therefore differ from the actual net import data provided by Eurostat, which incorporate the effect of changes in stocks. It is more properly thought of as the implied import *requirement* in the years listed.
The broken lines indicate a discontinuity in the series—German unification means that the 1991 data are not comparable with the data for earlier years.

EC were relying on imports for more than 80% of their requirements, and one of these countries (Portugal) was a new entrant which had not been a party to policy initiatives taken in the 1970s and the early 1980s.

That these changes occurred does not tell us why or how they occurred. Were they simply a function of EC policy responses to the oil crises of 1973 and 1979—which had exposed the EC's economy as overly dependent on imported oil? Or were there other factors at work?

Role of policy in reducing import dependency

An initial impression of the policy responses to the two oil crises is that EC energy policy did

Table 3. *Contributions to the increase in primary energy production by fuel 1974–90 (10 member state EC). Millions of tonnes of oil equivalent. Source: Eurostat*

Overall increase in primary production	195.2
Decrease in production from particular sources (coal)	38.9
Therefore overall expansion from other primary sources	234.1
Expansion contributed by	
Oil	105.0
Nuclear power	129.3
Expansion in oil production contributed by UK	91.8 (87.4%)
Expansion in nuclear power contributed by	
France	75.5 (58.4%)
(W)Germany	33.1 (25.6%)
Belgium	10.7 (8.3%)
UK	9.6 (7.4%)

play an important part in reducing the overall level of import dependency. In the aftermath of the 1973 oil crisis the exposure of the EC's vulnerability to external shocks led to a communiqué entitled *Towards a New Energy Policy Strategy for the Community*, which contained the expressed objective of reducing the degree of import dependency down to 50% by 1985 (European Commission 1983). In the wake of the second oil crisis of 1979–80, further Commission documents followed: *Energy—A Community Initiative, Energy Objectives for 1990* and *Energy Strategy for the Community* (European Commission 1983). These reiterated the desirability of improvements in self-sufficiency, but also stressed improving the security of supply by diversifying import sources and breaking the link between economic growth and energy consumption through improvements in energy efficiency. It would therefore seem that the success of the EC in reducing import dependency was related to a clearly articulated policy framework.

Table 3, however, which addresses how the reduction in import dependency came about, reveals a rather different picture. Firstly, it should be noted that the overall reduction in import dependency which occurred between 1974 and 1990 was not achieved through curbs on consumption, but rather because there was an increase in primary production amounting to 195.2 Mtoe. Once the decrease in the contribution by coal is added to this, the overall expansion in the contribution of other primary energy sources indigenous to the EC is shown to have been 234.1 Mtoe, and of this total 105 Mt was contributed by increasing oil production and 129.3 Mtoe by the increasing contribution of nuclear-generated electricity. These latter data in turn expose the true extent to which the reduction in import dependency was driven by EC policy.

First of all, the vast majority (87.4%) of the increase in oil production came from the UK. Secondly, the sharp increase in the production of nuclear electricity was driven by France in particular, which contributed 58.4% of the increase as its electricity supply industry was turned over almost entirely to nuclear power. In 1991 nuclear power contributed 86% of the fuel used by the French electricity supply industry (Eurostat; monthly and yearly energy statistics published by the Statistic Office of the European Communities, Luxembourg).

This examination of how the reduction in import dependency was achieved casts a different light on why it happened. Firstly, there was an element of good fortune contributed by the development of the North Sea's oil resources. This was also clearly related to national policy rather than EC policy. Secondly, the contribution of nuclear power, although it may have been sanctioned by EC policy, was driven by overriding national imperatives as France in particular sought to counter the competitive disadvantage represented by a large oil import bill. Certainly, this turn to nuclear power cannot in any way be described as an EC-wide policy because half of the 12 member EC do not have any nuclear power stations at all, despite the fact that they are among the most import-dependent members (Table 4). Moreover, those members which have developed nuclear capacity rely on it to widely different extents.

Contrasting energy economies and their implications for policy

If the broad conclusion from the previous section is that EC energy policy up until the

Table 4. *Composition of primary energy consumption in European Community countries: percentage shares in 1991. Source: Eurostat*

	Coal*	Oil	Gas	Nuclear
Belgium	20.1	40.6	17.5	21.5
Denmark	44.0	45.3	10.8	—
Germany	33.7	38.0	16.9	10.6
Greece	36.0	61.9	0.6	—
Spain	22.0	53.6	6.2	15.5
France	9.4	40.8	12.7	36.9
Ireland	33.9	46.9	18.6	—
Italy	9.1	58.1	27.0	—
Luxembourg	28.2	49.8	11.9	—
Netherlands	11.6	36.1	49.6	1.2
Portugal	19.0	74.9	—	—
United Kingdom	29.5	37.9	23.5	8.0

* 'Coal' includes brown coal as well as hard coal.

Table 5. *Composition of net energy imports in 1991 by country. Minus = net imports; plus = net exports. Source: Eurostat*

	Coal		Petroleum		Natural gas		Electricity	
	Amount (Mtoe)	Percentage of net imports	Amount (Mtoe)	Percentage of net imports	Amount (Mtoe)	Percentage of net imports	Amount (Mtoe)	Percentage of net imports
Belgium	−9.2	21.6	−24.6	58.0	−8.7	24.0	+0.2	—
Denmark	−7.8	79.5	−2.0	20.5	+1.3	—	+0.2	—
Germany	−7.8	4.3	−129.3	71.6	−43.4	24.1	+0.05	—
Greece	−0.9	6.0	−14.6	93.7	—	—	−0.06	0.4
Spain	−8.5	13.3	−50.8	79.7	−4.4	6.9	+0.06	—
France	−14.4	10.9	−91.3	69.5	−25.7	19.5	+4.5	—
Ireland	−2.1	29.7	−4.9	70.3	—	—	—	—
Italy	−13.7	10.6	−84.7	65.7	−27.5	21.3	−3.0	2.3
Luxembourg	−1.1	28.6	−1.8	50.0	−0.4	12.1	−0.3	9.3
Netherlands	−8.3	19.9	−32.7	78.2	+27.3	—	−0.8	1.9
Portugal	−2.7	18.1	−12.3	81.8	—	—	−0.008	0.1
U.K.	−11.5	62.3	+7.3	—	−5.6	30.1	−1.4	7.6
Totals	−88.0	—	−441.9	—	−81.7	—	−5.8	—

Net imports were calculated by deducting exports and re-exports from imports. Coal refers principally to hard coal, but does include some lignite, peat and brown coal briquettes. Petroleum includes petroleum products as well as crude oil.

mid-1980s reflected what would have happened anyway as individual members pursued their national interests, there are still further points to be made which relate to a more specific analysis of energy economy schisms within the EC and how they have affected the development of policy since the mid-1980s. These schisms still stem from the different relationships between energy production and consumption confronting member states. They are as follows.

Firstly, and despite the overall reduction in import dependency, there are still very substantial differences in the degree of self-sufficiency achieved by member states. Thus the 1991 position detailed in Table 1 identifies two energy-rich members, the UK and the Netherlands. At the other extreme, Belgium, Italy, Luxembourg and Portugal are still over 75% dependent on imports for meeting their energy requirements. France and Germany are in between, but the size of their economies means that their energy import requirements are still very large in absolute terms. These differences imply that one group of member states is likely to be more concerned about conservation, price and security of supply issues, while those countries with a more advantageous resource endowment, and thereby a less onerous import bill, are likely to be keener on exploiting these resources to the full. They are also less likely to welcome the emergence of an internal market in energy under which cross-border competition increases their import dependency.

Secondly, there are substantial differences in the way that different members of the EC satisfy their import requirements. Although Table 5 shows that for six members of the EC more than 70% of net energy imports by volume consisted of petroleum, at the other extreme Denmark relies on coal for almost 80% of its import requirements, whereas Belgium and Luxembourg pursue more balanced, diversified strategies. Moreover, those members which do rely heavily on oil imports meet the remainder of their requirements in very different ways. Thus Germany, France and Italy turn predominantly to natural gas rather than coal, whereas Greece, Spain, Ireland, the Netherlands and Portugal turn to coal. These differences imply that there will be varied emphases in positions taken about the security of supply issues, both in so far as these concern the price and availability of supplies indigenous to the EC and in so far as they concern policy towards third-party imports.

Thirdly, 70% of the EC's net imports of petroleum are imported by just three countries: Germany, France and Italy. This implies not only that these countries will be particularly sensitive about security of supply issues, but also that this sensitivity will vary with the price of oil.

Finally, the process of reducing the EC's import dependency between 1974 and 1990 divided the EC in two as far as nuclear power is concerned. Indeed, discounting the meagre production of nuclear energy by the Netherlands, it is only a minority of EC members which is truly committed to nuclear power.

Evolution of policy since the mid-1980s

Having now established the factors which are likely to be decisive in fashioning EC energy policy, or in paralysing it, we can now proceed to assess how far it may be inferred that they actually were decisive in the evolution of energy policy since the mid-1980s.

Security of supply

The policy of security of supply for the EC first began to shift in emphasis away from fostering supply sources within the EC towards seeking diversity of supply from outside. It then began to fade altogether as an issue. In February 1985, the EC issued *Energy 2000*, which contained a set of projections for EC energy demand and supply up until the turn of the century (DGXV11 1985). This envisaged that the bulk of the increase in the overall energy demand would be met by expanded nuclear output, whereas the increased requirements for coal would be met by imports. Overall import dependency was projected to increase slightly, to 46% by 2000.

Also in 1985, the Commission was elaborating new energy policy objectives for 1995. After a year of negotiations these were approved in September 1986. The general objectives were as follows (DGXV11 1986: 10–12): the development of the EC's own energy resources under satisfactory economic conditions; geographical diversification of the EC's external sources of supply; appropriate flexibility of energy systems and, *inter alia*, the development, as necessary, of network link-ups; effective crisis measures, particularly in the oil sector; vigorous policy for energy-saving and the rational use of energy; and diversification between different forms of energy.

The specific objectives corresponding to this general framework were as follows: the efficiency of final energy demand should be increased by at least 20% by 1995; oil consumption should be kept down to around 40% of energy consumption and net oil imports thus maintained at less than one-third of total energy consumption in the EC in 1995; to maintain the share of natural gas in the energy balance; the share of solid fuels in energy consumption should be increased; the proportion of electricity generated from hydrocarbons should be reduced to less than 15% in 1995; and to maintain the development of new and renewable energy.

Although this agenda clearly reflected the quintessence of the argument in the previous section as it catered for almost every conceivable national strategy, it did not encompass the implications of two momentous events which took place in 1986. The first of these was the dramatic decrease in oil prices, which fell to below $9/barrel in July before recovering to around $15/barrel after OPEC's short-term production-sharing agreement in August. Together with the sharp revaluation of the ECU against the dollar, this dramatically reduced the EC's oil import bill: by 1988 it stood at only 35% of its 1985 value (Eurostat). The second, in April 1986, was the accident at the Chernobyl nuclear power plant in the Ukraine.

The subsequent impact of the former was simply to reduce the force of the security of supply argument for the EC's large oil importers—this concern was shown to be price sensitive. As McGowan (1990: 52) puts it

> The psychological effect of this shift in energy prices and supply positions was a weakening of the scarcity culture which had prevailed among suppliers, consumers, governments and agencies such as the Commission. As prices fell and markets appeared well supplied so the concerns of policy focused less on supply at any costs and more on the price of supply and existence of obstacles to the lowest price.

By 1992, once low oil prices had proved durable, the much reduced concern about security of supply was sharply reflected in the most recent energy policy statement, *A View to the Future* (DGXV11 1992a). This envisages that 'Overall dependency on imported fuels could rise from 50% to just under 60% by 2005' (p. 16)—quite a change from the outlook which was espoused as recently as the mid-1980s. Moreover, the role of the European Energy Charter, if it ever gets off the ground, should be seen in this context: accepting that import dependency will increase again driven by increased oil, coal and natural gas imports, the objective of the Charter is to foster an increase in the number of sources of supply so that security will be sustained by the volume of supplies being introduced on to the international market, and prices will remain low in real terms (DGXV11 1993a).

This same strategy has already been possible with respect to the increasing extent to which the EC has come to rely on coal imports rather than indigenous production. Although this was envisaged in *Energy 2000* and reflected in the 1986 objectives, the circumstances for its successful execution were principally created by oil company investment in export projects round the world, rather than by anything linked to EC policy (see Rutledge & Wright 1992).

On the other hand, and perhaps surprisingly, the accident at Chernobyl does not appear to have had any practical effect whatsoever on the attitude of the EC towards nuclear power. In particular, the adjustments made to *Energy 2000* in the aftermath of the decrease in oil prices suggested a continuing increase in the contribution of nuclear power (DGXV11 1986: 17).

This response is, however, readily predictable from the argument in the previous section. Firstly, the nuclear club within the EC closed ranks. Secondly, the position was reinforced by the response to lower oil prices and lower energy prices generally: if there was to be a relaxation of security supply concerns on oil, gas and coal fronts, accompanied by an increase in import-dependency, then indigenous nuclear power became that much more important as a security bulwark, particularly for certain countries. Although *A View to the Future* does offer a less sanguine outlook for nuclear power, this is principally because its contribution to primary energy consumption may be squeezed slightly by the rapidly increasing contribution of natural gas—its absolute contribution is still expected to increase (DGXV11 1992a: 41).

Energy and the environment

It is now almost two years since the European Commission announced its proposal for a 'carbon tax', described as one of the most ambitious fiscal and environmental moves that has yet been contemplated within the EC' (Lascelles 1992). The proposal has two main aspects: (1) the tax would be modulated so that 50% fell on the energy content of a fuel and 50% on its carbon content; (2) it would be 'fiscally neutral' so that the massive revenues expected from the tax would be recycled to reduce other taxes.

The purpose is to try and stabilize CO_2 emissions at their 1990 level by the year 2000—in line with the UN's *Framework Convention on Climate Change*. That the first stage of the graduated introduction of this new tax was not implemented in January 1993 as first proposed, in spite of the two years of debate which it had taken to elaborate the proposal, is due to two sets of factors. Firstly, the now familiar energy-related conflicts of interest came to the fore. The element of the tax related to carbon content induces the following differential impact on different fuels (Table 6).

Table 6. *Impact of the proposed carbon tax on the price of different fuels. ECUs per tonne of oil equivalent. Source: DGXV11 1992b: 8.*

	1993	2000
Lignite	21.1	70.5
Hard coal	19.9	66.2
Residual fuel oil	18.0	59.9
Diesel/heating oil	17.5	58.3
Petrol	17.3	57.5
Natural gas	15.4	51.3

The fact that the burden is heaviest on solid fuels and lightest on gas immediately opens up a divide between the big coal producers and users, namely Spain (particularly concerned about the implications for its already uncompetitive coal industry), the UK, Germany, Denmark, Greece and Ireland on the one hand, and the Netherlands (big producer and consumer of gas) and France (big user of gas with nuclear power escaping the tax altogether).

Secondly, there was a general concern about the implication of the tax for EC competitiveness, about its effects on industrial costs and even about the impact on island economies such as the Canaries (DGXV11 1993*b*: 80).

Thus the schedule for the tax has been adjusted to 'somewhat later given the condition that other OECD countries also implement similar measures' (DGXV11 1992*b*: 8). Given the problems which the Clinton administration has had with its own energy tax proposal, this implies an indefinite postponement. The Clinton proposal started life as a radical energy consumption tax based on the thermal content of fuels (the so-called BTU tax), but in August 1992 it was finally transmuted into a straightforward increase in gasoline and diesel fuel tax, which is essentially addressed at the federal deficit rather than the environment (Crow 1993*a*, *b*).

The other main environmental measure which has had wide repercussions within the EC has been the Large Combustion Plant Directive of 1988. Building on an earlier Directive of 1984 which required that best practice anti-pollution technology should be deployed before authorization for substantial industrial plant would be granted, the Large Combustion Plant Directive obliged member states to comply with ceilings for emissions of SO_2 and NO_x from plants with a heat input of at least 50 MW. These ceilings are to be approached in three stages between 1993 and 2003, with each member state negotiating reduction targets. The latter again emphasizes the difficulty of achieving uniformity between member states. A more specific aspect of this is that there is the possibility of derogation for new plants burning indigenous solid fuel. These plants would be allowed to exceed the SO_2 limit, but by definition could now only be built by the UK, Germany, France, Spain, Greece and Ireland.

Competition and deregulation

Extending the Single Market to energy utilities is a difficult and controversial exercise, but one which nevertheless is being attempted. The Internal Energy Market is to be approached using the following building blocks (DGXV11 1992*c*: 10): opening of electricity generation to competition; liberalization of the construction of transmission lines and gas pipelines; freedom of purchases and sales transactions, through a limited scheme of third party access to electricity and gas networks; a safety net to ensure that the safe operation of electricity and gas networks is not endangered; protection of consumers against the risk of cross-subsidies; and transparency of production, transmission and distribution activities of integrated electricity and gas utilities.

Already in force are two Directives, one concerning the arrangements for the reciprocal use of gas and electricity transmission networks by transmission utilities and the other concerning the achievement of 'price transparency'. However, the key to the whole exercise is whether third party access (TPA) to transmission networks will be achieved. This is not such a problem in the UK where there is an ideological and political commitment to this kind of competition, and where it has already been made possible both in the process of privatising the electricity supply industry and in placing TPA obligations on British Gas. However, in the rest of the EC these proposals have met fierce opposition such that the Energy Commissioner has proposed a revised plan which settles for voluntary, negotiated TPA under subsidiarity arrangements (Mollet 1993).

Yet again the large differences in the extent to which the EC members are self-sufficient has made itself felt. For some member states, already suffering from the lack of control arising out of a major dependency on imports, the prospect of surrendering control over their transmission networks is too much to accommodate easily.

Consequences for coal

In line with the overall thrust of EC energy policy, and driven by the same factors, policy towards both the coal industry and the use of coal within the EC has been permissive rather than interventionist to a specific end.

Thus state aid to the industry, which had been sanctioned since 1951 under the auspices of the European Coal and Steel Community (Parker 1992: 6), was given an extended lease of life until the end of 1993 in the context of the 1986 energy policy objectives. The specific objective concerned was 'the development of the Community's own energy resources under satisfactory economic conditions'. Thus governments were allowed to continue to subsidize their coal industries as they saw fit, only limited by the

Table 7. *Production and consumption of hard coal in the European Community: selected years 1973–91. Millions of tonnes: t = t. Source: Eurostat*

		Production	Gross consumption	Production consumption (%)
1973	(EC10)	270.2	311.0	86.9
1980	(EC12)	260.3	330.9	78.7
1985	(EC12)	217.4	322.1	67.5
1988	(EC12)	214.7	311.7	68.9
1989	(EC12)	208.7	315.3	66.2
1990	(EC12)	197.2	320.1	61.6
1991	(EC12)	193.7	328.8	58.9

EC10 means ten member states; EC12 twelve member states
't = t' signifies that production and consumption have not been adjusted to reflect differences in calorific value. The broken line indicates a discontinuity in the series arising from the unification of Germany.

Table 8. *Production (Mt) and underground output per manhour (kg) in the hard coal industries of the European Community. Source: Eurostat*

	Belgium	Germany	Spain	France	UK	Portugal
Production						
1992	0.2	72.2	18.5	9.5	82.8	0.2
Productivity						
1990	361	673	341	634	704	NA
1991	355	695	354	728	794	NA
June 1992	Closed	703	373	662	922	NA
Percentage change, June 1992/90	—	4.5	9.4	4.4	31.0	—
Percentage of UK productivity, June 1992	—	76.2	40.5	71.8	—	—

rather general prescription that this aid should promote the competitiveness of the industry and enhance security of supply, or that they facilitate restructuring. The proposal under discussion to replace this policy from the end of 1993 involves a tightening of its terms by setting a 'reference cost' above which state aid will not be sanctioned (Directorate General for Research 1993: 92). However, the early signs are that it would be difficult to make recalcitrant member states (i.e. Germany and Spain) adhere to this policy because state aids can be rendered 'opaque' to Commission scrutiny. For example, in its 1992 interim report the Commission identified three difficulties in this respect (DGXV11 1992d: 15)

> ... firstly, the difficulty of defining what payments are actually aid; secondly, to put an end value on such aid; and, thirdly, the delayed and often incomplete notification of aid by Member States which, in turn, makes it impossible for the Commission to adopt the necessary decisions promptly.

In addition to the permissiveness of EC policy with respect to the use or misuse of coal resources indigenous to the EC, this approach has also been applied to coal imports. Under the auspices of 'diversifying the Community's external sources of supply', another objective from 1986, coal imports have been allowed to surge into EC markets. Table 7 shows the resulting decrease in coal self-sufficiency from 86.9% in 1973 down to 58.9% in 1991 (and much below this by the present day).

The overall consequence of this permissiveness has been entirely perverse as far as the rational use of the EC's coal reserves are concerned. Table 8 shows how far ahead of other EC coal industries the UK industry was mid-way through 1992, and how fast it was improving that performance. And yet it is the UK industry which has been closed down

Table 9. *Energy policy agenda in the European Community. DGXV11 (1992a)*

Past	Building blocks	Future
Diversity →	Economic and social situation →	Greater cohesion
Fragmentation →	Energy markets →	Internal Energy Market
Extensive use →	Environment →	More rational use
Important →	Research and development →	Major driving force

wholesale and not the extremely high cost producers in Germany and Spain. EC policy has been unable to restrain the UK government even though, on the Commission's own admission, the UK's average production costs were substantially below the proposed reference price of 110 ECU/tonne, whereas those of Spain and Germany were substantially above (DGXV11 1992d: 14–15).

'A view to the future'

This paper has given empirical form to the divergent energy economies of EC member states and argued that the EC has been unable to prevent the force of national interests which they foster from dominating the evolution of an energy policy. This is most sharply shown in the latest incarnation of energy policy *A View to the Future*, which contains the diagram in Table 9 (DGXV11 1992a: 11):

Apart from purveying only a dubious coherence, this agenda is notable both for retreating from the 'what' to emphasize the 'how', and for expecting cohesion to be produced by the 'economic and social situation' and the operation of 'energy markets'. It should be seen as the culmination of an impotence which has already contributed to the demise of the EC's coal industry, and which does not bode well for the grander projects of political and economic union which are on the EC's agenda.

References

Crow, P. 1993a. Marching toward a tax increase. *Oil & Gas Journal*, July 26, 30.
—— 1993b. Lessons from the tax bill. *Oil & Gas Journal*, August 16, 32.
DGXV11 (Directorate General for Energy) 1985. Energy 2000: the long-term outlook for the European Community. *Energy in Europe*, 1/**1985** (May) 19–24.
—— 1986. New Community Energy policy objectives for 1995. *Energy in Europe*, 6/**1986** (December), 10–12.
—— 1992a. A view to the future. *Energy in Europe*, Special Issue (September).
—— 1992b. CO_2 emissions stabilization: the Community strategy. *Energy in Europe*, **20**/1992 (December), 7–9.
—— 1992c. Completion of the Internal Market for electricity and gas. *Energy in Europe*, **19**/1992 (July), 9–13.
—— 1992d. Community aid to the coal mining industry. *Energy in Europe*, **19**/1992 (July), 14–15.
—— 1993a. The European Energy Charter: the road to the 'Lisbon Treaty'. *Energy in Europe*, **21**/1993 (July), 22–24.
—— 1993b. Energy and island development in the Community. *Energy in Europe*, **21**/1993 (July), 80–81.
Directorate General For Research 1993. *The Situation of the Coal Mining Industry in the European Community.* Working Paper, Energy and Research Series, 1-1993.
EC Inform-Energy 1993. *Member States' Strategies and the Difficulties for an EC Energy Policy*, London, p. 7.
European Commission 1983. *The European Community and the Energy Problem.* Office for Official Publications of European Communities, Luxembourg.
Lascelles, D. 1992. A mission to make polluters pay. *Financial Times*, January 28.
McGowan, F. 1990. European energy policy: 1992 vs 1995. *ENER Bulletin*, **7/90** (March), 47–64.
Mollet, P. 1993. Energy and environment debates take time out for Summer. *Petroleum Economist*, **August**, 33.
Parker, M. J. 1992. European Community coal policy: the subsidy issue. *ENER Bulletin*, **12/92** (July), 5–14.
Rutledge, I. D. & Wright, P. W. 1992. A company-level profile of the international coal trade. *Economia delle Fonti di Energia*, **47**/1992 (August), 33–53.
Treaty on European Union 1992. Office for Official Publications of the European Communities, Luxembourg.

Case histories

Exploration and exploitation of the East Pennine Coalfield

M. J. ALLEN

British Coal Corporation, Eastwood Hall, Eastwood, Nottinghamshire, NG16 3EB, UK

Abstract: The East Pennine Coalfield is the most important of all British Coal's resources. Exploration and exploitation are closely related and studies of the coalfield's past history, present development and future potential are relevant to the whole coal industry in the UK. At a local level the geology is often variable with uncertainty attached to reserves assessments. However, the overall development of the coalfield is substantially controlled by large-scale and easily determined geological features such as the size of the deposit, seam thickness and depth of cover. Studies on a coalfield scale can therefore contribute to assessments of the future of the coal industry.
The future production from this coalfield depends on both the size of the resource and on the proportion that can be recovered. Historical recovery data are discussed and some of the geological features of the deposit that affect coal recovery are reviewed.
Although the East Pennine Coalfield is at a mature stage of exploitation, with output from deep mines currently at about 54% of its peak level in the 1950s and 1960s, the resource has not yet been fully developed, even after major investments at Selby and Asfordby. Recent exploration results from the undeveloped eastern margin of the coalfield are briefly reviewed.

As an introduction to this paper, reference is made to the typical exploitation cycle of any large mineral resource, in which production builds up to a peak or plateau before declining and eventually returning to zero. For metalliferous minerals an analogous process has been discussed by Hewett (1929) and Schmitz (1976). British Coal deposits conform to this general production pattern, as shown in Fig. 1. The East Pennine Coalfield, covering parts of Yorkshire, Nottinghamshire and North Debyshire, is the most important of all British coal deposits. Its output peaked during the 1950s and 1960s at over $80 \times 10^6 \, t \, a^{-1}$, exceeding the maximum production achieved in the South Wales and Northumberland/Durham Coalfields, each of which peaked at just under $60 \times 10^6 \, t \, a^{-1}$ before World War I. In 1992/3 its deep mines produced $45.7 \times 10^6 \, t$, representing 72.7% of all the UK deep-mined output and 56.6% of total British coal production.

The detailed shape of a deposit's production

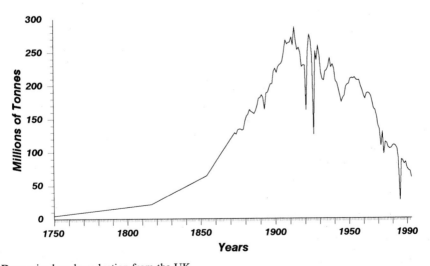

Fig. 1. Deep-mined coal production from the UK.

From Whateley, M. K. G. & Spears, D. A. (eds), 1995, *European Coal Geology*, Geological Society Special Publication No. 82, pp. 207–214

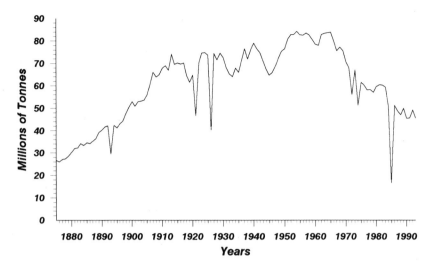

Fig. 2. Deep-mined coal production from the East Pennines.

curve and the amount of mineral recovered are affected by many external factors, including variable market demand, competition from other sources of coal and from substitute energy minerals, the development of new technology and conflict with surface environmental interests. All of these are now being experienced by British coal mines, and the inevitable uncertainty about future business and technical conditions makes it impossible to predict with confidence the future of coal recovery. However, under any set of external conditions, the geology of the deposit is one of the most important features determining the amount recoverable. A review of regional and local variations in the geology of the East Pennine Coalfield can indicate the sensitivity of future production to changes in external factors.

Boundaries of the coalfield

As a result of the advance of mining and the extensive use of new surface exploration technology since about 1970, most of the boundary of this coal resource has been well defined. It is formed by the outcrop of coal seams, or by their subcrop under younger rocks in the concealed part of the coalfield, and increasing depth imposes a practical limit of mining in the east (Fig. 3). Until the mid-1970s the concealed incrop around York and southeast of Nottingham was not proved, and recent exploration has yielded some surprising results in these areas; the construction of British Coal's newest colliery (Asfordby) is based on new reserves along the concealed edge of the coalfield proved by boreholes and seismic exploration since 1973. This exploration programme was part of a national search for potential new deep mine sites, which included nearly 700 boreholes drilled since 1973. As a result of this exploration it is considered unlikely that any further significant extensions of the East Pennine Coalfield remain undiscovered.

In general, the coal seams dip gently towards the east and have now been proved to extend to the North Sea coast and beyond. At the coast the top of the Coal Measures is around 2 km deep, but a practical mining limit is currently regarded as about 1.0–1.2 km.

The most extensively exploited seam is the Barnsley (known as Top Hard in the south), and the eastern edge of the coalfield is normally taken at the 1200 m Barnsley depth contour (Fig. 3). Locally the practical mining limit will vary, depending on the thickness of this seam and of others. In the east of the coalfield, the Barnsley is still being worked, notably at Selby where it yielded over 10×10^6 t in 1992/3, and there are large unexploited areas remaining along the eastern margin of the coalfield, as proved in recent boreholes.

Additional economic boundaries of the coalfield are defined by the top and bottom workable seams. Many geological, economic and mining criteria are used to determine whether or not a seam is workable, but coal thickness is very important. In the past, seams of 60 cm or less have been recovered from underground mines, but in the East Pennine Coalfield the thinnest

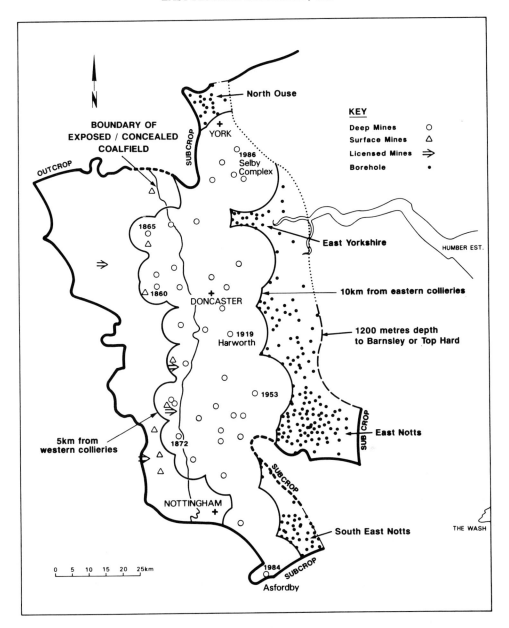

Fig. 3. East Pennine Coalfield exploitation as at August 1993.

seam now being worked in British Coal's deep mines is about 90 cm and the most successful collieries generally work seams over 150 cm thick.

One particularly important boundary within the coalfield is defined by the outcrop of the Permian and Triassic rocks which cover the Coal Measures. To the west of this line the coal seams are much more accessible; to the east the coal resource has been concealed and protected by a substantial thickness of unproductive overburden which includes major aquifers.

Other internal boundaries are defined by the extent of exploitation and divide the coalfield into three zones: western, central and eastern (Fig. 3). The boundaries between these zones are defined by circles centred on the shafts of existing deep mines; the radius used is 5 km for

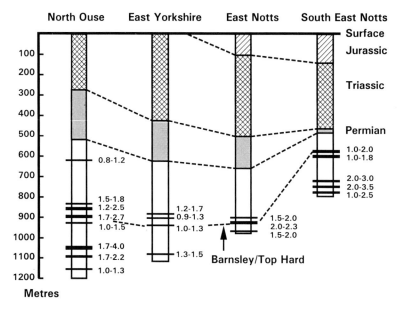

Fig. 4. Typical sections in unexploited areas.

the older collieries in the west and 10 km for the newer mines in the east. These lines are notional and do not define accessible reserves, which are determined by many factors.

Western zone

The first mines were developed on seam outcrops and exploitation has tended to follow the seams down-dip and towards the east. One example of this easterly advance of mining was mapped in a government assessment of coal reserves in the 1940s (Ministry of Fuel and Power 1945), and further detailed information is available in Wilcockson (1950). Similarly, exhaustion of underground mines generally started in the west, and a 'line of exhaustion' has advanced down-dip. The western zone is here defined as the area behind this line, which has been abandoned by large-scale underground mining.

For a variety of technical and economic reasons not all the coal was recovered by these old mines. Pillars of support were left to protect the mine or the surface; geology, access costs or market requirements have resulted in other areas being abandoned. In a few instances it has been possible for the private sector to establish small underground mines in these areas of unworked coal, but the pillars of coal among old workings and the thin seams that were not workable from underground collieries can now be economically recovered by opencast mining down to an average depth of 40 m.

Surface and licensed underground mines are not entirely confined to this western zone of the coalfield, but they are currently limited to the exposed Coal Measures. In some places it is possible for these mines to operate close to the large deep mines, depending on the disposition of reserves. There are at present four underground licensed mines, eight British Coal surface mines and a number of private opencast operations in this coalfield. Including output from licensed surface mines, the total production in 1991–2 from all these mines was less than 10% of the total from the large underground collieries in this coalfield.

Central zone

This is essentially the active part of the coalfield, containing all the deep underground mines now operating, and some of the surface mines. In general, the most eastern are the newest (including Selby), whereas some of those in the west are over 100 years old. The depth of working extends down to about 1000 m, and in recent years there has been considerable investment in this zone. Most of the capital spending in the last 15 years has been concentrated in reserves at the advancing eastern edge of the zone, in new mines at Selby and Asfordby and at

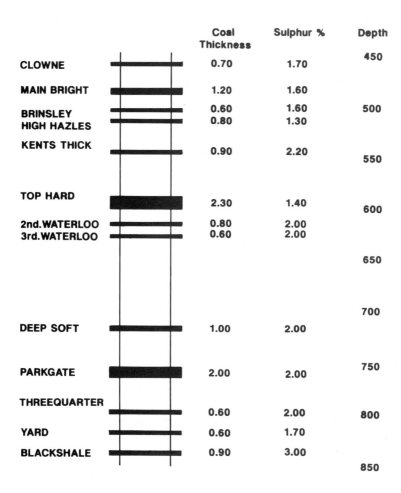

Fig. 5. Typical stratigraphic sequence: productive coal measures.

older mines with large reserves to the east, such as Harworth.

Eastern zone

This part of the coalfield is entirely concealed and unexploited and is generally the deepest part. It represents the potential for continuing the historical development of the coal resource to the east. Since 1974, after large increases in oil prices, 247 deep boreholes and 1100 km of seismic survey have been completed in the search for further coal reserves within the unexploited fringe of the coalfield. This is additional to the exploration programmes undertaken for Selby and Asfordby.

The sedimentary and tectonic geological environments are variable; however, the area remaining is substantial and the total coal resource is very large, with individual prospects containing of the order of 1000×10^6 t (Fig. 4). The extent to which any of this resource could be recovered depends not only on the geology, but on the many external factors noted previously.

Recovery of the coal resource

As with other minerals, it is not practical to achieve 100% recovery of the coal resource. The 1905 report of the Royal Commission on Coal

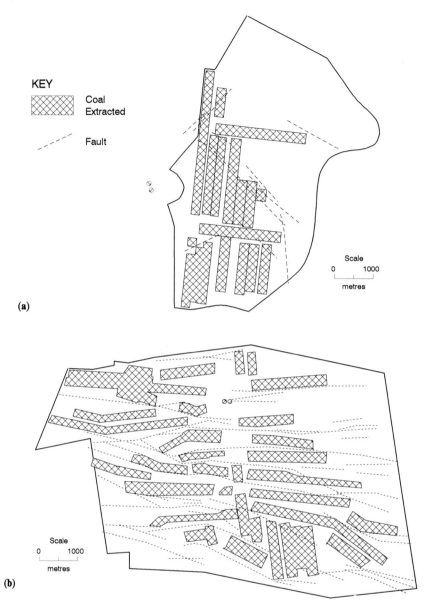

Fig. 6. Effect on coal recovery of different tectonic environments. (**a**) Colliery workings in a 'good' tectonic environment. (**b**) Colliery workings in a 'poor' tectonic environment.

Supplies indicated a practical expected recovery of just under 40% of the total resource in seams at least 0.6 m thick and not more than 1200 m deep.

In 1978 a National Coal Board report concluded that 52% recovery was being achieved, but this did not take into account areas and seams rejected as unworkable for a variety of reasons, so the real recovery was much lower. At Selby there is a planning permission to work only one seam, so on this basis the recovery will be relatively low and probably less than 10% of the total resources in all seams down to 0.6 m thick and 1200 m deep.

Unless new mines are constructed in the eastern zone, the future deep-mined output of

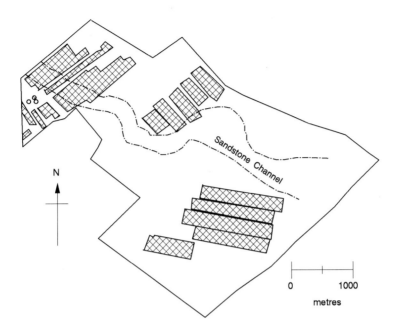

Fig. 7. Coal extraction restricted in a 'poor' sedimentary environment.

this coalfield will depend on the recovery of coal within the currently accessed central zone. The geological conditions that affect the percentage recovery vary considerably, both horizontally and vertically, and some of these are outlined in the following sections.

Seam thickness and quality

A typical seam succession from the southern part of the coalfield (Fig. 5) includes many named seams that locally reach workable thickness, but of which perhaps only two or three are sufficiently well developed at any one colliery to be economically attractive at present. The distribution of coal among seams at least 60 cm thick in Fig. 5 indicates how sensitive the recoverable reserves are to the minimum economic seam thickness. From the total coal thickness in this succession of 13 m, only 4.3 m could be recovered if the minimum workable section is 2 m, but 9.9 m might be exploited with an economic cutoff at 0.8 m thick. During the period from 1971 to 1993 the average working height of coalfaces in British mines has increased by over 40% from 1.37 to 1.97 m. Although this is partly due to the discovery of new reserves in thick seams (for example at Selby) it is also the result of concentrating production in thicker seams and writing off reserves in thin seams, in response to economic pressures.

This trend contrasts sharply with that forecast by the UK Government in 1945 (Ministry of Fuel and Power 1945), when the proportion of output from seams over 1.22 m thick was expected to decrease from 37 to 28% in 1994. In fact, large scale deep mining in seams of less than 1.22 m in the East Pennine Coalfield is now rare and restricted to one colliery.

As coal quality affects coal prices, this can also affect recovery. For example, a high sulphur content can make the coal less attractive to customers. In the UK, 3% sulphur is considered to be 'high' and 1% sulphur is considered to be 'low'. Figure 5 indicates the potential sensitivity of recoverable reserves to sulphur content.

Geological environment

Modern underground mining, with high initial costs for each production unit, is sensitive to the continuity of the coal seam. Figure 6 illustrates the variation in structural continuity between two collieries in the south of the coalfield. Workings in the 'poor' environment have now been abandoned, but those in the 'good' structure are continuing and their more regular pattern is reflected in lower mining costs and a higher rate of recovery.

Relatively small faults (less than 5 m vertical displacement) seriously affect coal production in this coalfield, and they can only be identified as the first seam is exploited. Even at a modern mine such as Selby, the intensity of 'small' faults was not known until mining started, as their size is below the resolution capability (around 7 m) of surface seismic surveys in that area (Houghton 1991). Modern three-dimensional seismic surveys might be able to reduce the structural uncertainty, but at present it is still difficult to estimate recoverable reserves in an area of first seam working.

Sedimentary continuity can also vary significantly, even within a single colliery. Figure 7 illustrates the severe effect of a roof channel on the recovery of coal at a colliery in Yorkshire. This seam has now been abandoned, and the relatively low percentage extraction is due to this and other sedimentary changes, together with faulting and surface environmental constraints.

Internal boundaries

The active central zone of the coalfield should not be regarded as a continuous area of exploitable coal. Within the current economic constraints, recoverable reserves are limited by local geological boundaries such as major faults and seam splits. The coalfield in practice consists of a series of discontinuous areas in each seam that are locally economic. The Selby complex is exploiting an isolated area of the Barnsley seam that is determined mainly by the geological barriers of sub-crop, seam splitting and increasing depth (Houghton 1991). This important deposit has its own production curve, which has now built up to its peak level of around $10-10^6 \, t \, a^{-1}$. The future production curve of the whole coalfield is the combination of all such curves for the individual economic deposits, some of which are worked by single collieries, whereas others are sufficiently large to support several mines. In view of this complexity and local uncertainties, any conclusions about the future of the coalfield are likely to be more reliable if they are based on a study of the development history and trends of the whole coalfield.

Conclusion

The East Pennine Coalfield can be regarded as being at a mature stage in its life cycle of exploitation, with deep-mined production in 1992–3 having decreased to about 54% of its peak level achieved in the 1950s and 1960s (Fig. 2). With the physical boundaries of the coalfield by now well proved, most of the uncertainty about future production is related to the percentage recovery that will be achieved, mainly from existing mines, but also from the unexploited area along the margin of the coalfield. Recent mining investments are based on substantial recoverable reserves in the central zone of this coalfield, but further progress in developing new exploration and mining technology will be one of the factors that will determine how much of the total resource will eventually be recovered.

This paper could not have been produced without the research and assistance provided by geological staff of the Technical Services and Research Executive at Bretby, especially Phil Eaton.

Thanks are also due to W. E. Hindmarsh, Head of Operations Department at Eastwood Hall, for permission to publish this paper. Any views expressed are those of the author and not necessarily those of British Coal Corporation.

References

Hewett, D. F. 1929. Cycles in metal production. *Transactions of the American Institute of Mining and Metallurgical Engineers, Yearbook.*

Houghton, A. 1991. The Selby Coalfield—problems and potential. *The Mining Engineer*, November, 135–144.

Ministry of Fuel and Power 1945. *North Eastern Coalfield*. Regional Survey Report. HMSO, London.

National Coal Board 1978. *Recovery of Reserves.* Mining Department Bulletin.

Schmitz, C. J. 1976. *World Non-Ferrous Metal Production and Prices 1700–1976.* Frank Cass.

Wilcockson, W. H. 1950. *Sections of Strata of the Coal Measures of Yorkshire.* 3rd Edn. The Midlands Institute of Mining Engineers, University of Sheffield.

Overview of the influence of syn-sedimentary tectonics and palaeo-fluvial systems on coal seam and sand body characteristics in the Westphalian C strata, Campine Basin, Belgium

ROLAND DREESEN,[1] DOMINIQUE BOSSIROY,[1] MICHIEL DUSAR[2] ROMEO M. FLORES,[3] & PAUL VERKAEREN[4]

[1] *Institut Scientifique de Service Public, 200 Rue du Chéra, B-4000 Liège, Belgium*
[2] *Belgian Geological Survey, 13 Jennerstraat, B-1040 Brussels, Belgium*
[3] *US Geological Survey, Federal Center, Branch of Coal Geology, Denver, CO 80225, USA*
[4] *NV Kempense Steenkolenmijnen, Aardkundige Dienst, 351 Koolmijnlaan, B-3540 Heusden-Zolder, Belgium*

Abstract: The Westphalian C strata found in the northeastern part of the former Belgian coal district (Campine Basin), which is part of an extensive northwest European paralic coal basin, are considered. The thickness and lateral continuity of the Westphalian C coal seams vary considerably stratigraphically and areally. Sedimentological facies analysis of borehole cores indicates that the deposition of Westphalian C coal-bearing strata was controlled by fluvial depositional systems whose architectures were ruled by local subsidence rates. The local subsidence rates may be related to major faults, which were intermittently reactivated during deposition. Lateral changes in coal seam groups are also reflected by marked variations of their seismic signatures. Westphalian C fluvial depositional systems include moderate to low sinuosity braided and anastomosed river systems. Stable tectonic conditions on upthrown, fault-bounded platforms favoured deposition by braided rivers and the associated development of relatively thick, laterally continuous coal seams in raised mires. In contrast, rapidly subsiding downthrown fault blocks favoured aggradation, probably by anastomosed rivers and the development of relatively thin, highly discontinuous coal seams in topogenous mires.

The application of sedimentology to coal mining in the Campine mining district (northern Belgium) is a new approach. In the past, sedimentological analysis was not utilized in understanding the causes of geological hazards and mining impediments such as wash-outs, seam splits and roof falls, which were responsible for important production losses. Presently, reconstruction of the palaeoenvironments of the coal-bearing strata has proved to be helpful in locating potential hazardous areas which may affect longwall mining in the Campine collieries (Dreesen 1991, 1993).

Ninety years of deep-mine coal extraction and the more recent seismic and borehole exploration between 1979 and 1988 in the north of the Campine mine district have provided a good insight into the subsurface structural setting of the Campine basin (Bouckaert & Dusar 1987). Deep-mine coal extraction has been strongly influenced by the presence of a grid of NW–SE trending faults. These faults with throws of up to several hundred metres, parallel the Rur Valley Graben margin (Fig. 1). Variscan (Late Carboniferous) and Kimmerian (Middle to Late Jurassic) movements along these faults follow the same direction (Rossa 1987). Moreover, seismic interpretation indicates that contemporary movement on some of these faults affected Late Carboniferous sedimentation, with notable differences in basin subsidence (Bouckaert & Dusar 1987; Vandenberghe 1984; Dreesen et al. 1987).

This study is based on a series of cored boreholes covering the Westphalian C–D stratigraphic interval in the eastern part of the Campine Coalfield (Figs 2 and 3). Although the boreholes are less than 1–4 km apart, correlation of the study interval is by thick, persistent coal beds that contain tonstein units. These marker beds are more laterally continuous than the interbedded sandstones. Where possible, correlation of rocks between these marker beds is based on brackish marine bands (e.g.

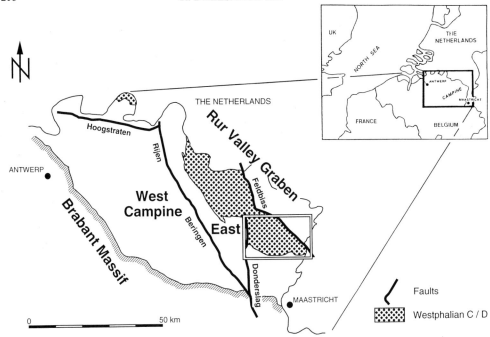

Fig. 1. Distribution of Westphalian C/D deposits in the Campine Basin (modified after Dusar 1989a; Langenaeker & Dusar 1992). Westphalian C/D coal deposits with a total preserved thickness of 1.425 m occur in the northeastern part of the basin. The productive Coal Measures are confined to the east of the Beringen-Reijen fault and to the north of the Hoogstraten fault. The study area is located to the east of the Donderslag transpressional fault, which is of Asturian age (Dusar & Langenaeker 1992). Inset refers to Fig. 2.

bivalves, ostracods, ichnofossils), similarity of physical properties and position in sequence. The studied stratigraphic interval includes the coal-bearing strata above the Maurage Marine Band (= Aegiranum Marine Band) (Paproth *et al.* 1983), with special emphasis on a sequence within the Lower Westphalian C. The sequence is defined by the Erda tonstein at the base and the Hagen 1 tonstein at the top (Burger 1985) (Figs 4 and 5). This sequence is part of the youngest Coal Measures (Westphalian C and D) in the Campine Coalfield, which have never been mined, although the Westphalian C contains large reserves of economic to subeconomic bituminous coals. The thickness and lateral continuity of these Westphalian C coals vary considerably. This is due to the combined effects of syn-sedimentary subsidence and contemporary fluvial processes; both phenomena will be discussed in this paper. Similar variations in thickness and continuity of coal seams have also been recorded in the Upper Westphalian C coal seams (Dusar 1989b). The varying seismic signature for the different Westphalian C-D coal-bearing sequences probably relates to similar mechanisms of combined syn-sedimentary subsidence and related changes in the fluvial processes. These processes were probably caused by intra-Westphalian C tectonic events.

Previous studies of depositional environments

In the past, the emphasis of exploration has been to identify tectonic effects on coal deposits and their lateral extent, to carry out structural reconstructions and to estimate coal reserves north of the working collieries. Furthermore, more attention has been paid during longwall mining to unexpected adverse conditions of tectonic origin (e.g. faults) rather than to geological hazards and mining impediments of sedimentary origin (Dreesen 1991, 1993). Washouts, seam splits and seam deterioration (increase in dirt bands or partings; pinch-outs) have long been recognized as having adverse effects on longwall coal mining in the Limburg–Campine–Ruhr mining district (e.g. Thiadens & Haites 1944; Stassen 1948, 1949; Stoppel & Bless 1981). However, although a direct or indirect link of these features with fluvial processes has generally been accepted, the prediction of seam discontinuities and geological hazards often failed due to a lack of a proper understanding of the fluvial origin of these coal-bearing rocks.

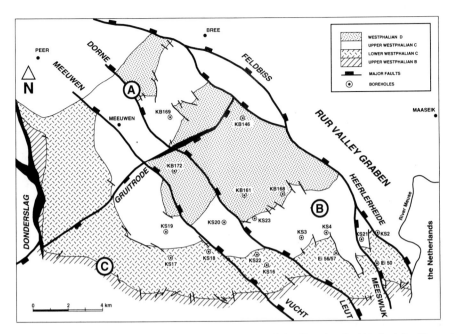

Fig. 2. Simplified subsurface geological map of the eastern part of the Campine Coal Basin (after Dusar 1989a) with location of studied exploration coreholes, including surface (KB and KS boreholes) and underground coreholes (Ei boreholes). In the northeastern Campine Coal Basin, the Gruitrode lineament separates the Meeuwen-Bree (A) and the Neeroeteren-Rotem (B) coalfields, whereas the Dorne-Leut fault separates the latter from the Zwartberg-Opglabbeek (C) Coalfield.

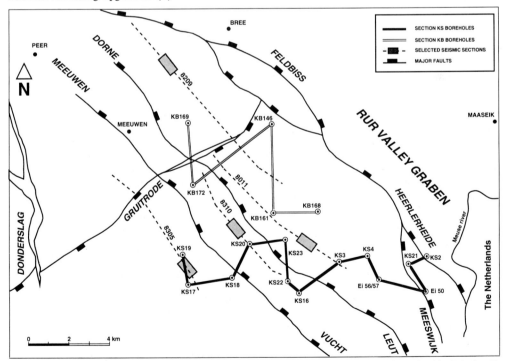

Fig. 3. Location of studied cored borehole sections and seismic sections with respect to the major block faults. For section along KB boreholes, see Fig. 3; section along KS boreholes, see Figs 5–8; parts of seismic sections (shaded blocks) are depicted in Figs 10 and 11.

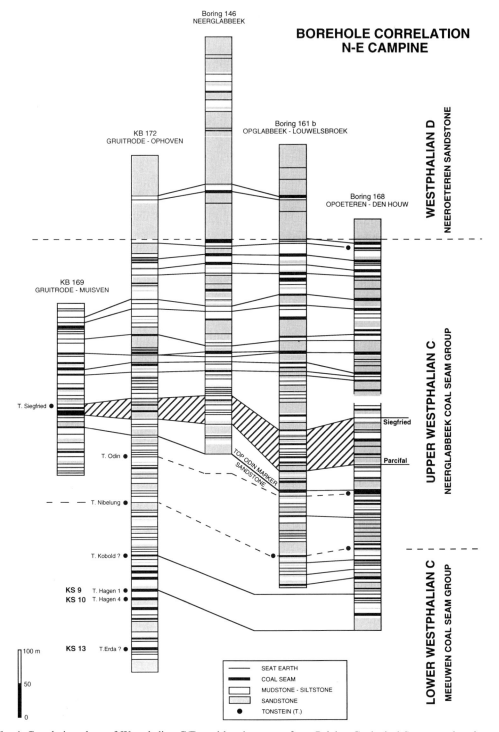

Fig. 4. Correlation chart of Westphalian C/D coal-bearing strata from Belgian Geological Survey exploration coreholes (after Dusar 1989a). The shaded area corresponds to the Parcifal–Rubezahl–Siegfried stratigraphic interval. Correlations are based on geophysical logs, combined with lithofacies characteristics, tonstein occurrences and palynological data. See Fig. 3 for location of KB boreholes.

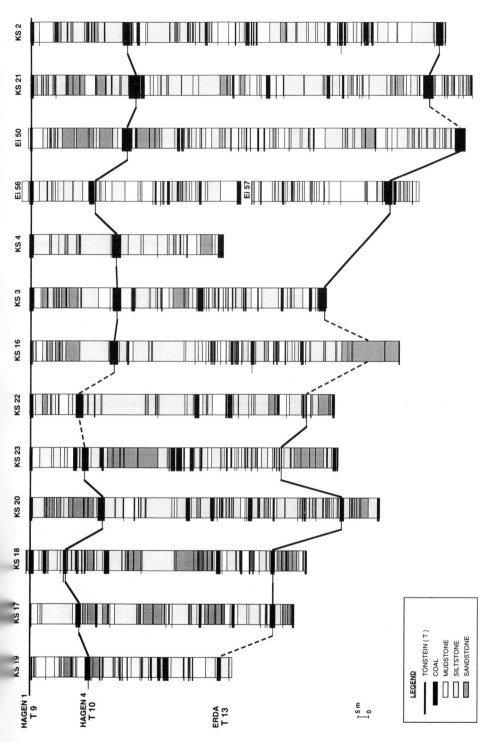

Fig. 5. Correlation of KS surface and subsurface cored exploration boreholes (Ei 56-57-50). The basic framework for stratigraphic correlation within the Lower Westphalian C is provided by tonstein marker beds T9, T10 and T13. The latter tonsteins have been correlated with the 'Kaolin-Kohlentonstein' horizons Hagen 1, Hagen 4 and Erda, first established in the Ruhr Coal District (Burger 1985). Detailed seam by seam correlation is hampered by multiple splitting in this particular interval. See Fig. 3 for location of boreholes.

Recognition of channel and floodplain deposits is an important step towards understanding the alluvial sedimentary environment and its influence on the formation of coal deposits (Rahmani & Flores 1984; Flores et al. 1985, 1993; McCabe 1984, 1987). Coal seam geometry, lateral continuity and quality depend to a high degree on the type of fluvial environment in which the coals were deposited. Thus applied sedimentology should be an important tool in coal exploration and development. Although it is commonly accepted that sedimentological analysis has an important part to play in all stages of coal exploration and mine planning, its usage has not been widely reported in northwest Europe. The UK is an exception and examples of studies describing the applications of sedimentology to deep-coal mining have been reported by Guion & Fulton (1993).

In the western part of the Campine Basin, geophysical wireline logs have been used to interpret the sedimentary history of the Namurian and earliest Westphalian A deposits. The subsurface studies show an evolution of the depositional settings of these strata, from turbidite-fronted delta complexes in the Upper Namurian to a lower delta plain ('sheet delta') in the earliest Westphalian A (Langenaeker & Dusar 1992). Similar delta plain–delta mouth facies have also been suggested by Bless (1973) for early Westphalian A deposits in the South Limburg Coalfield adjacent to the Campine Basin. Van Amerom & Pagnier (1990) studied the relationship of Westphalian B–C plant macrofossils and fluvio-deltaic deposits in a borehole located in the South Limburg Coalfield. These workers were able to identify deposits of floodplain, lake floor and filling, distributary channel and crevasse subenvironments.

A sedimentological analysis of a series of cores of boreholes covering the late Westphalian A to early Westphalian B stratigraphic interval in the western part of the Campine Basin (Beringen and Zolder collieries) indicated deposition in a fluvial-dominated delta plain complex that shifted from a lower to upper delta plain (Lorenzi et al. 1992). Apparently this shift took place around the Westphalian A–B boundary (= Quaregnon or Vanderbeckei Marine Band). Several subenvironments in this deltaic complex were identified based on a combination of lithological, sedimentological and palaeoecological criteria: low sinuosity migrating fluvial channels, interdistributary channels in fluvial-dominated delta plains, proximal and distal crevasse splays, interdistributary lakes and bays, lake margins and swamps (Dreesen 1993). Analogous depositional environments have been described from time-equivalent strata in Britain (Scott & Collinson 1983; Fielding 1984; Guion & Fielding 1988), the Netherlands (Van Amerom & Pagnier 1990), the Aachen-Erkelenz Coal District (Müller & Steingrobe 1991) and the Ruhr area, NW Germany (Strehlau 1990).

The general environmental settings of the Westphalian C–D Coal Measures are poorly studied. However, investigations by Wouters & Gullentops (1988) and Wouters et al. (1989) of a 130 m thick multi-storey sandstone complex (the Neeroeteren Sandstone) cored in the lower Westphalian D strata of the eastern Campine Basin, indicate deposition in a large, fan-shaped, gravelly braided river. Microresistivity (dip meter) measurements carried out in two boreholes (KB 161 and KB 172) suggest a palaeocurrent direction from the SE (average N61°W) for the Neeroeteren Sandstone (Dusar et al. 1987).

Van Wijhe & Bless (1974) proposed a general model for the Westphalian of northwest Europe, which was based on the idealized relationship between major sedimentary facies and vegetational pattern based on miospores. The depositional model of these workers suggested an evolution from coastal marshes (Lower Westphalian A), through backswamps (Westphalian B) and fluvial plains (Westphalian C) to a 'hinterland' setting (Westphalian D). A progressive withdrawal of the marine environment was accompanied by a gradual change of the palaeoclimate that became more arid in late Westphalian times. However, some marine-influenced fauna horizons still occur in the Westphalian D of Germany.

Fluvial depositional environments in the Lower Westphalian C

A detailed sedimentological facies analysis of the lower Westphalian C Coal Measures, between the Erda and Hagen 1 tonsteins, was carried out on cores of boreholes in the eastern Campine Basin (Fig. 4). An important method of sedimentological analysis is the recognition of coarsening upward (CU) and fining upward (FU) sequences. The combination of CU and FU sequences with their lithology, thickness, nature of lithological contact (e.g. gradational, erosional), sedimentary structures and flora and fauna (e.g. body and trace fossils) lead to the diagnosis of depositional environments. The vertical succession of these environments can be expressed as a vertical profile, which illustrates the evolution in space and time

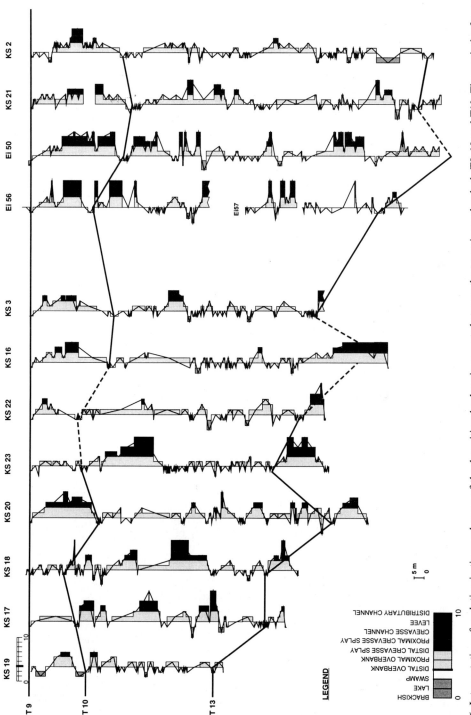

Fig. 6. Interpretation of evolution in time and space of the depositional subenvironments between the tonstein marker beds T9–10 and T13. The vertical reference line corresponds to the limit between distal and proximal overbank (floodplain) settings. The environmental settings at the right-hand side of the column show, from left to right, the increasing proximity to a fluvial channel complex. The settings to the left indicate peatlands and increasingly lacustrine to brackish conditions. See Fig. 3 for location of boreholes.

(switching) of the depositional setting at each borehole locality (Fig. 6). The changes in the 'blocks' reflect variations in lithology, or more specifically, grain size. The plots resulting from the connection of midpoints in the 'blocks' are the reflection of the deposition of various grain sizes during different energy levels of the fluvial system (e.g. channels, crevasse, floodplains). Fluvial environments affected by channelized flow and floods are believed to have resulted in the deposition of coarser sediments (e.g. sands and gravels). In contrast, areas only intermittently affected by channelized flow were dominated by the deposition of fine sediments (e.g. muds and silts). Sand and gravel formed mainly in channels and crevasses, whereas mud and silt accumulated on distal floodplains and levees (McLean & Jerzykizwicz 1978; Ethridge et al. 1981; Frostick et al. 1983; Flores & Pillmore 1987). The FU channel sandstones are interpreted to reflect deposition by channelized flows from high to low discharges. The CU sequences may represent either overbank or crevasse splay deposits formed during floods.

When comparing the plots in Fig. 6, the following observations may be made: (1) there is a high concentration of thick, fluvial channels and associated facies (levees, proximal crevasse deposits) in the upper and lowermost parts of the intervals, compared with a lack of thick fluvial channel sandstones in the middle part of the interval; (2) conspicuous vertical stacking of dominantly distal floodplain facies with abundant lacustrine deposits occurs in the central and eastern parts of the middle interval, as opposed to much thicker fluvial channel complexes in the western part of the same interval.

The interpretation of the sandstone bodies as channel or as crevasse deposits was made by studying the vertical change in grain size (FU versus CU). This genetic classification was of assistance in estimating the lateral correlation potential of the sandstone bodies, which is of great importance in the construction of lithostratigraphic panel or fence diagrams (Figs 7 and 8). The combination of lateral extent, thickness, FU or CU trends, sediment type, internal architecture (e.g. single or multi-scour, multi-storey), sedimentary structures and lithologies of channel sandstones, should eventually result in the identification of the prevailing fluvial facies.

Large sandstone belts (Fig. 7) with thick multi-storey FU channel sandstone bodies and extensive sandstone–siltstone sheets, recurrent coarse or gravelly sandstone beds, as well as the associated poor development of overbank deposits, and more especially that of flood basin sediments, have been interpreted as the deposit of a sandy braid plain fluvial environment (Ramos & Sopena 1983). On the other hand, thick sequences of interbedded thin coals, thin channel sandstones with limited lateral extent, common freshwater fossiliferous (lacustrine) mudstones and common crevasse sheet sandstones (CU sequences) (Fig. 8) might suggest deposition in an anastomosed fluvial depositional system (Smith 1983).

Although the continuity and contemporaneity of the channel sandstones are difficult to demonstrate given the spacing of the boreholes, the common association of these discontinuous sandstones with freshwater lacustrine mudstones and crevasse sandstones deposited in mud-rich lacustrine floodplains may indicate a relationship with fine-grained, anastomosing low sinuosity channels described by Schumm (1968).

Examples of coal-bearing braid plain environments are known in the Namurian of the Pennine Basin of northern England (Bristow 1988), in the Lower Permian Coal Measures of India (Casshyap & Tewari 1984), in the Late Permian Newcastle Coalfield of New South Wales, Australia (Diessel 1992), in Early Tertiary coal basins of northern Alaska (Merritt 1986) and in Early Tertiary intermontane coal basins of the Rocky Mountain region, USA (Flores et al. 1985; Flores 1989). Case histories of anastomosed coal-forming fluvial environments have been described by Flores (1981, 1983, 1986, 1993) and Flores & Hanley (1984) from the Palaeocene–Eocene (Fort Union Formation) in the Powder River Basin, Montana and Wyoming, USA, by Putnam & Oliver (1980) from the Upper Mannville (Albian) in east-central Alberta, Canada and by Rust et al. (1984) from the Pennsylvanian Cumberland Group of Nova Scotia, Canada.

It is suggested that the abrupt lateral change from braided (high gradient) to anastomosed (low gradient) fluvial systems (Fig. 8) is related to the effects of differential subsidence during deposition (Fig. 9). It is also suggested that this differential subsidence was related to the reactivation of some of the major fault blocks in the northeastern part of the Campine Basin. This interpreted change from a braided to an anastomosed river system coincides with the location of two major faults which delimit tectonic blocks, namely the Meeuwen-Vucht and the Leut-Dorne faults (Fig. 2). The anastomosed fluvial system is believed to have developed on the gradually subsiding eastern fault block (the so-called Neeroeteren-Rotem Coalfield), whereas the braid plain setting developed during the more stable tectonic conditions of the western block (the Zwartberg-

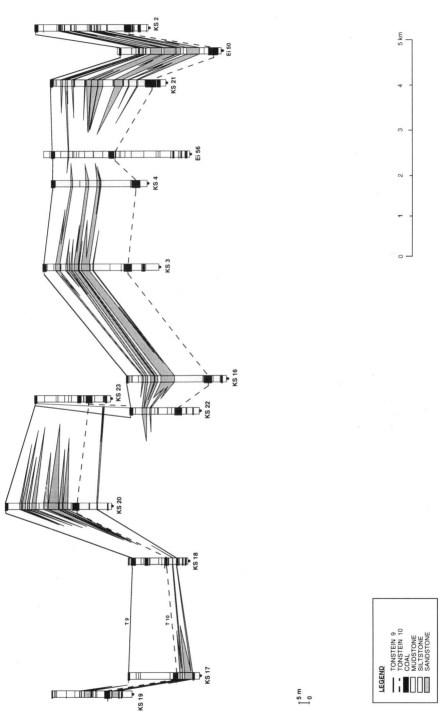

Fig 7. Lithostratigraphic fence diagram for the interval between tonsteins Hagen 1 (T9) and Hagen 4 (T10). Note presence of large (up to 4 km wide) belts of multi-storey channel sandstones and associated siltstones, accompanied by relatively thick coal seams. See Fig. 3 for location of boreholes.

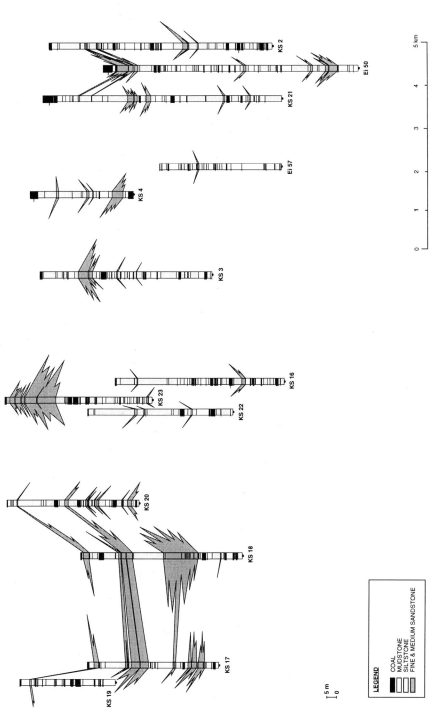

Fig. 8. Lithostratigraphic fence diagram for the interval between Hagen 4 (T10) and Erda (T13) tonsteins. Note sharp contrast in channel density and channel type between west and east, with lack of thick channel sandstone bodies and abundance of mudstone–siltstone suites containing a multitude of thin coal seams in the east. Isolated thin sandstone or siltstone bodies have been interpreted as distal crevasse splay deposits. See Fig. 3 for location of boreholes.

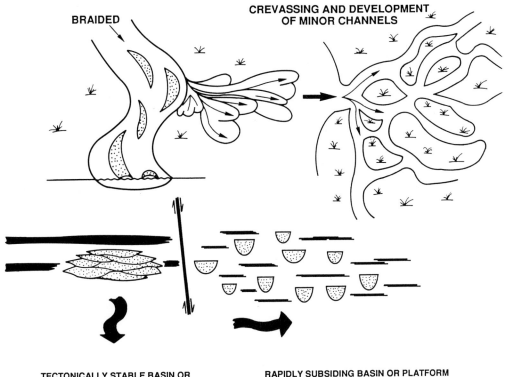

Fig. 9. Diagrammatic representation of the possible mechanism which controlled the distribution of fluvial depositional systems during the Early Westphalian C in the northeastern Campine Coalfield. Contemporary tectonic control had an important impact on the thickness, areal distribution and quality of the peats.

Opglabbeek Coalfield). However, tectonic-induced subsidence was episodic in this area, as shown by the widespread development, seen in the next higher stratigraphic interval, of a braid plain fluvial depositional environment on both sides of the Leut-Dorne lineament (compare Figs 7 and 8).

Upper Westphalian C sediment distribution

The Upper Westphalian C strata in the northeastern Campine Basin could be correlated by means of geophysical well logs, supported by tonstein units (identified by the method of Hedemann et al. 1984), despite considerable

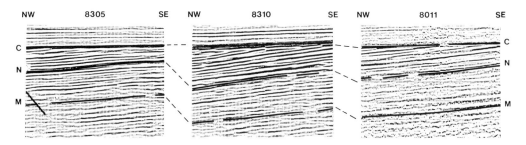

Fig. 10. Different seismic facies types of the Westphalian C deposits of the northeastern Campine Coalfields. 'Reverberative' seismic facies are characterized by persistent medium to high amplitude reflections, developed on sequences with high frequency thin coals (Upper Westphalian C, above horizon N on sections 8305 and 8310 of the Zwartberg-Opglabbeek Coalfield). 'Transparent' to 'quiet' seismic facies are characterized by low amplitude 'wormy' reflection configurations with fewer continuous reflections, developed on sequences with some thick coal seams and thick sandstones (Lower Westphalian C, between horizons N and M; these reflections are most pronounced on section 8011 of the Neeroeteren-Rotem Coalfield). Note also that section 8011 was obtained with a lower frequency signal and after filtering. C, Unconformity at the base of the Cretaceous; N, tonstein Nibelung, base of Upper Westphalian C; and M, Maurage = Aegiranum Marine Band, base of Lower Westphalian C. See Fig. 3 for location of seismic sections.

differences in sedimentary facies distribution (Fig. 4). During the Late Westphalian C thick, continuous coal seams associated with sand-rich braid plain deposits prevailed on the more stable structural block (Neeroeteren-Rotem), whereas a succession of thinner seams and abundant lake deposits was deposited on the active structural blocks (Meeuwen-Bree, or, transitionally, Zwartberg-Opglabbeek blocks) (Dusar 1989b). Although thickness variation is prominent, the present thickness may be mostly due to compaction differences in the variable sand–mud–peat sequences. Therefore it is not recommended to generalize the interpretation of sedimentary facies from sediment thickness distribution maps alone (Strack & Freudenberg 1984). The contrast between the KB161–168 boreholes (Neeroeteren-Rotem coalfield) and the KB169 borehole (Meeuwen-Bree coalfield) is noteworthy (Fig. 3). The thickness reduction almost attains 50% for borehole KB169. This reduction is not entirely due to compaction differences, however. The so-called Parcifal-Rubezahl interval, between the Odin and Siegfried coal seam groups (see Fig. 3) near the base of the Upper Westphalian C (for coal denominations, see Fiebig & Groscurth 1984), shows a thickness difference of 31 m (from 43 to 12 m) or about 100 m after decompaction, within similar sedimentary environments, which can be accommodated only by differential subsidence. Moreover, the Meeuwen-Bree block, north of the Gruitrode lineament (Fig. 2) apparently underwent not only subsidence standstills, but also tilting during the Late Westphalian C, deduced from seismic evidence.

Seismic evidence of changing depositional systems

Seismic exploration in the northeastern Campine Coalfield has allowed the structural reconstruction of the concealed basin and, to a lesser degree, the recognition of seismic facies and, hence, lithological successions and coal seam distribution. In the Campine Basin coal seam thickness variations are such that of the 62 potentially exploitable coal seams, only three were mined throughout the coal basin (Delmer 1963). Major fault zones or structural discontinuities often limit the lateral extent of mineable seams, implying either post-depositional or syn-sedimentary dislocations; the latter may have affected the sediment distribution and architecture of the palaeofluvial systems. The north–south oriented Donderslag transpressional fault (Fig. 1) has largely affected sedimentary facies distribution and coal seam thicknesses for the Westphalian A/B (Dusar & Langenaeker 1992; Delmer 1963). The NE–SW Gruitrode lineament in the north of the prospect area (Fig. 2) shows a change in seismic character for the Westphalian D, which is characterized by seismically 'transparent to chaotic' coarse-grained stacked sandstones south of the lineament and a seismically more 'reverberative' succession of thin sandstones and coals north, of the lineament (Bouckaert & Dusar 1987). Similar 'reverbera-

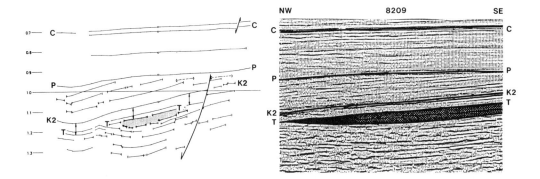

Fig. 11. Seismic section 8209 in the Meeuwen-Bree Coalfield, north of the Gruitrode lineament, with interpretation on a compressed (left) section (by H. G. Rossa, Geco-Prakla). This seismic section shows an unconformity at horizon T, with a sagging structure (stippled area) in the lower sequence. C, Unconformity at base of Cretaceous; P, unconformity at base of Permo-Triassic; K2, marker horizon near the Westphalian C/D boundary; and T, unconformity horizon. For location of seismic sections, see Fig. 3.

tive and transparant' seismic facies types have also been recognized in the Westphalian A/B and in the Westphalian C/D, respectively, of the southern North Sea (Evans et al. 1992). Tectonic instability and uplift of the Variscan front in the south of the prospect area could have provided sources for the seismically transparent coarse clastics in the Westphalian C/D of the eastern Campine Coal Basin (Thorez & Bless 1977).

Similar, though less abrupt, vertical and lateral changes in seismic facies can be observed for the Westphalian C (Fig. 2). The Upper (400–475 m) and Lower (300–400 m) Westphalian C strata contain high frequency occurrences of coal seams of uneven thickness (Fig. 4). Sequences containing thick seams and/or thick sandstones (= braid plain deposits?) display a more 'quiet' seismic character, with few continuous reflections. In contrast, sequences with many, but consistently thinner, unexploitable coals display a larger number of medium to high amplitude reflections (Fig. 10).

Occasional angular unconformities are discernible on the seismic sections (Fig. 11). These may either represent major channel complexes or regional unconformities, possibly related to one of the syn-depositional deformation phases in the Upper Westphalian C. In the latter instance, a correlation with the Symon unconformity of intra-Westphalian C age (Tubb et al. 1986) is tentatively suggested. The onset of the Upper Westphalian C sequence is well marked in the Zwartberg-Opglabbeek coalfield (Fig. 11), and the associated fault block tilting could have been triggered by the same tectonic event that produced the Symon unconformity in the southern North Sea. This event is attributed to a peripheral shift in thermal subsidence during the Late Carboniferous in the Variscan foreland basin (Leeder 1982; Tubb et al. 1986).

Westphalian palaeogeography of the Campine Basin and conclusions

A combination of allocyclic (eustatic basin–subsidence–palaeoclimatic) and autocyclic (fluvial avulsion–abandonment–crevassing) mechanisms are responsible for the palaegeographical evolution and alluvial architecture of the Campine Basin during the Westphalian times (Fig. 12). A gradual shift from more marine-influenced settings (in the Late Namurian–Early Westphalian A) to more continental ('hinterland') settings (in the Late Westphalian) is evidenced by the decreased importance of marine or brackish incursions. Moreover, pedological–sedimentological evidence (Besly 1988) indicates increased aridity towards the end of the Westphalian. This is also supported by the occurrence of red beds in alluvial fan-type coarse sandstones (the Neeroeteren Sandstone) of the Early Westphalian D in the eastern Campine area (Wouters & Gullentops 1987).

In the late Westphalian A and earliest Westphalian B, lower delta plain settings prevailed (Dreesen 1993), with large interdistributary areas enclosing lakes and bays. Low gradient meandering channels interrupted extensive delta plain swamps. Excess discharge was diverted during flood periods from the distribu-

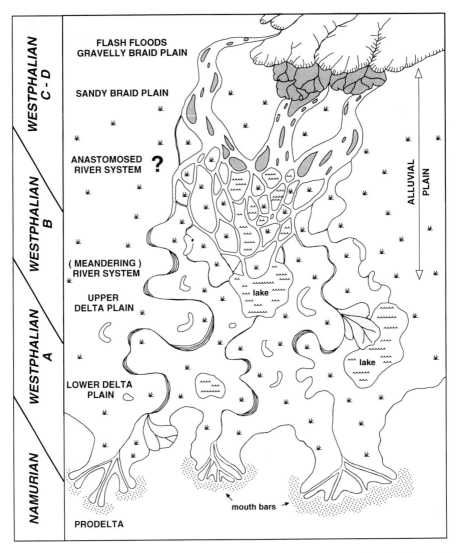

Fig. 12. Diagrammatic representation of suggested temporal palaeogeographical evolution of the Campine Basin during the Late Carboniferous (not to scale). Based on data from (in stratigraphically ascending order). Dreesen (1991, 1993); Langenaeker & Dusar (1991); Lorenzi et al. (1992); and Wouters et al. (1988).

tary channels into the floodplains, or into the interdistributary bays and lakes. Minor incursions of sediment-laden waters produced dirt bands or ash partings in the peat, whereas important inundations due to crevasses in the channel levees during floods generated crevasse splays that contributed to the drowning of swamps and infilling of lakes and bays. Crevasse splays may have prograded into lakes or bays, which enhanced the effects of coal splitting by post-depositional compaction (Guion 1984). Lateral channel migration and avulsion were a common cause of wash-outs in economic coals. Both fluvial processes have accounted for important production losses in the deep coal mines of the Campine area (Dreesen 1991, 1993).

A gradual shift of deposition to more upper delta plain settings is inferred during the Late Westphalian A–Early Westphalian B. This change is accompanied by a greater proportion of siltstones and sandstones (high channel sandstone density) in Westphalian B strata, compared with the mudstone-dominated Westphalian A strata (Lorenzi et al. 1992). In addition, the scarcity or even absence of marine beds in the Late Westphalian A–Early Westphalian B reflects deposition in an upper delta plain. A high channel sandstone density for upper delta plain/fluvial plain settings as opposed to the lower delta plain setting has been suggested by Fielding (1984).

An important feature of sandy braid plains is the lateral relationship of relatively thick peat (economic coals) and braid plain channels (large belts of thick multi-scour, multi-storey sandstone bodies). Although individual channels had some degree of freedom to move within the braid plain, the latter remained confined to more or less the same position by the surrounding peat, possibly accumulated for thousands of years, and by cohesive floodplain muds. Where channels were juxtaposed with peat, this occurred when the rates of peat accumulation exceeded the rates of inorganic overbank sedimentation, thus maintaining a swamp topography well above the local drainage level. The relatively low ash (average 8%) content of the thick Westphalian C coals probably reflects a swamp setting evolving towards a 'raised bog' environment (the ash content has been calculated according to the ISO 1171(81) standard procedures).

Another aspect of the Westphalian C coal-bearing strata is the common occurrence of freshwater floodplain lacustrine mudstones associated with CU crevasse splay sandstones. These deposits are interbedded with thin, discontinuous ashy coals (average 18% ash) and minor channel sandstones of restricted extent (see Fig. 6 between T10 and T13). Facies associations of ashy coals, discontinuous channel sandstones, floodplain lacustrine mudstones and crevasse splay sandstones have been observed in other coal-bearing intervals (e.g. Flores & Hanley 1984; Flores 1986, 1993) and peat-bearing sediments (e.g. Smith & Smith 1980; Smith 1983; Smith et al. 1989), which were interpreted to form in anastomosed river systems.

During the Westphalian D, more gravelly braid plains developed with thick, coarse sandstone deposits and few thin, non-mineable coals. The early Westphalian D Neeroeteren Sandstone (studied in detail from borehole KB 161) has been interpreted as the result of flash floods (gravelly braided alluvial system) under supposedly less humid climatological conditions (Wouters & Gullentops 1988; Wouters et al. 1989).

Anastomosed river belts are associated with rapid vertical aggradation occurring during periods when differential compaction of sediments or basin subsidence have lowered the base level of the alluvial plain below (Smith 1983). Vertical aggradation in the alluvial plain might have been initiated by crevassing (Fig. 9). The levees of fluvial channels were breached during floods; subsequent progradation, bifurcation and reunification of crevasse channels produced anastomosis (Flores 1986, 1993; Flores & Pillmore 1987).

The development of braided streams may be attributed to a variety of mechanisms, including the increased influx of bedload sediments due to increased sediment input, proximity to the source area, the high gradient imposed by local tectonism or a combination of these. In the Campine Basin, this may be related to the proximity of an emerging Variscan mountain belt to the south ('the hinterland') and/or to the effect of local syn-sedimentary block faulting. Our case study in the Lower Westphalian C indicates that thick and laterally more continuous coals accumulated on more tectonically stable, upthrown fault-bounded platforms where thick, long-lived major channel systems developed, whereas subeconomic, highly discontinuous and ashy coals accumulated in more rapidly subsiding downthrown fault blocks where thin, short-lived minor channel systems were formed.

Our coal facies model contrasts with that of Fielding et al. (1988), which studied the effect of contemporaneous local tectonic activity on sedimentation and peat accumulation in Early Carboniferous lower delta plain settings in Scotland. Here, coals were best developed and least affected by oxidation in an elongate, fault-bounded zone of enhanced subsidence (graben). In an earlier paper, Fielding (1984) also emphasized the structural control on coal seam distribution for the Westphalian Coal Measures of the Durham Coalfield (NE England). Here, the deposition of the coal-bearing sequences was controlled on a medium scale (several hundreds of square kilometres) by a subsiding basin, characterized by the occasional 'vertical stacking' pattern of upper delta plain channel sandstone bodies. The occurrence of expanded sedimentary 'cycles' in elongate belts aligned parallel to major faults and the presence of thin discontinuous coals were interpreted as remnants of subtle graben or half-graben topographic lows, which were created by the reactivation of structural weaknesses (possible Caledonian structures). These observations are in contrast with those of Weisenfluh & Ferm (1984), who reported accumulations of thick coals in upthrown fault blocks where peats were formed in swamps associated with a meandering fluvial system in the Black Warrior Basin (USA).

Thus although in some instances subsiding basins may promote the accumulation of thick coals, our study, as well as others, indicates that relatively thick, laterally extensive coal seams also form on upthrown, stable fault blocks. These coal seams are associated with thick multistorey channel sandstones. In adjacent subsiding downthrown fault blocks, coal seams tend to be thin and discontinuous. These coals are associated with thin, discontinuous channel sandstones commonly interbedded with floodplain lacustrine mudstones and crevasse splay sandstones. We therefore suggest this model as an alternative tool in promoting the exploration and development of coal seams, particularly in coal basins influenced by syn-depositional tectonics.

This project has been partially conducted with the financial support of the European Commission for Steel and Coal (convention 7220-AF/211). H. G. Rossa and M. Hemmerich (formerly Geco-Prakla, Hanover) gave valuable assistance in seismic interpretation. D. Schmitz (DMT, Bochum) and A. Schuster (Neuenhaus) helped with geophysical well log correlation. Seismic and borehole data have been provided courtesy of the Director of the Belgian Geological Survey (Brussels) and of the staff of the NV Kempense Steenkolenmijnen (Zolder). The manuscript benefited from the valuable criticism, corrections and helpful comments of B. R. Turner and P. D. Guion.

References

BESLY, B. M. 1988. Palaeogeographic implications of late Westphalian through early Permian red-beds, central England. *In:* BESLY, B. M. & KELLING, G. (eds) *Sedimentation in a Synorogenic Basin Complex. The Upper Carboniferous of NW Europe.* Blackie, Glasgow, 200–221.

BLESS, M. J. M. 1973. The history of the Finefrau Nebenband Marine Band (Lower Westphalian) in South Limburg (The Netherlands). A case of interaction between paleogeography, paleotectonics and paleoecology. *Mededelingen Rijks geologische Dienst Nieuwe Series*, **24**, 57–103.

BOUCKAERT, J. & DUSAR, M. 1987. Arguments geophysiques pour une tectonique cassante en Campine (Belgique), active au Paléozoique supérieur et réactivée depuis le Jurassique supérieur. *Annales de la Société Géologique du Nord*, **CVI**, 201–208.

BRISTOW, C. S. 1988. Controls on the sedimentation of the Rough Rock Group (Namurian) from the Pennine Basin of northern England. *In:* BESLY, B. M. & KELLING, G. (eds) *Sedimentation in a Synorogenic Basin Complex. The Upper Carboniferous of NW Europe.* Blackie, Glasgow, 114–131.

BURGER, K. 1985. Kohlentonsteine im Oberkarbon NW-Europas. Ein Beitrag zur Geochronologie. *Dixième Congrès International de Stratigraphie et de Géologie du Carbonifère*, **8**, 433–447.

CASSHYAP, S. M. & TEWARI, R. C. 1984. Fluvial models of the Lower Permian coal measures of Son-Mahanadi and Koel-Damodar Valley basins, India. *In:* RAHMANI, R. A. & FLORES, R. M. (eds) *Sedimentology of Coal and Coal-bearing Sequences.* International Association of Sedimentologists, Special Publication, **6**, 121–148.

DELMER, A. 1963. Carte des mines du bassin houiller de la Campine. *Annales des Mines de Belgique*, **1963/6**, 739–754.

DIESSEL, C. F. K. 1992. *Coal-bearing Depositional Systems.* Springer-Verlag, Berlin, 721pp.

DREESEN, R. 1991. Planification et conduite optimisées des exploitations fondées sur l'interpretation des données géologiques et leur traitement automatique en cours de l'exploitation. *Commission des Communautés Europeennes. Recherche Technique Charbon.* Rapport EUR 13401 FR. Contrat 7220-AF/205, 138pp.

—— 1993. Seam thickness and geological hazards forecasting in deep coal mining: a feasibility study from the Campine Collieries (N-Belgium). *Bulletin de la Société belge de Géologie*, **101**, 209–254.

——, BOUCKAERT, J., DUSAR, M., SOILLE, J. & VANDENBERGHE, N. 1987. Subsurface structural analysis of the late-Dinantian carbonate shelf at the northern flank of the Brabant Massif (Campine Basin, N-Belgium). *Service Géologique de Belgique Mémoires Explicatives pour les Cartes Géologiques et Minières de la Belgique*, **21**, 1–37.

DUSAR, M. 1989a. Non-marine lamellibranchs in the Westphalian C/D of the Campine coalfield. *Bulletin de la Société belge de Géologie*, **98**, 483–493.

—— 1989b. The Westphalian C in the Campine Basin: coal content influenced by tectonics. *Annales de la Société géologique de Belgique*, **112**, 248–249.

—— & LANGENAEKER, V. 1992. De Oostrand van het Massief van Brabant, met beschrijving van de geologische verkenningsboring te Martenslinde. Belgische Geologische Dienst, Professional Paper, 1992/5, **255**, 22pp.

——, BLESS, M. J. M. & 18 others 1987. *De steenkoolverkeningsboring Gruitrode-Ophovenderheide (Boring 172 van het Kempens Bekken).* Belgian Geological Survey, Professional Paper 1987/3, **230**, 235pp.

ETHRIDGE, F. G., JACKSON, T. J. & YOUNGBERG, A. D. 1981. Floodplain sequence of fine-grained meander belt subsystem: the coal-bearing Lower Wasatch and Upper Fort Union Formations, Southern Powder River Basin, Wyoming. *In:* ETHRIDGE, F. G. & FLORES, R. M. (eds) *Recent and Ancient Nonmarine Depositional Environments: Models for Exploration.* Society of Economic Paleontologists and Mineralogists, Special Publication, **31**, 191–209.

EVANS, D. J., MENEILLY, A. & BROWN, G. 1992. Seismic facies analysis of Westphalian sequences of the southern North Sea. *Marine and Petroleum Geology*, **9**, 578–589.

FIEBIG, H. & GROSCURTH, J. 1984. Das Westfal C im nördlichen Ruhrgebiet. *Fortschritte Geologie Rheinland Westfalen*, **32**, 257–267.

FIELDING, C. R. 1984. A coal depositional model for the Durham Coal Measures of NE England. *Journal of the Geological Society, London*, **141**, 919–931.

——, AL-RUBALL, M. & WALTON, E. K. 1988. Deltaic sedimentation in an unstable tectonic environment—the Lower Limestone Group (Lower Carboniferous) of East Fife, Scotland. *Geological Magazine*, **125**, 241–255.

FLORES, R. M. 1981. Coal deposition in fluvial paleoenvironments of the Paleocene Tongue River Member of the Fort Union Formation, Powder River area, Powder River Basin, Wyoming and Montana. *In:* ETHRIDGE, F. G. & FLORES, R. M. (eds) *Recent and Ancient Nonmarine Depositional Environments—Models for Exploration*. Society of Economic Paleontologists and Mineralogists, Special Publication, **31**, 169–190.

—— 1983. Basin facies analysis of coal-rich Tertiary fluvial deposits, northern Powder River Basin, Montana and Wyoming. *In:* COLLINSON, D. J. & LEWIN, J. (eds) *Modern and Ancient Fluvial Systems*. International Association of Sedimentologists, Special Publication, **6**, 501–516.

—— 1986. Styles of coal deposition in Tertiary alluvial deposits, Powder River Basin, Montana and Wyoming. *In:* LYONS, P. C. & RICE, C. L. (eds) *Palaeoenvironmental and Tectonic Controls in Coal-forming Basins of the United States*. Geological Society of America, Special Paper, **210**, 79–104.

—— 1989. Rocky Mountain Tertiary coal-basin models and their applicability to some world basins. *In:* LYONS, P. C. & ALPERN, B. (eds) *Peat and Coal: Origin, Facies, and Depositional Models*. International Journal of Coal Geology, **17**, 767–798.

—— 1993. Coal-bed and related depositional environments in methane gas-producing sequences, chapter 2. *In:* LAW, B. E. & RICE, D. D. (eds) *Hydrocarbons from Coals*, American Association of Petroleum Geologists, Studies in Geology Series, **38**, 13–37.

—— & HANLEY, J. H. 1984. Anastomosed and associated coal-bearing fluvial deposits: Upper Tongue River Member, Palaeocene Fort Union Formation, northern Powder River Basin, Wyoming, USA. *In:* RAHMANI, R. A. & FLORES, R. M. (eds) *Sedimentology of Coal and Coal-bearing Sequences*. International Association of Sedimentologists, Special Publication, **7**, 85–103.

—— & PILLMORE, C. L. 1987. Tectonic control on alluvial paleoarchitecture of the Cretaceous and Tertiary Raton Basin, Colorado and New Mexico. *In:* ETHRIDGE, F. G., FLORES, R. M. & HARVEY M. D. (eds) *Recent Developments in Fluvial Sedimentology*. Society of Economic Paleontologists and Mineralogists, **39**, 311–320.

——, ETHRIDGE, F. G., MIALL, A. D., GALLOWAY, W. E. & FOUCH, T. D. 1985. *Recognition of Fluvial Depositional Systems and their Resource Potential*. Society of Economic Paleontologists and Mineralogists, Short Course, **19**, 290pp.

FROSTICK, L. E., REID, I. & LAYMAN, J. T. 1983. Changing size distribution of suspended sediment in arid-zone flash floods. *In:* COLLINSON, J. D. & LEWIN, J. (eds) *Modern and Ancient Fluvial Systems*. International Association of Sedimentologists, Special Publication, **6**, 97–106.

GUION, P. 1984. Crevasse splay deposits and roof-rock quality in the Threequarters Seam (Carboniferous) in the East Midlands Coalfield, U.K. *In:* RAHMANI, R. A. & FLORES, R. M. (eds) *Sedimentology of Coal and Coal-bearing Sequences*. International Association of Sedimentologists, Special Publication, **7**, 291–308.

—— & FIELDING, C. 1988. Westphalian A and B sedimentation in the Pennine Basin, UK. *In:* BESLY, B. M. & KELLING, G. (eds) *Sedimentation in a Synorogenic Basin Complex. The Upper Carboniferous of NW Europe*. Blackie, Glasgow, 153–177.

—— & FULTON, I. M. 1993. The importance of sedimentology in deep-mined coal extraction. *Geoscientist*, **3**, (2), 25–33.

HEDEMANN, H. A., SCHUSTER, A., STANCU-KRISTOFF, G. & LÖSCH, J. 1984. Die Verbreitung der Kohlenflöze des Oberkarbons in Nordwestdeutschland und ihre stratigraphische Einstufung. *Fortschritte Geologie Rheinland Westfalen*, **32**, 39–88.

LANGENAEKER, V. & DUSAR, M. 1992. Subsurface facies analysis of the Namurian and earliest Westphalian in the western part of the Campine Basin. *Geologie en Mijnbouw*, **71**, 161–172.

LEEDER, M. R. 1982. Upper Palaeozoic basins of the British Isles—Caledonide inheritance versus Hercynian plate margin processes. *Journal of the Geological Society, London*, **139**, 479–491.

LORENZI, G., BOSSIROY, D. & DREESEN, R. 1992. Les minéraux argileux au service des correlations stratigraphiques des formations houillères du Carbonifère. *Commission des Communautés Européennes. Recherche Technique Charbon. Rapport EUR 14024 FR.* Contrat 7220-AF/206, 164pp.

MCCABE, P. J. 1984. Depositional environments of coal and coal-bearing strata. *In:* RAHMANI, R. A. & FLORES, R. M. (eds) *Sedimentology of Coal and Coal-bearing Sequences*. International Association of Sedimentologists, Special Publication, **7**, 13–42.

—— 1987. Facies studies of coal and coal-bearing strata. *In:* SCOTT, A. C. (ed.) *Coal and Coal-bearing Strata: Recent Advances*. Geological Society, Special Publication, **32**, 51–66.

MCLEAN, J. R. & JERZYKIEWICZ, T. 1978. Cyclicity, tectonics, and coal: some aspects of fluvial sedimentology in the Brazeau-Puskapoo Formations, Coal Valley area, Alberta, Canada. *In:* MIALL, A. D. (ed.) *Fluvial Sedimentology*. Canadian Society of Petroleum Geologists, **5**, 441–468.

MERITT, R. D. 1986. Paleoenvironmental and tectonic controls in major coal basins of Alaska. *In:* LYONS, P. C. & RICE, C. L. (eds) *Paleoenviron-*

mental and Tectonic Controls in Coal-forming Basins of the United States. Geological Society of America, Special Paper, **210**, 173–200.

MÜLLER, A. & STEINGROBE, B. 1991. Sedimentologie der oberkarbonischen Schichtenfolge in der Forschungsbohrung Frenzer Staffel 1 (1985). Aachen-Erkelenzer Steinkohlenrevier—Deutung der vertikalen und laterlalen Trendentwicklungen. *Geologisches Jahrbuch*, A **116**, 87–127.

PAPROTH, E., DUSAR, M. & 11 others 1983. Bio- and lithostratigraphic subdivisions of the Silesian in Belgium. A review. *Annales de la Société géologique de Belgique*, **106**, 241–283.

PUTNAM, P. E. & OLIVER, T. A. 1980. Stratigraphic traps in channel sandstones in the Upper Mannville (Albian) of east-central Alberta. *Canadian Petroleum Geologists Bulletin*, **28**, 489–508.

RAHMANI, R. A. & FLORES, R. M. (eds) 1984. *Sedimentology of Coal and Coal-bearing Sequences*. International Association of Sedimentologists, Special Publication, **6**, 412.

RAMOS, A. & SOPENA, A. 1983. Gravel bars in low-sinuosity streams (Permian and Triassic, Central Spain). *In:* COLLINSON, J. D. & LEWIN, J. (eds) *Modern and Ancient Fluvial Systems*. International Association of Sedimentologists, Special Publication, **6**, 301–312.

ROSSA, H. G. 1987. Upper Cretaceous and Tertiary inversion tectonics in the Western part of the Rhenish-Westphalian coal district (FRG) and in the Campine area (N. Belgium). *Annales de la Société géologique de Belgique*, **109**, 367–410.

RUST, B. R., GIBLING, M. R. & LEGUN, A. S. 1984. Coal deposition in an anastomising – fluvial system: the Pennsylvanian Cumberland Group south of Joggins, Nova Scotia, Canada. *In:* RAHMANI, R. A. & FLORES, R. M. (eds) *Sedimentology of Coal and Coal-bearing Sequences*. International Association of Sedimentologists, Special Publication, **7**, 105–120.

SCHUMM, S. A. 1968. *River Adjustment to Altered Hydrologic Regime, Murrumbidgee River and Paleochannels, Australia*. United States Geological Survey, Professional Paper, **598**, 65pp.

SCOTT, A. C. & COLLINSON, M. 1983. Investigating fossil plant beds. *Geology Teaching*, **7**, 114–122.

SMITH, D. G. 1983. Anastomosed fluvial deposits: modern examples from Western Canada. *In:* COLLINSON, J. D. & LEWIN, J. (ed.) *Modern and Ancient Fluvial Systems*, International Association of Sedimentologists, Special Publication, **6**, 155–168.

—— & SMITH, N. O. 1980. Sedimentation in anastomosed river systems: examples from alluvial valleys near Banff, Alberta. *Journal of Sedimentary Petrology*, **50**, 151–164.

SMITH, N. D., CROSS, T. A., DUFFICY, J. P. & CLOUGH, S. R. 1989. Anatomy of an avulsion. *Sedimentology*, **36**, 1–23.

STASSEN, P. 1948. Wash-out et dédoublement de couches aux charbonnages de Houthalen. *Annales de la Société géologique de Beligique*, **71**, B101–114.

—— 1949. Quelques wash-outs et dédoublements de couches dans le terrain houiller de la Campine et enseignements que l'on peut en tirer. *Annales de la Société géologique de Belgique*, **72**, 389–420.

STOPPEL, D. & BLESS, M. J. M. 1981. Flözunregelmässigkeiten im Oberkarbon. *Mededelingen Rijks Geologische Dienst*, **35-8/14**, 269–332.

STREHLAU, K. 1990. Facies and genesis of Carboniferous coal seams in Northwest Germany. *International Journal of Coal Geology*, **51**, 245–292.

STRACK, A. & FREUDENBERG, U. 1984. Schichtenmächtigkeiten und Kohleninhalte im Westfal des Niederrheinisch-Westfälischen Steinkohlenreviers. *Fortschritte Geologie Rheinland Westfalen*, **32**, 243–256.

THIADENS, A. A. & HAITES, T. D. 1944. Splits and wash-outs in the Netherlands Coal Measures. *Mededelingen Geologische Stichting, Serie C-11*, **1**, 1–51.

THOREZ, J. & BLESS, M. J. M. 1977. On the possible origin of the Lower Westphalian D Neeroeteren sandstone (Campine, Belgium). *Mededelingen Rijks Geologische Dienst*, **28**, 128–132.

TUBB, S. R., SOULSBY, A. & LAWRENCE, S. R. 1986. Palaeozoic prospects on the northern flanks of the London–Brabant Massif. *In:* BROOKS, J., GOFF, J. C. & VAN HOORN, B. (eds) *Habitat of Palaeozoic gas in NW Europe*. Geological Society, Special Publication, **23**, 55–72.

VAN AMEROM, H. W. J. & PAGNIER, H. J. M. 1990. Palaeoecological studies of the late Carboniferous plant macrofossils from borehole Kemperkoul-1 (Sittard, The Netherlands). *Mededelingen Rijks Geologische Dienst*, **44**, 1–19.

VAN WIJHE, D. H. & BLESS, M. J. M. 1974. The Westphalian of the Netherlands with special reference to miospore assemblages. *Geologie en Mijnbouw*, **53**, 295–328.

VANDENBERGHE, N. 1984. The subsurface geology of the Meer area in North Belgium, and its significance for the occurrence of hydrocarbons. *Journal of Petroleum Geology*, **7**, 56–66.

WEISENFLUH, G. A. & FERM, J. C. 1984. Geologic controls on deposition of the Pratt seam, Black Warrior Basin, Alabama, USA. *In:* RAHMANI, R. A. & FLORES, R. M. (eds) *Sedimentology of Coal and Coal-bearing Sequences*. International Association of Sedimentologists, Special Publication, **7**, 317–332.

WOUTERS, L. & GULLENTOPS, F. 1988. The sedimentology of the Westphalian D Neeroetern sandstone, Kempen (Belgium). *Annales de la Société Géologique du Nord*, **CVII**, 191–202.

——, ——, BOLLE, L., DE LOOSE, J., VAN LISHOUT, S. & DUSAR, M. 1989. The sedimentology of the Neeroeteren Sandstone, Upper Westphalian, Kempen (Well GD161 & 161b). *Aardkundige Mededelingen (Leuven)*, **4**, 47–101.

Structural geological factors in open pit coal mine design, with special reference to thrusting: case study from the Ffyndaff sites in the South Wales Coalfield

ROD GAYER, TANYA HATHAWAY & JOHN DAVIS

Laboratory for Strain Analysis, Department of Earth Sciences, University of Wales Cardiff, PO Box 914, Cardiff CF1 3YE, UK

Abstract: A review of the shear strength characteristics of Coal Measures lithologies suggests that major structurally controlled discontinuities in mudrock or coal are likely to show near-residual shear strength values, with minimal cohesion and a friction angle of approximately 12°. The principal structurally controlled discontinuities are those associated with tilting, folding and faulting. The former result from bedding plane slip during flexural folding that smoothes the bedding planes and reduces cohesion. The latter are produced by fault movements that weaken the rock and produce a structural anisotropy in the fault surface. Both are likely to confine the movement of water through the rock, enhancing weathering and further weakening the surface. In open pit design, the stability of the highwall is discussed in relation to the orientation of the various types of structurally controlled discontinuity.

A case study from Ffyndaff Opencast Coal Site in the South Wales Coalfield is analysed, and shows that the major normal faults that cut the site intersect the highwall with a stable orientation. Regional dip in the site also has a stable orientation, but the site is affected by several major north-verging thrusts that have been investigated by computer analysis of borehole and coal extraction data. The results indicate that the thrusts have a ramp-flat geometry and are likely to imbricate upwards from a bed-parallel detachment in a coal seam. The thrusts have generally developed in a piggy-back sequence, although at least one formed as a break-back thrust. The thrust displacement (slip separation) variation along the thrusts is described and suggests a normal displacement gradient of 0.16–0.18. These geometrical characteristics are used to interpret the thrust structure of a potential site from exploratory borehole logs, where the predicted geometry limits the location of the bounding highwall.

With the progressively improved performance of earth-moving machinery, it has become viable to excavate open pits for the recovery of coal to depths in excess of 200 m. The resultant greater heights of rock faces involved has required close scrutiny of slope stability and the need to design the open pit in such a way as to minimize the risk of slope failure by taking into account all relevant geotechnical parameters. In many instances, environmental considerations or local planning regulations restrict the possible options, making it difficult to utilize the optimum design. Under these circumstances it is of paramount importance to have an accurate knowledge of the geological structure of the proposed site, as many of the potential discontinuities that might give rise to slope failure are a direct consequence of tectonic structures (Fig. 1). In addition, these discontinuities are made weaker by slip along the discontinuity induced by the structure, or they may be rotated into an unfavourable orientation by a structure.

In this account the geotechnical properties of the coal measure sequence are discussed and the structures that can affect open pit design. The general principles are illustrated in the context of a major opencast site in the South Wales Coalfield.

Geological aspects of stable slope design

Coal Measures geotechnics

Strata containing productive coal seams are typically developed in a cyclic sequence within coastal or alluvial plain environments. The cyclicity is attributable to a complex interplay between intra- and extra-basinal controls, including: variations in subsidence rates resulting from tectonism and/or sediment compaction; variations in sediment supply due to tectonism and climatic change in the source region; and eustatic sea-level movements. These controls act on the distribution of fluvial systems and on the interface between fluvial and marine environments in such a way as to produce the

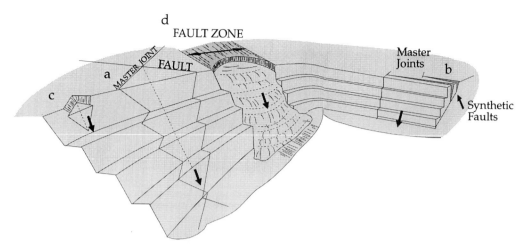

Fig. 1. Types of slope failure in open pit mines. Modified from Patton & Deare (1970). (**a**) Large-scale wedge failure with sliding on fault and master joint. Plunge of line of intersection of discontinuities dips at a sufficiently high angle for sliding to occur and daylights in bench of the open pit. (**b**) Large-scale plane failure on major normal fault. Fault plane dips out of slope at a high angle and daylights in the lower bench level. A block of rock is released laterally along the master joint set, and tension cracks at the rear of the block are produced by synthetic and antithetic normal faults. (**c**) Small-scale local failures caused by intersection of discontinuities such that failure is restricted to one or two benches of the open pit. (**d**) Circular failure of intensely fractured and weathered rock along a major fault zone. Weakened rock has the properties of soil.

characteristic coal measure sequence (Fielding 1987; Hartley 1993).

It is probable that the coals, formed by the accumulation and compaction of peat in a raised mire, although normally forming less than 10% of the sequence, represent the greatest time interval of the cycle. Coals are normally underlain by a rooted seat earth or underclay. The most abundant lithologies, often representing more than 80% of the cycle, are mudrocks. These claystones, shales and siltstones are usually well laminated, contain abundant plant debris and are often interbedded with thin bands or nodules of early diagenetic sideritic ironstones. They commonly represent overbank deposition of fluvial floodplains and in interdistributory lakes. Rooted units within the mudstone sequence represent periods of non-deposition and soil development, but with no accumulation of coal-forming peat. Mudstones, containing marine benthonic or freshwater bivalve faunas, form relatively thin but laterally persistent units within the sequence, characteristically forming the immediate roof of a coal seam. They represent the breakdown of the peat-forming rain forest environment as a result of eustatically rising sea levels (Hartley 1993). Thin to thick beds of sandstone, either of wide lateral extent with only minor thickness variations, or forming discontinuous lenticular bodies showing diagnostic internal bedforms of deposition in fluvial channels, usually form less than 10% of the sequence. Comparatively rare conglomeratic units, normally associated with channel sandstones, represent lag deposits.

The geotechnical parameters that are most relevant to rock slope stability are: (1) the persistence, attitude and nature of discontinuities within the rock mass that could form potential failure or release surfaces; (2) the shear strength characteristics both within the rock mass and along discontinuities, allowing the determination of friction angle and cohesive force across the discontinuity; (3) the rock density; and (4) the potential for build-up of water pressure within the rock mass and in tension cracks in the rock slope. Together with the geometry of the designed slope, these parameters allow the calculation of forces acting both to promote and to prevent sliding (or toppling) along the discontinuity (see Hoek & Bray 1981 for a full discussion of such analyses). For most coal measure lithologies these basic geotechnical parameters are well known; however, considerable variation exists and for most potential opencast sites a set of simple laboratory tests should be carried out to determine the precise characteristics of the principal lithologies (Taylor & Spears 1981).

The shear strength properties of rock along an established discontinuity are strongly dependent on the nature of the surface, including its

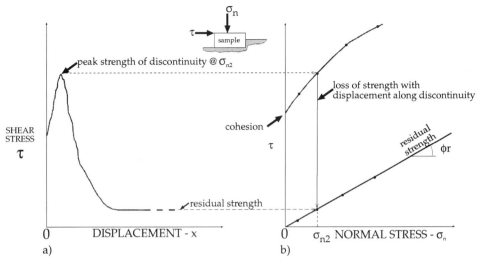

Fig. 2. Modification of shear strength following movement along a discontinuity. After Patton & Deare (1970). (**a**) Relation between shear stress (τ) and displacement (x) along a discontinuity at constant confining (normal) stress (σ_{n2}) to show peak and residual values of shear strength. (**b**) Relation between shear stress and normal stress to show decrease in strength after movement along a discontinuity. ϕ_r = residual friction angle.

roughness, its weathering characteristics and the strength of any infilling. If the discontinuity is rough, the peak shear strength is increased, particularly at low confining stress, according to the amplitude and wavelength of asperities causing its roughness. With extreme roughness in stronger rocks the shear strength approaches that of intact rock. The effect of increased roughness is to increase the stability of the discontinuity (Patton 1966). Conversely, the effect of smoothness is to decrease the stability, particularly if the discontinuity is infilled with poorly cemented material with a lower shear strength. In addition, the shear strength along a smooth discontinuity will decrease to residual values after relatively small displacements along the discontinuity (Fig. 2). In open pit environments, where rock movement during blasting may be sufficient to overcome the initial peak strength resistance to shearing, smooth discontinuities may realize only residual strength friction angles and zero cohesion. However, blasting commonly causes a shuffling movement along a range of discontinuities, producing an effective increase in roughness where the smoothed discontinuity surfaces are slightly offset.

Principal geological structures affecting slope stability

Bedding planes are the most common throughgoing discontinuity in Coal Measures sequences. They are most obvious when they separate beds of different lithologies, representing distinct changes in depositional conditions. They also often occur between beds of the same lithology, where they probably reflect periods of non-deposition or post-depositional diagenesis. Bedding planes form initially as relatively rough surfaces, affected by such sedimentary structures as ripple marks, groove casts or cross-bedding although in mudstones bedding planes are often smooth. The most likely processes of diagenesis are compaction consolidation and cementation, which produce irregular concretions (iron carbonate, etc.) or zones of irregular cohesion along flatter bedding contacts. Bedding planes normally initially form as subhorizontal surfaces and as such are unlikely to cause stability problems. However, regional tilting can increase the dip significantly and, at the same time, reduce the resistance to shearing by bed-parallel movements as a result of flexural slip. In coal measure mudrocks, flexural slip during folding is concentrated along master bedding planes and causes any original roughness to be broken down, producing a new foliated fabric parallel to the discontinuity, described as clay mylonite by Stimpson & Walton (1970). These weakened bedding planes can be of very wide lateral extent and commonly have near-residual shear strength characteristics, with friction angles of $\approx 11°$ and extremely low values of cohesion (Stimpson & Walton 1970; Walton & Coates 1980). Bedding plane discontinuities

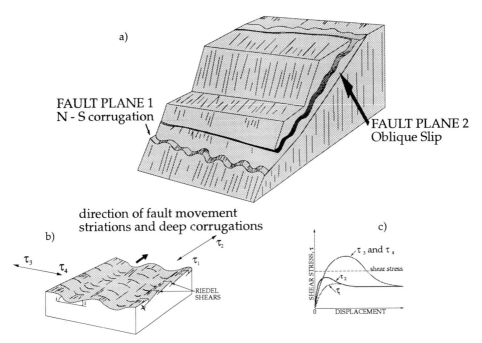

Fig. 3. Effects of fault plane anisotropy on peak shear strength. Modified from Patton & Deare (1970). (a) Block diagram showing two faults. Fault 1 is a dip-slip fault with slickenside corrugations plunging down-dip of the fault, and fault 2 is an oblique-slip fault with slickenside corrugations plunging obliquely down the fault plane. (b) Details of fault plane anisotropy for dip-slip fault 1 of (a). i, Angle of roughness; τ_1, τ_2, τ_3 and τ_4, orientations of shear values in directions indicated and shown in (c). (c) Graph of shear stress against displacement showing peak shear strengths for four directions across the fault plane in (b). Note that for displacements perpendicular to the fault plane anisotropy, the peak shear strength is greater than the shear stress generated along the discontinuity and the block is stable.

containing clay mylonite seams in mudrocks are a likely occurrence in opencast sites with bedding dips in excess of ≈15° and should be carefully monitored. Where possible, the site should be designed to develop high walls either parallel to the dip of the strata or with the bedding dipping directly into the slope.

Joints form the next most abundant category of discontinuity. Closely spaced joint discontinuities tend to be restricted to the more competent lithologies. Master joints, however, with wide lateral extent and cutting through several bed units, may also be developed, although they are commonly more widely spaced. These may form in almost undeformed strata and are thought to result from far-field stresses operating in over-pressured sequences (Lorenz et al. 1991). Cleat development in coal is an example of this process of joint formation (Gayer & Pesek 1992). In more tectonized regions a number of joint sets with systematic orientations may be developed, requiring a detailed analysis to assess their effect on slope stability. Discontinuities produced by joints usually have rough surfaces and display a high resistance to shearing, in some instances approaching the shear strength of intact rock (Patton & Deare 1970). The main stability problems relating to joints are likely to arise from water pressure buildup and weathering of the rock adjacent to the joint surface. Joint systems are often activated as release surfaces or tension cracks at the side and rear of major slip failures.

The discontinuities likely to pose the most serious stability problems in opencast sites are those associated with faulting. The three principal reasons for this are: (1) faults are normally laterally extensive, near-planar surfaces—consequently, they form major through-going discontinuities; (2) tectonic displacement along the fault will cut through any original asperities on the initially rough surface, thus reducing the resistance to shearing to values close to residual—the fault movement may also result in a weak fault gouge separating the hanging wall and footwall, reducing the shear strength still further; and (3) the dip of faults is variable, but they are characteristically inclined at angles between 30° and 90°. The discontinuities produced by major faults are therefore not only large, through-going surfaces with little resistance to shearing, but they also dip at angles which are likely to exceed the angle of friction along the discontinuity. In addition, major faults tend to develop as zones of closely spaced, subparallel surfaces, complicating any remedial stability treatment. The fault surface is commonly grooved in the direction of slip of the fault. These grooves may either plunge down the dip of the fault, in which case they have little effect on the strength of the discontinuity, or they may be parallel or oblique to the strike of the fault, in which case they will act as asperities, increasing the resistance to shearing (Fig. 3).

The three main categories of faults, normal, thrust and transcurrent, pose different stability problems. Normal faults have been extensively studied in coal measure strata (Watterson 1986; Walsh & Watterson 1988, 1989, 1990). The faults are dip-slip structures and, consequently, the slickenside grooves plunge down the dip of the fault surface and do not add significantly to the strength of the discontinuity (Fig. 3). They form isolated, planar, elliptical surfaces with a length to width ratio of approximately 2:1. Displacement varies across the fault, from a maximum at the centre to zero around the tip-line loop. The area of the fault surface depends on the magnitude of the maximum displacement, with a power law relationship. The fault plane dips at an average angle of 69°. Major normal faults form conjugate systems, with one set (synthetic) dipping subparallel to the master fault and the other (antithetic) dipping in the opposite direction. Normal faults form as a result of crustal extension. They are commonly associated with extensional veins that dip in the same direction, but more steeply, as the master fault. The fault surface is commonly filled with poorly compacted fault breccia, formed by fragmentation of the wallrock during movement. The shear strength of this fault gouge is critical in assessing the stability of the discontinuity. The steep dip of a normal fault, although almost always greater than the angle of friction for the discontinuity, will make it less likely to daylight in a rock face. In a site containing a major normal fault zone, the open pit should be designed so that the fault zone either dips directly into a highwall, or preferably cuts across the centre of the site, striking at a high angle into the highwall. In the former instance, care should be taken to ensure there is no connected discontinuity dipping out of the face in such a manner as to develop a biplanar failure (Boyd et al. 1978).

Thrust faults are also dip-slip structures, with slickenside grooves plunging down the fault surface. Although they can occur as isolated structures, comparable with normal faults, they commonly occur as a linked, imbricate system with individual thrusts branching upwards from a flat-lying detachment. Displacement varies along the thrust surface to a tip-line, but the details of this variation are not so well understood as for normal faults and will be discussed in the following. The shape of the thrust surface is variable. Classically, it has been described as a ramp-flat geometry (Boyer & Elliott 1982), with the ramp dipping at angles between 30° and 90° and cutting up through the stratigraphy, and the flat dipping parallel to the associated strata. The geometry of thrusts in coal measure strata is very complex, due to easy slip thrusting (Frodsham et al. 1993). Thrusts commonly form extensive flats within either coal seams or their seat earths, with ramps imbricating upwards from the flat detachments. Fault propagation folds and tip folds commonly develop above and in front of the detachments such that, as the detachment propagates forwards, the thrust cuts through the fold limbs giving the appearance of a ramp. Thrusts form as the result of crustal shortening and cause repetitions of strata, including coal seams. The thrust surfaces are characteristically polished, often forming a mirror-like surface in coals and carbonaceous shales. Where the detachments pass through mudrocks and seat earths a new foliated fabric is developed, with an identical appearance and strength characteristics to those of the clay mylonites associated with flexural slip. Thrusting therefore generates discontinuities with a very low resistance to shearing and with complex geometries, often associated with folding, that are liable to give rise to stability problems (Fig. 4). The design of opencast coal sites in areas of thrusting requires a detailed knowledge of the thrust system to minimize the risk of slope failure along the thrusts or weakened bedding planes. Some of the aspects of the geometrical analysis of thrusts and their recognition and characterization from borehole logs are described in the following case study.

Transcurrent faults are strike-slip structures with the slickenside grooving oriented parallel to the strike of the fault surface. These grooves therefore lie at right angles to the direction of potential sliding along the fault discontinuity

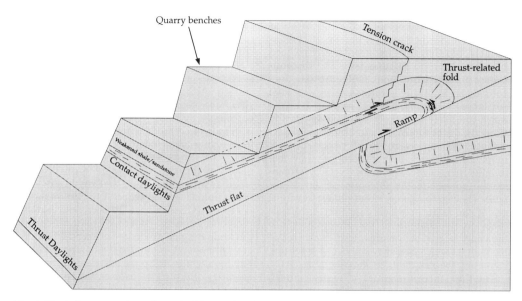

Fig. 4. Block diagram to show discontinuities produced by thrusting. Hanging wall and footwall folds, generated by normal drag along the thrust plane, develop flexural slip discontinuities along master bedding planes.

and increase the resistance to shearing. Strike-slip faults are normally subvertical and thus seldom daylight in a rock face. They may, however, give rise to toppling instability, and care should be taken to avoid designing an opencast site with a transcurrent fault parallel to a highwall. As with normal faults, transcurrent faults should pass through the centre of the site, striking into the highwall at a high angle.

Case studies from the South Wales Coalfield

Regional tectonics

Coalfields developed as foreland basins at the margins of orogenic belts [such as the South Wales, Ruhr and Upper Silesian coalfields along the northern Variscan margin (e.g. Havlena 1963; Gayer et al. 1993; Jankowski et al. 1993); the Pennsylvanian and Black Warrior basins along the western Appalachian margin (e.g. Leach & Rowan 1986; Clendenin & Duane 1990; Daniels et al. 1990); and the Bowen Basin in Queensland (e.g. Fielding 1991), are commonly strongly deformed by folds and thrusts propagating into the basin towards the orogenic foreland. The South Wales Coalfield is a good example of such a basin (Kelling 1988; Gayer & Jones 1989) and has been chosen to illustrate the effects of geological structure on open pit design because the entire remnant basin crops out in a major east–west synform, with existing opencast coal operations distributed along the northern and southern limbs of the fold structure (Fig. 5). The coal basin contains about 3 km of Silesian Coal Measures, overlying Dinantian platform carbonates, and non-productive, largely marine, Namurian siliciclastic sediments. The main sequence of productive Coal Measures occurs in the Upper Westphalian A to Lower Westphalian C and is dominated by lower to upper coastal plain mudrocks with thin sandstones and abundant coals. Marine band mudstones overlie coals in the lower and upper parts of the sequence and form distinctive marker beds at the Westphalian A/B boundary (Vanderbeckei Marine Band) and the Westphalian B/C boundary (Aegiranum Marine Band). A thick sequence (>1 km) of medium- to coarse-grained sandstones with thin interbedded coals (the Pennant Measures) overlie the productive measures. The coals vary in rank from high volatile bituminous coal ($>33\%$ volatile matter dry ash free (d.a.f.)) in the east and south of the coalfield, to anthracite ($<5\%$ volatile matter d.a.f.) in the northwest of the coalfield. The coals were ranked before the main folding affected the basin-fill (White 1991).

Variscan deformation has resulted from a generally south to north compression, with the development of the north verging main basin syn-form giving steep north dips (30–90°N)

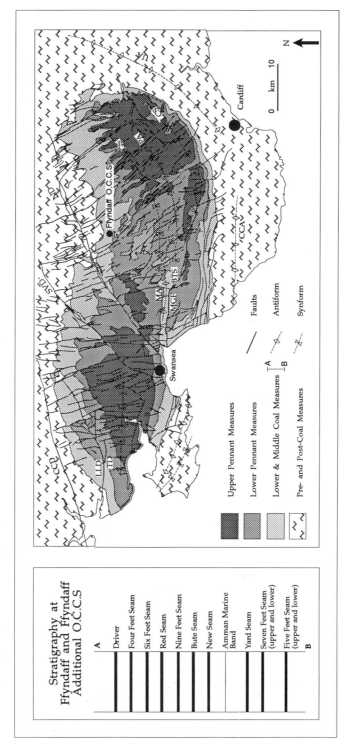

Fig. 5. Geological map of the South Wales Coalfield showing major tectonic elements and the location of Ffyndaff OCCS. BTS, Betws–Tonyrefail Syn-form; CCA, Cardiff–Cowbridge Antiform; CCD, Careg Cennen Disturbance; GS, Gelligaer Syn-form; LLD, Llanon Disturbance; LCS, Llantwit–Caerphilly Syn-form; MA, Maesteg Antiform; MGF, Moel Gilau Fault; ND, Vale of Neath Disturbance; PA, Pontypridd Antiform; SVD, Swansea Valley Disturbance; TD, Trimsaron Disturbance; and UA, Usk Antiform.

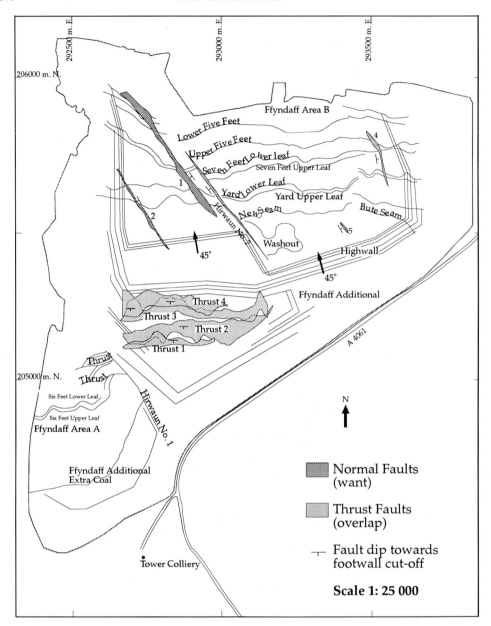

Fig. 6. Plan of Ffyndaff OCCS, located in Fig. 5, showing positions of principal highwalls during site development, outcrop of extracted coal seams and locations of major normal faults and thrusts described in the text. Shaded areas associated with normal faults and thrusts indicate zones of want and overlap, respectively.

along the south limb and shallow south dips (8–10°S) along the north limb. The syn-form hinge plunges inwards to give a basin structure and an elliptically shaped outcrop. Parasitic folds occur on the main syn-form, usually associated with thrusts. The thrusting generally verges northwards to the north of the coalfield, but to the south along the southern margin. This has been attributed to back-thrusting in the south above a major thrust wedge of Lower Carboniferous and older rocks, driven northwards beneath the southern limb (the passive roof duplex model of Jones 1991).

In detail, the thrusting is complex. Thrust

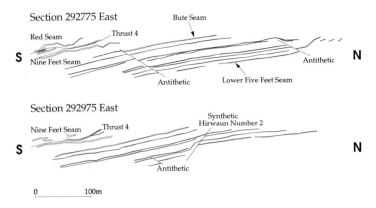

Fig. 7. Digitized, north–south, computer-modified extraction sections across Ffyndaff area B, locatd in Fig. 6. Sections show three normal faults, drag in the hanging wall of Hirwaun No. 2 fault, and thrust 4. See text for explanation.

detachments occur along many of the coal seams, which acted as weak, easy slip horizons. The thrusts develop ductile strains in the volume around the slip surface, the most prevalent of which are asymmetrical folds, verging with the same sense as the associated thrust. Along the southern margin and in the west of the coalfield the thrust structures indicate that several detachments were activated simultaneously. In these instances, complex break-back thrust geometries are developed, following the lock-up of higher detachments deformed by folds in the hanging wall of lower detachments. Frodsham et al. (1993) described these structures from Ffos Llas Opencast Coal Site (OCCS) in the west of the coalfield, where between 55 and 65% tectonic shortening can be demonstrated. Further east in the coalfield the recorded strains are less (35–45%) and the resultant thrust geometry less complex (Jones 1991).

Associated with the east–west trending folds and thrusts are a set of NW–SE cross-faults (e.g. Owen & Weaver 1983). Some of these faults show evidence of early, syn-thrusting dextral strike-slip displacement (Trotter 1947; Archer 1968; Gayer et al. 1973) but, in all the instances where these structures have been exposed, they have been shown to be pure dip-slip normal faults, post-dating the thrust deformation and recording late Variscan orogen-parallel extension (Cole et al. 1991). A second set of major oblique faults, striking ENE–WSW, traverse the coalfield and demonstrate at least two stages of sinistral strike-slip movement (Owen 1974). They are thought to represent reactivated Caledonian basement faults. Many of the major faults in the coalfield have probably been reactivated during the Mesozoic and Tertiary evolution of the Bristol Channel basin (e.g. Brooks et al. 1988, in press; Nemcok et al. in press).

Ffyndaff Opencast Coal Site

Geology of site. Ffyndaff OCCS is a complex of open pit workings situated on the northern margin of the South Wales Coalfield, approximately 3.5 km SE of the major sinistral strike-slip, ENE–WSW trending Neath Disturbance (Fig. 5). The site operates to a working depth of 70 m, exploiting eight major coal seams within about 180 m of the main productive coal measures, spanning the Westphalian A–B boundary (Fig. 6). The dominant inter-seam strata are mudstones with numerous thin bands of ironstone and ironstone concretions. Several thin, <1 m, fine-grained quartz sandstones are also present. All the coal seams overlie muddy seat earths and several seams are divided into distinct leaves by up to 6 m of mudrock. Seams often contain thin, laterally persistent mud partings, <0.05 m thick.

The regional dip is 6–8° to 170–190°, although this is considerably affected by the various faults cutting the site. The most prominent of these is the Hirwaun No. 2 cross-fault, a NW–SE trending normal fault, dipping ≈55°SW, with up to 28 m downthrow to the southwest (Fig. 6). The displacement along the fault decreases both up-dip and towards the northwest, where major normal drag is present in the hanging wall, producing local dips of up to 50° SW (Fig. 7). At

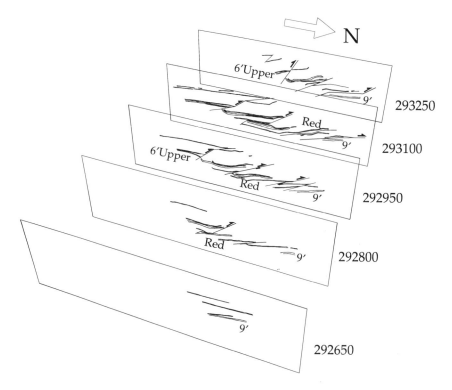

Fig. 8. Block diagram generated in AutoCAD, displaying series of north–south sections at 150 m intervals through Ffyndaff Additional Site, looking westwards, and showing the four thrusts described in the text.

least two antithetic normal faults occur in the hanging wall of the Hirwaun No. 2 fault. These dip 45–60° NE with displacements of <10 m (Fig. 7). Displacement decreases both towards the southeast, in the opposite direction to the main synthetic structure, and upwards, producing tiplines, with zero displacement, that plunge gently southwest. Dip-slip slickenside corrugations and fibres are commonly developed. The normal faults displace earlier thrusts.

At least five major thrusts are developed in the south of the site, where they dip between 7° and 40° S. The shallower dips reflect bedding-parallel flats, and the steeper dips thrust-ramps. The thrust surfaces are highly polished and often show dip-slip corrugations. The thrust flats commonly follow the roof or floor of a coal seam and are associated with a zone of shearing in which a new fabric is formed. The shear zones may be <2 m thick and show many features characteristic of ductile deformation, such as an oblique, sigmoidally curved foliation and asymmetrical folds (Frodsham *et al.* 1993). These strongly deformed zones in mudrocks are thought to possess very low residual shear strengths, with almost zero cohesion and friction angles $\approx 12°$ (e.g. Walton & Coates 1980). Meso-scale northward verging folds, with sheared bedding surfaces in the steep limbs dipping up to 90°, are often developed in the hanging wall of the thrusts. These sheared bedding surfaces also show low resistance to shearing and are potential slip surfaces.

Analysis of thrust structures: methods. An analysis of the thrust structures exposed progressively in the site has been undertaken in an attempt to define the geometrical properties of thrusts and thrust-related structures in coal measure lithologies. This has been achieved with the aid of AutoCAD software, customized using Lisp programs. The data used in the computer programs were available in the form of: (a) contractor surveyor's coal extraction sections in the form of line drawings, at 25 m intervals across the site; (b) British Coal surveyor's seam plans; and (c) British Coal exploration borehole logs (the latter two in the form of computer files). In none of these data sets were the thrusts identified and it was

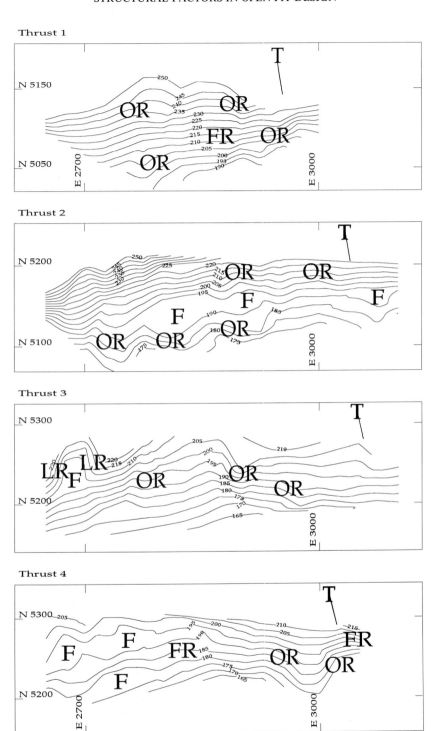

Fig. 9. Contoured plans, generated in AutoCAD and SURPAC, of four thrusts in Ffyndaff Additional Site, located in Fig. 6. Contours allow the definition of thrust transport direction (T), thrust flats (F), frontal ramps (FR), oblique ramps (OR), lateral ramps (LR) and their dip characteristics

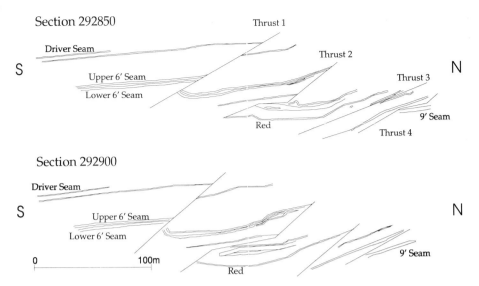

Fig. 10. Digitized, north–south extraction sections, produced in AutoCAD, of thrusts in Ffyndaff Additional Site. Thrust 2 shows ramp-flat geometry, with seams folded in the hanging wall and an antiform in the footwall immediately north of the ramp-flat fault bend (discussed in text). Thurst 1 has planar geometry, unaffected by, and post-dating, fault bend fold associated with thrust 2.

necessary to interpret the information by inspection. In the case of the extraction sections the data were digitized to produce a set of AutoCAD drawings, one for each section, to which the interpreted thrusts were added by inspection. These drawings were then combined to give three-dimensional displays of the coal seams and thrusts. For the seam plans, which were already in the form of a set of AutoCAD drawings, the data were easily accessible for the insertion of the thrusts. The borehole data, however, had to be downloaded from a mainframe computer to give a text file of borehole logs. The depths to the base of the relevant seams were then extracted from the logs, converted to an AutoCAD DXF file and imported to the drawings.

The analysis of thrust displacement (dip separation) was carried out using the digitized extraction section drawings. For each thrust on successive sections the displacements of the seams were calculated using an interactive AutoLisp program that allowed the normal drag components of displacement to be incorporated. Displacement values were recorded at points along the thrust mid-way between seam hanging-wall and footwall cutoff points. The distances along the thrust to a fixed reference point common to each section, e.g. a specific seam cutoff point, or a particular elevation, were also recorded by the program. The displacement data were contoured and displayed either as maps or in three dimensions.

Analysis of thrust structures: results. Four thrusts, numbered one to four, from south to north, have been analysed. Figure 8 shows a set of faces representing one of the thrusts created interactively using a customized program, whereas Fig. 9 is an annotated plan view of structure contours for the same thrust. The contoured plans of the four thrusts allowed the following to be determined: (a) the average dip and dip direction; (b) the lateral extent of thrust flats; and (c) identification of frontal, lateral and oblique ramps, together with their lateral extent, dip and dip direction. The average dip of the thrusts shows a systematic increase from 26° for thrust 4 to 38° for thrust 2. Thrust 1 has an average dip of 39°, the same as for thrust 3. Thrust flats are developed on all but the most southerly thrust, which shows a smooth frontal ramp geometry varying in dip along-strike from 27° to 38° (Fig. 9). The flats are invariably associated with coal seams, although they are limited to a strike length of ≈40 m, being bounded along-strike by oblique or lateral ramps, e.g. thrust 2 (Fig. 9). The dip of thrust ramps varies from 21° for thrust 4 to 41° for thrust 2, and shows the same systematic southward increase as in the average dip. Together, this southward increase in dip of the thrusts and their ramps suggests a piggy-back sequence of

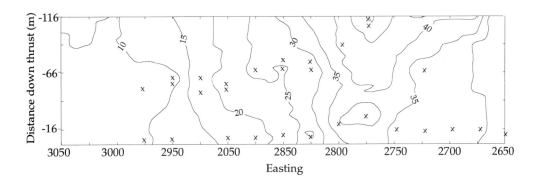

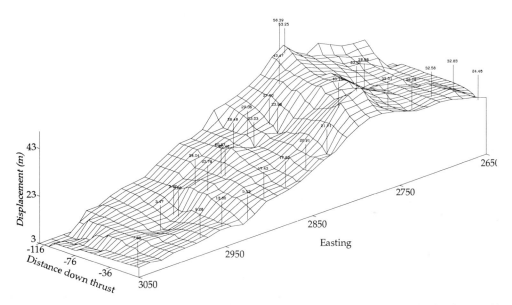

Fig. 11. Displacement diagrams, generated from extraction sections using AutoCAD and Surfer, for thrust 1. (**a**) Contoured displacement map and (**b**) perspective view. The thrust displacement decreases towards the east, but increases continually down-dip. Average displacement gradient parallel to strike is 0.18.

thrust evolution, with thrust propagation towards the north. Earlier formed thrusts in the south have been passively rotated in the hanging wall of later thrusts. Thrust 1 is anomalous in having an average dip of 30°, a shallower dip than thrust 2 and the same dip as that of thrust 3. A possible explanation is that this thrust developed out of sequence and after the formation of thrust 2. The fact that thrust 1 is not affected by folds in the hanging wall of thrust 2 supports this sequence of thrust development (Fig. 10). Another example of out of sequence thrusting is developed in the hanging wall of thrust 4 (Fig. 7) and was described from the surveyor's extraction sections by Jones (1991).

Thrust 2 develops a complex footwall structure at the bend between the thrust flat and frontal ramp (Fig. 10). Inspection of this structure, on-site, revealed an antiformal stack in which the strata between the Lower and Upper Six Foot Seam are repeated by duplexing thrusts; the floor thrust lies in the roof of the Lower Six Foot Seam and the roof thrust in the floor of the Upper Six Foot Seam. The structure suggests that the flat in thrust 2 extends along the roof of the Lower Six Foot Seam beyond the point where the thrust bends into the overlying

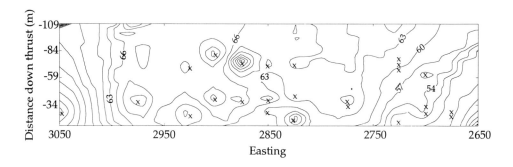

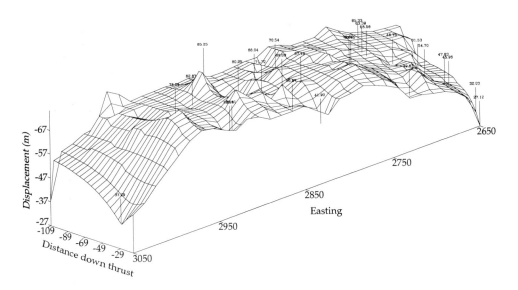

Fig. 12. Displacement diagrams, generated from extraction sections using AutoCAD and Surfer, for thrust 2. (**a**) Contoured displacement map and (**b**) perspective view. The thrust displacement shows distinctive maxima and minima related to thrust-flats. Average displacement gradient parallel to strike is 0.16.

ramp. It is not clear whether the extension of the flat occurred after the formation of the overlying ramp, as a footwall collapse structure, or whether bed-parallel thrusting was an early, pre-ramp event in which an antiformal stack developed, which effectively locked-up further bed-parallel movements and triggered the formation of the ramp.

Analysis of thrust displacement shows significant variations within the thrust surface. Thrust 1 has a maximum displacement of ≈50 m at Easting 2775, which decreases eastwards to zero over 275 m, giving a strike-parallel displacement gradient of 0.18 (Fig. 11). This value is similar to that of normal faults (Walsh & Watterson 1988).

The displacement increases steadily down-dip, with no indication of an approach to a maximum. Thrust 2 displacement variation is far more complex than in thrust 1, with distinctive maxima and minima (Fig. 12). The maxima are associated with thrust-flats and clearly reflect the low angle between the bedding and thrust surface. The region of maximum displacement is approximately elliptical and the displacement decreases from ≈78 to 26 m in the west, with an average strike-parallel gradient of 0.16. The relatively high displacement gradients between the maxima and minima may reflect linkages between discrete, isolated thrust surfaces, or they may be the effect of sharp bends in

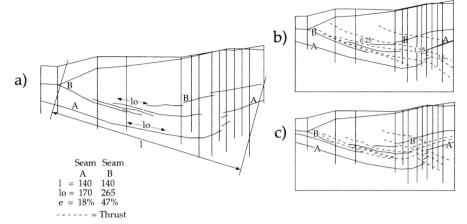

Fig. 13. North–south cross-section generated from exploratory borehole data from a potential site in the South Wales Coalfield. (**a**) Principal seams deformed by major fold, and location of boreholes. (**b**) Thrust interpretation based on single thrusting event. (**c**) Thrust interpretation based on early bed-parallel thrust detachment and later thrust ramping event. Numbers in (**b**) indicate displacement gradients for segments of thrusts indicated. Inset shows values of bed length shortening for seams A and B; stratigraphic level of change in shortening coincides with high displacement gradient and projected level of thrust detachment.

the thrust surface. Additional work on the geometry and displacement variation at bends is required to understand their relationships.

Thrusts and slope stability. The analysis of thrust geometry shows that the thrusts invariably dip south with angles varying from 7°, subparallel to regional bedding, to 40°. The high wall in the site is aligned parallel to the strike of both bedding and thrusts and has a batter of 45°N. Thus the thrusts, dipping directly into the highwall, do not pose any stability problem. Normal drag in the volume around the thrust, however, produces asymmetrical folds verging northwards. Flexural slip along the bedding, associated with the folding, has reduced the resistance to shearing of the mudrocks along these discontinuities to very low values. Thus the steeply dipping north limbs of these folds create localized regions where slip out of the highwall is possible.

A situation where thrusting has a more direct adverse effect on stability is illustrated from another South Wales site. Here an exploratory borehole investigation has produced a structural interpretation suggesting the presence of a set of major thrusts dipping consistently southwards (Fig. 13a), and thus posing no obvious threat to stability in any north-dipping high face bounding the site to the south. However, the thrusts cut a large-scale fold in which bedding dips at up to 30° in the opposite direction to the thrusts, and thus out of any bounding face. If the folding was associated with flexural slip mechanisms, bedding discontinuities in the mudrocks would be expected to possess a low resistance to shearing, with a consequent risk of slope failure of a southern boundary face. A more serious threat, however, would be posed by the thrusts. Inspection of the thrusts predicted in the structural interpretation indicates abrupt changes in displacement gradient down the constantly dipping thrust ramp. Displacement gradients vary from values of ≈0.25, close to those observed in the Ffyndaff thrusts, to much higher gradients up to 1.35, that appear to be unrealistically high (Fig. 13b). The effect of these high displacement gradients is to give markedly disparate values for bed-length shortening for adjacent seams in the section. Across the length of the section, seam A has been shortened by 18%, whereas seam B has been shortened by 47%. The changes occur at the same stratigraphic level and an alternative interpretation of the structure is that an early bed-parallel thrust system imbricated through the stratigraphy to give the large displacements in seam B. A later set of thrusts then cut through the earlier thrusts to produce the structure shown in the second interpretation (Fig. 13c). These later thrusts have much lower displacement gradients, comparable with those described from Ffyndaff. As the bed-parallel thrusts follow the stratigraphy around the fold, the latter must have occurred between the two thrusting events. The implications for the stability of any north-dipping boundary face are severe. The presence of a major thrust zone inclined at angles up to 30°

out of the face would lead to almost certain slope failure. Thus the structural investigation of this proposed open pit mine, together with other considerations, required that the southern boundary face of the site was located further north in the region where both the bedding and thrust planes dip southwards.

Conclusions

1. Significant slope failure in open pit coal mines can occur as a result of structurally controlled discontinuities within mudrock and/or coal that daylight in the highwall. Such discontinuities will contain structurally degraded rock and have near-residual shear strengths, being almost cohesionless and with friction angles $\approx 12°$.
2. The major structurally controlled discontinuities are: (i) weakened bedding planes in mudrock and seat earth produced by flexural-slip mechanisms during tilting/folding; (ii) normal faults with dips between 65° and 75° and with dip-slip slickenside corrugations; (iii) thrust faults, commonly with dips <45° and often bed-parallel, and with shear zone fabrics parallel to dip-slip slickenside lineations; and (iv) both normal and thrust faults characteristically develop normal drag structures with increased bedding dips and associated bed weakening.
3. Transcurrent (strike-slip) faults, unless major, do not normally lead to slope instability.
4. Computer analysis of borehole and coal extraction data shows that thrusts in coal measure strata develop ramp-flat geometries that imbricate from detachments within the floor or roof of coals, normally show a piggy-back sequence of thrust propagation and have displacement gradients of 0.16–0.18. These characteristics help to interpret thrust structures from borehole data and to assess potential slope stability.
5. Open pit designs should attempt to avoid the presence of adversely oriented structurally controlled discontinuities in the highwall.

Field work for the South Wales case study was partly funded by a NERC Special Topic research grant to R.A.G. T.H. is supported by a NERC/British Coal Opencast CASE studentship. The data used in the computer analysis of thrusts were provided by British Coal Opencast. We are grateful to K. Frodsham for allowing us to use the structural cross-section drawn from exploratory boreholes in Fig. 13, and to J. Argent for help with digitizing coal extraction sections. The manuscript was greatly improved following discussions with P. Norris.

References

ARCHER, A. A. 1968. *The Geology of the Gwendraeth Valley and Adjoining Areas.* Memoir, Geological Survey, Great Britain, 216.

BOYD, G. L., KOMDEUR, W. & RICHARDS, B. G. 1978. Open strip pitwall instability at Goonyella mine—causes and effects. *In: Proceedings of the Annual Conference, Australian Institute of Mining and Metallurgy*, 139–157.

BOYER, S. E. & ELLIOTT, D. 1982. Thrust systems. *American Association of Petroleum Geologists Bulletin*, **66**, 1196–1230.

BROOKS, M., HILLIER, B. V. & MILIORIZOS, M. New seismic evidence for a major geological boundary at shallow depth under north Devon. *Journal of the Geological Society, London*, in press.

——, TRAYNER, P. M. & TRIMBLE, T. J. 1988. Mesozoic reactivation of Variscan thrusting in the Bristol Channel area, UK. *Journal of the Geological Society, London*, **145**, 439–444.

CLENDENIN, C. W. & DUANE, M. J. 1990. Focused fluid flow and Ozark Mississippi Valley-type deposits. *Geology*, **18**, 116–119.

COLE, J. E., MILIORIZOS, M., FRODSHAM, K., GAYER, R. A., GILLESPIE, P. A., HARTLEY, A. J. & WHITE, S. C. 1991. Variscan structures in the opencast coal sites of the South Wales Coalfield. *Proceedings of the Ussher Society*, **7**, 375–379.

DANIELS, E. J., ALTANER, P. & MARSHAK, S. 1990. Hydrothermal alteration in anthracite from eastern Pennsylvania: implications for mechanisms of anthracite formation. *Geology*, **18**, 247–250.

FIELDING, C. R. 1987. Coal depositional models for deltaic and alluvial plain sequences. *Geology*, **15**, 661–664.

—— 1991. The geological setting of Queensland coal. *In: Queensland Coal Symposium, Brisbane*, Australian Institute of Mining and Metallurgy, 97–103.

FRODSHAM, K., GAYER, R. A., JAMES, J. E. & PRYCE, 1993. Variscan thrust deformation in the South Wales Coalfield—a case study from Ffos-Las Opencast Coal Site. *In:* GAYER, R. A., GREILING, R. O. & VOGEL, A. (eds) *The Rhenohercynian and Sub-Variscan Fold Belts.* Earth Evolution Science Series. Vieweg, Braunschweig, 315–348.

GAYER, R. A. & JONES, J. 1989. The Variscan foreland in South Wales. *Proceedings of the Ussher Society*, **7**, 177–179.

—— & PESEK, J. 1992. Cannibalisation of Coal Measures in the South Wales coalfield—significance for foreland basin evolution. *Proceedings of the Ussher Society*, **8**, 44–49.

——, ALLEN, K. C., BASSETT, M. G. & EDWARDS, D. 1973. The structure of the Taff Gorge area, Glamorgan, and the stratigraphy of the Old Red Sandstone–Carboniferous Limestone transition. *Geological Journal*, **8**, 345–375.

——, ——, GREILING, R. O., HECHT, C. & JONES, J. 1993. Comparative evolution of coal-bearing foreland basins along the Variscan northern margin in Europe. *In:* GAYER, R. A., GREILING, R. O. & VOGEL, A. (eds) *The Rhenohercynian and Sub-Variscan Fold Belts*. Earth Evolution Science Series. Vieweg, Braunschweig, 47–82.

HARTLEY, A. J. 1993. A depositional model for the Mid-Westphalian A to Late Westphalian B Coal Measures of South Wales. *Journal of the Geological Society, London*, **150**, 1121–1136.

HAVLENA, V. 1963. Geologie uhelnych lozisek 1. *Nakladatelstvi Ceskoslovenske adkademie ved.* Prague.

HOEK, E. & BRAY, J. W. 1981. *Rock Slope Engineering*, 3rd Edn. Institute of Mining and Metallurgy, London, 358.

JANKOWSKI, B., DAVID, F. & SELTER, V. 1993. Facies complexes of the Upper Carboniferous in Northwest Germany and their structural implications. *In:* GAYER, R. A., GREILING, R. O. & VOGEL, A. (eds) *The Rhenohercynian and Sub-Variscan Fold Belts*. Earth Evolution Science Series. Vieweg, Braunschweig, 137–158.

JONES, J. A. 1991. A mountain front model for the Variscan deformation of the South Wales coalfield. *Journal of the Geological Society, London*, **148**, 881–891.

KELLING, G. 1988. Silesian sedimentation and tectonics in the South Wales Basin: a brief review. *In:* BESLY, B. & KELLING, G. (eds) *Sedimentation in a Syn-orogenic Basin Complex: the Upper Carboniferous of NW Europe*. Blackie, Glasgow, 38–42.

LEACH, D. L. & ROWAN, E. L. 1986. Genetic link between Ouaticha foldbelt tectonism and the Mississippi Valley-type lead–zinc deposits of the Ozarks. *Geology*, **14**, 931–935.

LORENZ, J. C., TEUFEL, L. W. & WARPINSKI, N. R. 1991. Regional fractures I: a mechanism for the formation of regional fractures at depth in flat-laying reservoirs. *American Association of Petroleum Geologists Bulletin*, **75**, 1714–1737.

NEMCOCK, M., GAYER, R. A. & MILIORIZOS, M. Inversion of the Inner Bristol Channel basin: implications for fracture permeability. *In:* BUCHANAN, J. G. & BUCHANAN, P. G. (eds) *Basin Inversion*, Geological Society, London, Special Publication, in press.

OWEN, T. R. 1974. The Variscan orogeny in Wales. *In:* OWEN, T. R. (ed.) *The Upper Palaeozoic and Post-Palaeozoic Rocks of Wales*. University of Wales Press, Cardiff, 285–294.

—— & WEAVER, J. D. 1983. The structure of the main South Wales Coalfield and its margins. *In:* HANCOCK, P. L. (ed.) *The Variscan Foldbelt in the British Isles*. Adam Hilger, Bristol, 74–87.

PATTON, F. D. 1966. Multiple modes of shear failure in rock. *In: Proceedings of the 1st International Congress of Rock Mechanics, Lisbon,* Laboratoria Nazional de Engenharia Civil, 509–513.

—— & DEARE, D. U. 1970. Significant geologic factors in rock slope stability. *In:* VAN RENSBURG, P. W. J. (ed.) *Planning Open Pit Mines. Proceedings of the Symposium on the Theoretical Background to Planning of Open Pit Mines with Special Reference to Slope Stability*. South African Institute of Mining and Metallurgy, 143–151.

STIMPSON, B. & WALTON, G. 1970. Clay mylonites in English coal measures. *In: Proceedings of 1st Congress International Association of Engineering geology, Paris,* 1388–1393.

TAYLOR, R. K. & SPEARS, D. A. 1981. Laboratory investigation of mudrocks. *Quarterly Journal of Engineering Geology*, **14**, 291–309.

TROTTER, F. M. 1947. The structure of the Coal Measures in the Pontardawe-Ammanford area, South Wales. *Quarterly Journal of the Geological Society, London*, **103**, 89–133.

WALTON, G. & COATES, M. 1980. Some footwall failure modes in South Wales opencast workings. *In:* GEDDES, J. D. L. (ed.) *Proceedings of the 2nd International Conference on Ground Movement and Structures*, Cardiff, 1–17.

WALSH, J. J. & WATTERSON, J. 1988. Dips of normal faults in British Coal Measures and other sedimentary sequences. *Journal of the Geological Society, London*, **145**, 859–873.

—— & —— 1989. Displacement gradients on fault surfaces. *Journal of Structural Geology*, **11**, 307–316.

—— & —— 1990. New methods of fault projection for coal mine planning. *Proceedings of the Yorkshire Geological Society*, **48**, 209–219.

WATTERSON, J. 1986. Fault dimensions, displacements and growth. *Pure and Applied Geophysics*, **124**, 365–373.

WHITE, S. C. 1991. Palaeo-geothermal profiling across the South Wales Coalfield. *Proceedings of the Ussher Society*, **7**, 368–374.

Controls of coalbed methane prospectivity in Great Britain

H. E. BAILY, B. W. GLOVER, S. HOLLOWAY & S. R. YOUNG

British Geological Survey, Kingsley Dunham Centre, Keyworth, Nottingham NG12 5GG, UK

Abstract: Only in the last few years has there been a significant increase in exploration for coalbed methane (CBM) in Great Britain. There are several geological controls on British CBM prospectivity which combine with socio-political constraints in limiting the direct transfer of experience and technology from the USA where, until recently, CBM development has been concentrated. In many parts of the world, the depth of burial and rank of coals may be used as approximate indicators of CBM potential. However, in most British coal-bearing basins, one of the most important factors controlling the amount of preserved adsorbed methane in coals appears to be the degree of syn- and post-depositional basin inversion. Although enough methane to saturate the coals was probably generated during the formation of the Late Carboniferous basins, extensive degasification took place during the end-Carboniferous Variscan orogeny. Subsequent Permo-Triassic and later reburial of coals seems to have been insufficient to replenish adsorbed methane over much of the CBM target areas of Great Britain. Those British coal-bearing basins which can be identified as having been originally deeply buried, only mildly inverted during the Variscan orogeny and which have remained relatively deeply buried beneath Mesozoic cover until early Cenozoic times are likely to contain the most preserved adsorbed methane and consequently prove the best prospects. Identification of a consumer and an adequate infrastructure further limit the potential of CBM prospects. In addition, at present, geological factors such as *in situ* stress, Coal Measures sedimentology, coal cleat orientation, hydrogeology and hydrology, and planning and environmental issues are considered in more detail only at the well-siting stage. As a knowledge of British CBM grows, however, these factors will become more important in initial licence acquisition.

Gas inflow into British deep mines has been recorded since at least the 17th century (Robinson & Grayson 1990). Initially it was regarded solely as a hazard to mining; firedamp explosions have resulted in the loss of over 10 000 lives since 1850 (Turton 1981). The first exploitation of this gas occurred in the 1950s when the National Coal Board (now British Coal Corporation) began to use methane from the ventilation systems in deep mines to provide a source of energy at the pithead (Bromilow 1959). These schemes concentrated on enhancing safe working practice and were not designed to exploit coalbed methane (CBM) to its full potential.

Coalbed methane is becoming an increasingly important hydrocarbon resource. Rapid technological development of the CBM industry in the USA (Kelley 1989), coupled with the need to use CBM as a sustainable resource rather than allowing it to contribute to global warming (Mitchell 1990), has led to the initiation of exploration programmes for CBM world wide (Kuuskraa *et al.* 1992).

At the beginning of the 1990s, US hydrocarbon companies began to look abroad for areas with possible CBM potential. The vast coal reserves of northwest Europe were an obvious target. The first licences in Great Britain sought primarily for CBM exploration were issued late in 1991 and, although the presence of a considerable geological database played an important part in initiating British CBM exploration, it was the existence of a framework for the independent sale of gas which prompted the early acquisition of licences. The Department of Trade and Industry (DTI) estimate that current CBM reserves in Britain may represent £50 billion to the nation (Knott 1993).

This paper sets out the factors which have controlled and are likely to influence further CBM development in Britain. Geological issues are considered, together with geographical and currently topical political and legislative problems.

Age and depositional setting of British coalbed methane source rocks

Coal-bearing strata which form CBM exploration targets in Great Britain are Carboniferous in age (Glover *et al.* 1993a). They were deposited on the foreland north of the emerging Variscan

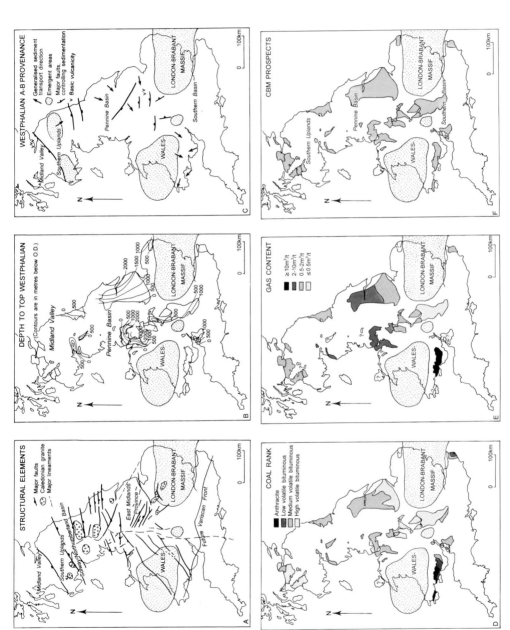

Fig. 1. (A) Main structural elements which influence the Carboniferous and Permian to Mesozoic basin formation in Great Britain (partly after Fraser et al. 1990). (B) Approximate depth in metres below OD to the top of Westphalian strata. (C) Provenance and structural influences during the deposition of Westphalian A–B strata (adapted from Guion & Fielding 1988; Kelling 1974). (D) Rank map for Carboniferous coals (after National Coal Board 1979). (E) Hypothetical methane gas content for coals from Carboniferous strata (partly after Creedy 1991, and references cited therein). (F) The main CBM prospects of Great Britain (after Glover et al. 1993a).

orogenic mountain belt now preserved in four regions, three of which lie to the north of a persistent Carboniferous land mass known as the Wales–London–Brabant Massif (WLBM) (Fig. 1). The fourth region lies to the south of the WLBM, in a belt stretching from South Wales to Kent. Leeder (1982, 1988), Leeder & McMahon (1988) and Besly (1988) have reviewed the geodynamics of British Carboniferous basins. Ramsbottom et al. (1978) established a comprehensive correlation of Silesian strata in Great Britain.

The area north of the Wales–London–Brabant Massif

In the northern regions, coal-bearing strata range in age from Visean to Westphalian C. They were formed in three main sedimentary basins: the Pennine Basin, the Northumberland–Solway Basin and the Midland Valley Basin of Scotland (Fig. 1C). Outcrops and subcrops of coal-bearing strata now form a series of structural sub-basins. Three main structural trends are recognized in the present day distribution of the Coal Measures in these basins: a NE to ENE trend in the north of the region, including the Northumberland–Solway Basin and the Midland Valley of Scotland; a mainly NE to N trend west and southwest of the Pennines, including the area of the Permo-Triassic Cheshire Basin, and a NW trend east of the Pennines (Fig. 1A and 1C).

The following reviews give further details: Read (1988) and Francis (1991) (Midland Valley of Scotland); Scott & Colter (1987) and Chadwick et al. (1993) (Northumberland–Solway Basin); Besly (1988) and Waters et al. (1994) (west and southwest of the Pennines); and Fraser et al. (1990) (east of the Pennines and general).

Strata containing economically significant amounts of coal were first laid down in the Midland Valley of Scotland and northern England during Visean times (Leeder 1987; Fraser et al. 1990). Coal-forming environments proliferated during the Namurian and, in the subsequent Westphalian A to early Westphalian C, covered much of Britain with Coal Measures. More than 3000 m of coal-bearing strata were deposited in the Pennine Basin, with subsidence centred in the southwest of the basin, near the present day position of the city of Manchester (Wills 1956). Deposition was largely upon poorly drained alluvial plains which were periodically inundated during basin-wide marine incursions (Guion & Fielding 1988). Peat deposits over 90 m thick before compaction formed along the southern margin of the basin, whereas thinner peats developed in the central areas where subsidence was relatively greater (Wills 1956) (Fig. 1B).

During Namurian and Westphalian A times, minor localized folding and fault inversion and uplift occurred, induced by the developing Variscan tectonism. This had little effect on the large-scale subsidence patterns, but locally, especially at basin margins, it resulted in erosion or stratigraphical condensation (Fraser et al. 1990; Glover et al. 1990; Waters et al. 1994). In late Westphalian B and early Westphalian C times, tectonic activity intensified, resulting in the northwards expansion of well drained alluvial plains and the consequent shrinkage of coal-forming environments (Besly 1988; Glover et al. 1993b). Variscan compressive deformation intensified further in late Westphalian C times, resulting in the development of a widespread unconformity along most of the southern margin of the Pennine Basin. This can be correlated with unconformities of similar age in the North Sea (Leeder & Hardman 1990) and the Northumberland–Solway Basin (Poole 1988).

Renewed subsidence in Westphalian D times was flexurally induced and caused by nappe emplacement to the south of the Pennine Basin (Kelling 1988). Westphalian D coal-bearing sequences occur along the southern margin of the basin, but these do not represent significant CBM targets as the coals are generally of low rank, thin and laterally impersistent. The remainder of Silesian sedimentation within the Pennine Basin was largely upon a well drained alluvial plain fed mainly by molasse from Variscan nappes to the south.

The area south of the Wales–London–Brabant Massif

A number of structural basins occur south of the WLBM. The age of CBM targets within these basins ranges from Westphalian A to D. Relevant reviews are provided by Shephard-Thorn (1988) (Kent) Dunham & Poole (1974) and Foster et al. (1989) (Oxfordshire–Berkshire), Kellaway & Welsh (1993) (Bristol–Somerset) and Cornelius et al. (1993), Jones (1991), Thomas (1974) and Owen & Weaver (1983) (South Wales).

Depositional environments of coal-bearing strata were comparable with those in the Pennine Basin. Being closer to the Variscan fold belt, the effects of Variscan compressive deformation were much greater than to the north of the WLBM, which effectively acted as a buttress, limiting the northwards propagation of thrusts. As with the Pennine Basin, minor Variscan tectonism occurred through much of Silesian times (e.g. Jones 1989).

As exemplified in South Wales, there are several structural trends, the most important of which appears to be E–W to ESE–WNW. This is the dominant fold axis and thrust plane strike developed during Variscan deformation (Coward & Smallwood 1984).

Subsequent to the Late Carboniferous–Permian Variscan basin inversion and erosion, there have been several basin-forming periods. These have been interrupted by phases of relative uplift and folding. A summary of the tectonic history and styles of basin-fill is given in Fig. 2.

Geological controls on coalbed methane formation

Coal volume

Coal volume is of prime importance in the quantitative assessment of CBM potential. The British Carboniferous basins commonly contain relatively high proportions of coal. Those formed along the margins of the Pennine Basin contain up to 20% coal within the Westphalian A–C interval; coal may total 120–200 m. In more central settings, coal percentages within this interval are still relatively high (6–9%). In addition, many coal seams persist laterally for thousands of square kilometres (e.g. the Top Hard, Warren or Barnsley Coal of eastern Yorkshire, Lincolnshire and Nottinghamshire), providing good predictability in the assessment of CBM potential.

Within the British Coal Measures the greatest potential for CBM lies in the coal-bearing sequence of Westphalian A–C age, considered here as the main target interval. This is composed mainly of low permeability mudstones and siltstones which have prevented the early escape of methane during coalification. Where sandstones form coal wash-outs or the principal roof lithology to coals, the coals may be undersaturated in methane (Ulery 1988).

Coal composition

Coal composition can control CBM sorption capacity (Creedy 1991), although Ulery (1988) concluded that such effects were negligible. Barker-Read (1984) and Barker-Read & Radchenko (1989) have shown that there is a clear link between ash content (i.e. the residual material of coal left after incineration, including mineral matter such as pyrite, clay compounds and ankerite) and the sorption capacity of coal. Commonly, coals formed near basin margins contain higher ash contents. This negative factor can, however, be compensated by the thicker coals that are present in these areas. Coal composition is likely to affect permeability as the main constituents of coal, i.e. the coal macerals, have differing microstructure (e.g. Stach et al. 1982, fig 27) and hence porosity and permeability. However, these effects on porosity and permeability are negligible in comparison with coal cleat, which is responsible for most of the permeability in coal.

Timing and degree of coalification

A knowledge of the timing and degree of coalification is crucial to CBM appraisal. All other factors being equal, the higher the coal rank, the more methane can be adsorbed onto maceral surfaces. Fundamental to coal rank attainment are temperature, and, to a lesser extent, pressure. These are dependent in turn on the depth of burial and heat flow. Time may also be an important contributor to the process of coalification (Bostick 1979), although its importance has been debated (Price 1983; Barker & Goldstein 1990).

The rank of coal may not be the product of a single burial event; it may have been subject to uplift and later reburial or thermal metamorphism. Suggate (1976) suggested that in the East Pennine Coalfield, peak coalification occurred during the deposition of the Permian and Mesozoic strata which unconformably overlie the Westphalian Coal Measures, citing the systematic increase in coal rank beneath the sub-Permian unconformity. In contrast, Creedy (1988) inferred coalification in this region to have occurred before Variscan uplift (latest Westphalian–?early Permian). Ayers et al. (1993) showed that, in the Pennine Basin, isopachs for the thickness of Westphalian A–B strata could be correlated with coal isorank lines, indicating that, in broad terms, the greatest degree of coalification took place during deposition of the Westphalian Coal Measures within the Pennine Basin.

Given the great thickness of strata in British Permian and Mesozoic basins (e.g. over 3000 m in the Cheshire Basin), it seems likely that Permian–Mesozoic deposition had an effect on coal rank at least locally where Permian and Mesozoic successions overlie thick Upper Carboniferous successions. Suggate (1976) published a plot showing the variation of calorific value and moisture content with depth in a cross-section of folded coal-bearing strata (Fig.

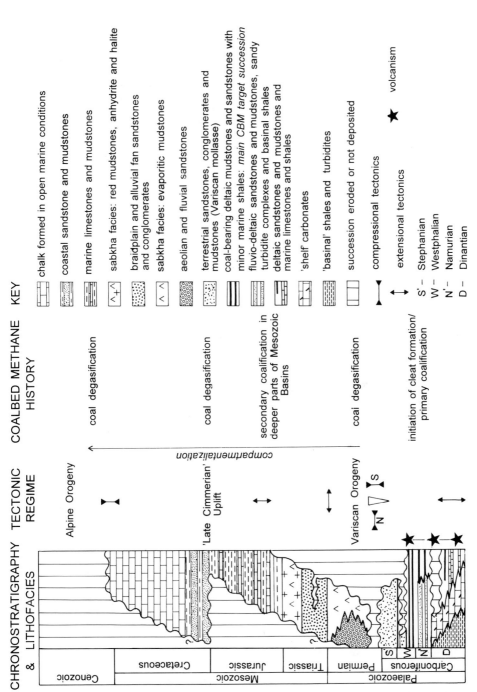

Fig. 2. Chronology, lithofacies, stress and methane gas history of British coal basins and Permian to Mesozoic overburden.

3). This clearly shows that variations in coal rank broadly mimic stratal folding. That the trends are not exactly coincident may well indicate partial secondary coalification rather than the complex localized variations in the geothermal gradient envisaged by Suggate (1976).

We suggest the following generalized, simplified coalification history for Britain (Fig. 2):

1. Syn-depositional coalification occurred during Westphalian A–Stephanian times. Minor early Variscan tectonism may have caused local exceptions to the normal increase in coalification with depth.
2. Variscan deformation folded coals such as to affect isoranks. In addition, if folding was initiated at depth, a pattern of calorific values, moisture contents and structure similar to that shown in Fig. 3 may have been produced. Uplift and peneplanation occurred, exposing high ranking coals in the cores of anticlines.
3. Secondary coalification took place during Permian–Mesozoic times where the combined influences of burial and temperature were greater than those acting during Silesian times. This may have served to broaden the cores of isorank synclines, producing a broad fit to the pattern of previously produced folds.

Estimation of extent, thickness and nature of former total overburden to target intervals in Britain has been made difficult by Permo-Carboniferous and subsequent phases of basin inversion and regional uplift. Recent interpretations suggest a significant Permian to Mesozoic cover to areas of currently exposed Palaeozoic strata (Holliday 1993; e.g. 700–1750 m on the northern Pennines and Lake District). Holliday (1993) suggested that earlier estimates of over 3000 m of strata covering such areas (e.g. Green 1986, 1989; Lewis et al. 1992) resulted from the incorrect application of apatite fission track-derived palaeotemperature data to former cover thicknesses.

Magmatic and metasomatic influences on coal rank

South of the WLBM, in South Wales, anomalously high coal ranks (low volatile bituminous and anthracitic coals) are present in the northwest of the coalfield. Trotter (1948) suggested that the pattern of coalification was related to tectonism. But, as noted by Stach et al. (1982: 58), pressure alone is unlikely to contribute significantly to coalification other than to produce frictional heating along fault and thrust planes. Geochemical studies have indicated that mineralization associated with the Variscan orogeny may have contributed to enhanced coal ranks (Davies & Bloxam 1974). Bloxam & Owen (1984) concluded that anthracitization may have resulted from magmatic heat from an igneous source at depths of no more than 3.5 km. In contrast, Gayer et al. (1991) suggested that a hot fluid flux was responsible for anthracitization.

Evidence from northeast England and Scotland has confirmed the influence of deep-seated plutons, such as the Weardale granite (Bott 1967; Creaney 1980; British Geological Survey 1987; Evans et al. 1988) and hypabyssal intrusions (Creaney 1980; Creedy 1985, 1988) on the degree of coalification.

Coal cleat

Coal cleat generally forms perpendicular to bedding planes, providing pathways for fluids and gases to migrate along and out of the coal as coalification proceeds. Indeed, it has been suggested by Patching (1965) that the movement of gas in coal is almost entirely by flow along fissures rather than diffusion. Cleat is formed in all types of humic coal from subbituminous to anthracitic ranks. The orientation and intensity of cleat development is related to the degree of coalification and the intensity of stress at the time of formation as well as the maceral and mineral content of the coal. The nature of coal cleat determines its permeability. To promote fluid conductivity and hence degasification in CBM production, coal cleats should be open, i.e. they should contain little or no mineralization.

Creedy (1988) suggested that coal cleats in the East Pennine Coalfield were mineralized during the Variscan orogeny and that subsequent gas migration was strongly retarded due to reduced permeability. However, the nationwide applicability of such claims remains to be tested. Indeed, our field observations indicate that cleat mineralization is generally only patchily developed in many British coalfields.

Importance of in-seam gas pressure

A simple correlation of increasing methane volumes with increasing rank cannot be applied comprehensively to British CBM targets. Although the rank of coal reflects the maximum

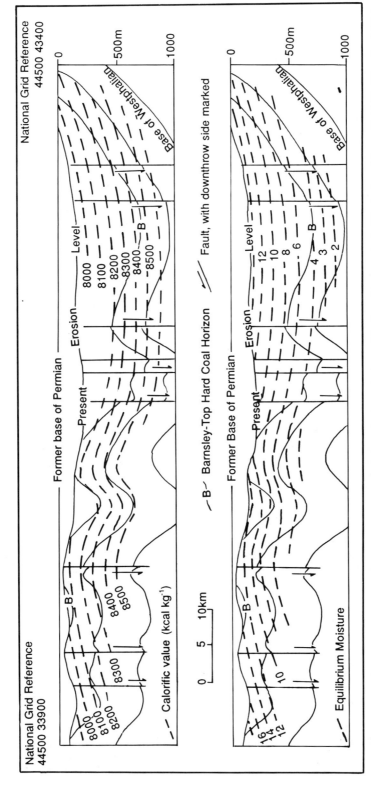

Fig. 3. Cross-sections showing structure, calorific values and equilibrium moisture to the base of the Permian strata in the Nottinghamshire coalfield (East Pennines). Note how both the calorific value and moisture content broadly mimic the structure (after Suggate 1976).

temperature attained during burial, it is the in-seam gas pressure during coalification and subsequent confining gas pressure up to the present day that dictate the amount of preserved methane. Kaiser (1993) suggested that CBM content of coals correlates better with reservoir pressure than with coal rank. In several US basins, artesian overpressuring is most conducive to high gas yields. Paradoxically, although low permeability, underpressured coals may produce less gas during depressurization, they also produce less water. This reduced cost of water disposal may offset the loss of gas revenue.

The apparent difficulties in measuring in-seam gas pressures have led to the use of indirect calculations (Creedy 1985, 1988). With the exception of the coalfield in the vicinity of Point of Ayr Colliery in North Wales, British coals are reported to be underpressured. This means that their gas sorption capacities are less than would normally be expected if the coal were permeable. Ayers *et al.* (1993) pointed out that incorrect gas pressure estimates may have resulted from inaccurate gas content measurements, the incorrect assessment of gasification history and, significantly, the assumption of there being no fluids other than gas present in the coal seam.

High heat flow in northern England and parts of the Midland Valley of Scotland resulted in the attainment of relatively high coal rank. However, the shallow burial depths, and hence lithostatic pressures, meant that coal gas sorption capacities were relatively low and therefore, much of the methane formed during coalification would have escaped, assuming a relatively permeable system.

In contrast, in South Wales, the high methane content indicates that coalification took place under significant confining pressures. Gayer *et al.* (1991) and White (1991) suggested that fluid overpressuring was required within coal seams to enable the style of thrusting to form. Overpressuring, and hence coalification and methane generation, must have occurred before the main folding, as coal isovols parallel the fold pattern of the coalfield (White 1991).

High coal ranks also occur in the Kent coalfield. However, these do not appear to be associated with any obvious structural complexity comparable with the South Wales Coalfield and also have significantly lower gas contents within the coals. Smith (1993) concluded that the over-all pattern of coalification south of the WLBM reflected differences in the Late Carboniferous geothermal gradient as the rapid increase in maturities with depth is unlikely to have been related to burial.

Post-depositional structural controls

Basin inversion

Basin inversion and uplift causes depressurization of coal and, assuming a relatively permeable system, accompanying degasification. The longer the period of time spent at shallower depths, the greater the amount of gas that will escape from the coals.

There have been at least three major erosional episodes in Britain since the deposition of the coal-bearing strata (Fig. 2). In the East Pennine Coalfield, significant degasification appears to have taken place before the deposition of Permian strata (Creedy 1988). The methane content of coals is generally close to or at zero directly beneath Permian strata and, in general, this Variscan phase of uplift and erosion was likely to have been responsible for the most significant loss of methane from coals. In Kent, where coals are mainly of low volatile bituminous rank, Variscan inversion led to the exposure of the coalfield throughout Permian and Triassic times and intermittent exposure throughout Jurassic times. This apparently caused significant degasification (Creedy 1988). Evidence for this gas loss is based on the lack of firedamp emissions in the Kent coalfield and measured gas contents taken from Westphalian D coals, which are generally of a much lower gas content than Westphalian A–B coals throughout Britain.

The extent of late Jurassic to early Cretaceous (late Cimmerian) uplift on the coalfields of Britain is hard to assess due to the paucity of Mesozoic cover. However, it is possible that significant degassing may have taken place at this time in some coalfields (Fig. 2).

Early Cenozoic conversion and uplift was responsible for the present day configuration of strata in Britain and resulted in further degasification of coals as seams were depressurized during uplift. The apparent impermeability of British coals (Creedy 1988) may have retarded rapid degasification at this stage.

Faulting

Significant normal, reverse and transcurrent faulting occurred subsequent to the deposition of coal-bearing strata. Fault movements have resulted in various degrees of compartmentalization of CBM target successions. With the exception of the eastern Pennines, hectametre-scale faulting is commonplace. Larger scale faults commonly delineate the edges of coalfields. Fault trends are generally controlled by the basement grain. Variscan thrusting south of the WLBM in South Wales, Bristol–Somerset, Oxford and possibly Kent coalfields, has also resulted in the compartmentalization of prospective CBM target sequences (Cornelius et al. 1993). In areas where the primary porosity in coal is low, it may be advantageous to identify areas where fracturing has produced secondary porosity.

Hydrogeological considerations

Experience from the US has shown the importance of a thorough understanding of the hydrogeology of a CBM target area (Kaiser et al. 1991). More permeable coals tend to produce better CBM reservoir performance (e.g. Sparks et al. 1993). A knowledge of Coal Measures water chemistry and measurement of total dissolved solids in a CBM target area will give vital evidence pertaining to the coal permeability and degree of recharge.

In CBM wells, it is often necessary to pump water from completed target intervals to reduce hydrostatic pressure within the coals and enable methane desorption. Several variables affect the volume and length of time of water production. Multiple zone completions will produce more water than single zones and permeable strata, such as sandstones, within zones will also increase the amount of water produced. Consequently, the hydrogeological regime within each zone and the degree of connection between different zones are most important in assessing pumping times and rates. Development of CBM in the USA has shown that it can be up to six months before methane is produced and up to three years before peak production occurs. Maximum methane production rates for wells in the USA are about $66\,000\,m^3$ each day in the San Juan Basin, although more typical rates are $6{-}17\,000\,m^3$ each day.

Unfortunately, there is a lack of hydrogeological data for Westphalian strata of Britain as they do not contain any major aquifers. However, hydrogeological maps covering most of the country have been produced by the British Geological Survey. These contain general data such as potentiometric levels and basic hydrochemistry. A review of groundwater conditions in the UK is provided by Downing et al. (1987).

The poorly permeable nature of most of the Westphalian strata in Great Britain and the retention of highly saline formation waters (e.g. Tonks et al. 1928) suggest that little modern day recharge has occurred. In such successions, pumped water volumes are likely to be small, reducing the costs of disposal. There are, however, notable exceptions to this, such as in parts of South Wales (Ineson 1967).

In situ stress

Measurement of the magnitude of in situ horizontal stress in the subsurface is important to CBM prospectivity in that it may give an indication of how well the natural fractures in coals are likely to transmit gases. Studies in the USA (e.g. Sparks et al. 1993) have highlighted the need for a knowledge of in situ stress to assess accurately CBM reservoir potential and to optimize hydrofracture design. The relationship between cleat orientation and in situ stress can be just as important, if not more so, than resource volume in locating the optimum reservoir potential.

Only sparse accounts of in situ crustal stress data are available for Great Britain. There appears to be a reduction in stress at depths of 700–800 m (Evans 1987; Evans & Brereton 1990). On a more local scale, the magnitude of horizontal stress varies greatly both laterally and vertically and may be significantly influenced by faulting.

Measurements of the in situ crustal stress direction derived from borehole breakout directions indicate a dominant minimum in situ crustal stress direction oriented NE–SW. Comparison of coal cleat trends suggests that the maximum horizontal stress orientations are commonly parallel to the dominant cleat over large parts of Britain (R. S. Ellison, pers. comm. 1993).

How much gas is there?

According to published data, the amount of gas (dominantly methane) within British coal seams is highly variable across the country and is generally lower than similar rank coals in the USA (Creedy 1986, 1988, 1991). Significantly, methods of measurement vary considerably between the USA and Britain (Diamond & Levine 1981; Creedy 1986). However, even if gas content figures have been underestimated, they serve to show the general trends across Britain.

As a general rule, except where thermal metamorphism due to igneous activity has enhanced coal rank, high methane values correspond to those areas of high rank. As has been outlined above, however, post-coalification history can significantly affect the volume of preserved methane while not affecting rank coal. In the Pennine Basin, high values correspond to areas of maximum thickness of Westphalian A–

B strata (e.g. Lancashire). The highest recorded methane values in Britain are found within the anthracitic coals of South Wales. The implications of these values are discussed in the following sections.

Political considerations controlling coalbed methane

According to the 1934 Petroleum (Production) Act of Great Britain, methane is petroleum and therefore the property of the Crown. The point was recently reiterated by the Minister for Energy in a written Commons reply (Hansard 27/7/93, col. 944–945). According to Moorhouse (1992), British Coal 'only has licence to extract methane necessary for the safety of the mine'. Primacy over coal remains unclear. Currently there are no legal precedents covering this issue. This may have increasing significance as the CBM industry grows in Great Britain.

British Coal currently own all of the coal in Great Britain and individual CBM operators negotiate with British Coal for data and permission to drill through the coal seams. With the forthcoming privatization of British Coal, this position may alter.

Legislation

As CBM development could be regarded as 'sustainable' on the grounds that it leaves coal available for future use and generally causes less disturbance to an area than conventional hydrocarbon exploration and production, it could warrant special attention with regard to legislation.

At present, however, there is no legislation specific to the CBM industry. Currently, the entire hydrocarbon industry is controlled by the Department of Trade and Industry under the Petroleum (Production) Act 1934 and the Petroleum (Production) (Landward) Regulations 1991. The regulations provide for separate licences for the exploration, appraisal, and production stages of resource exploitation. Initial exploration licences are granted for six years, during which time a maximum of 90 days may be spent testing. Activities carried out under the licences are subject to normal planning processes and access to land must be obtained from landowners or occupiers. Notification of any construction work lasting six weeks or more must be given to the Health and Safety Executive. To drill a well, the permission of the Secretary of State is required, along with evidence of planning permission. Regulations relating to onshore licensing in Great Britain are currently being reviewed and may help unify and clarify the position with regard to CBM development.

The drilling of boreholes for CBM exploration and production also requires the consent of the National Rivers Authority under S.32(3) of the Water Resources Act 1991 and a subsequent abstraction licence is required for the produced water (C. Thomas, pers. comm. 1993).

Development of CBM is controlled under the Town and Country Planning Act 1990, where the term 'development' includes building, engineering, mining or other operations in, over or under land. Planning permission is usually obtained through the relevant county council or metropolitan district council and consequently has to be dealt with on a site-specific basis. If successful, planning permission may be granted unconditionally, or with conditions attached pertaining to site access, landscaping or site aftercare and restoration. Such conditions may vary according to factors such as land use at the site or population density of the area.

Water disposal

The production phase of a CBM well involves pumping of water of variable quality and quantity, the disposal of which may prove problematic. This water is classified as controlled waste of industrial origin and its disposal is controlled by the Control of Pollution Act 1974 (and future regulations under the Environmental Protection Act 1990) and Collection and Disposal of Waste Regulation 1988. Produced water may be disposed of in several ways (most of which require a licence) depending on its salinity, pH, dissolved solids, etc. It may be added directly to ground or surface waters, which requires a discharge consent from the National Rivers Authority; this is tightly controlled and monitored and is unlikely to be an option for the disposal of CBM water. Alternatively, the water may be released into sewers, which requires a discharge consent from the local water company. Much of the water produced during CBM production in the USA is spread directly on the land. However, in Britain this option is only possible if the addition of the CBM water is proved to be beneficial to agriculture and does not pose an environmental hazard. To meet one of the aforementioned disposal options, the quality of CBM water may be improved by undergoing aeration and sedimentation treatment in ponds. This aids the removal of constituents such as iron or manganese and increases the dissolved oxygen content, whereas membrane desalting removes chlorides and other dissolved solids. Examples of these treatment techniques for CBM water in coal basins of Alabama have been outlined in Davis et al. (1993).

If the produced water is not of a suitable quality for discharge to sewers or ground or surface water it may be removed by tanker from the site by registered carrier and disposed of in, for example, a landfill site or water treatment plant. This option does not require the water producer to obtain a licence, but is self-regulatory and operates under a Duty of Care which requires that all reasonable prudence is taken by waste producers, carriers, treaters and disposers to ensure the legal disposal of this waste. It should be noted that water disposal by tankers during pre-production and production phases would not generally constitute an economically viable option and it is likely that many landfills would not accept large amounts on a regular basis.

In the USA, Ortiz et al. (1993) have studied the feasibility of reinjection into deep, naturally fractured formations as an alternative method of disposing of CBM water. However, deep injection of water is not encouraged in Britain as many aquifers are deep and their water quality may be adversely affected as a result of this process (C. Thomas, pers. comm. 1993).

Geographical and human constraints

Coalbed methane prospectivity may be controlled to some extent by human and geographical factors, many of which may manifest themselves during the planning permission application process and may be dealt with as the relevant county council sees fit. These human considerations may include, for example, noise pollution arising from drilling, the visual impact of the site and the effects of works traffic in populated areas. Other influences such as complex land ownership and high population densities over most of the country, together with 'green issues' may be important factors in curtailing the large-scale development of the CBM industry in Great Britain.

The past, present and future extent of coal mining is also an important factor in CBM prospectivity. Extensive mining in the past will have removed much of the coal that hosts the CBM, although some methane may be preserved in the collapse zone behind the working face. If coal extraction and CBM extraction proceed simultaneously, problems may arise from the spacing of gas drainage wells, with both parties requiring a different configuration of wells. Additionally, coal mining may require the removal of gas from mines for safety reasons, thus reducing the CBM available for the CBM operator. Finally, the fear of stimulation rendering tracts of coal unmineable cannot be substantiated (Lewin et al. 1993). Indeed, inspection of fractures induced by stimulation which have been later intercepted by underground mining have indicated no adverse impact on mining (Steidl 1993).

Geographical constraints on CBM prospectivity may vary from the current land use of the potential site to the relative proximity to the consumer of the produced gas. The former, again, would be provided for under the planning permission process and the importance of the latter depends on the destination of the produced gas. If the gas is to be fed into the national transmission system, proximity to a pipeline is important. However, such gas may require calorific upgrading, a process which may not be economically viable. Alternatively, the gas may be destined for direct use or for electricity generation at or near the site. Ultimately, the development of CBM at any site depends on the economies of the local energy demand being sufficient to make the venture viable.

Identifying coalbed methane prospects in Britain

Initial identification of CBM prospects has been governed largely by: (1) the extent of mine workings; (2) the depth of the target sequence, preferably less than 1500 m (Fig. 1C); (3) the volumes of coal; and (4) coal rank and measured gas content (Fig. 1D and 1E).

Thity-four CBM licences have been issued in Britain since September 1991. Essentially these cover the main areas of the occurrence of unmined coal-bearing strata at average depths of 800–1000 m, i.e. suitable depths for optimum CBM extraction. Several companies have taken the approach of acquiring licences in different geological settings, but the largest number of licences has been awarded east and west of the Pennines, coinciding with the greatest volumes of unmined coal and relatively high measured CBM values.

Currently an early success is needed in the industry to sustain interest and ensure research aimed at recognizing the full potential of CBM in Great Britain. As drilling and well testing begins, the initial approach outlined above will inevitably give way to more comprehensive CBM appraisal combining all the factors described in the previous sections. Indeed, well site appraisal for individual licence blocks already integrates all the available geological data in ensuring optimum positioning. There is still, however, a large degree of calculated guesswork required at this early stage in CBM development in Britain, largely because of the absence of permeability data. A cross-fertilization of current US and British knowledge is therefore vital to the industry to ensure efficient exploration.

Economic considerations

Low costs are essential to the formation of a sustainable industry. Initially, a lack of dedicated equipment and services in Britain has inflated exploration costs. Added to this is the cost of 'coal rental' from British Coal and the cost of data acquisition. Currently, the estimated cost for each exploration well is around £400 000–500 000. Experience from the US indicates that CBM exploitation is unlikely to produce enough profit to allow for substantial royalties (Knott 1993). Indeed, development in the US has slowed significantly after the removal of Section 29 tax credits, the mechanism by which the CBM industry was 'jump started' in the USA. The lack of similar tax credits in Britain during the infancy of the CBM industry may slow development considerably.

Conclusions

1. Coalbed methane prospectivity in Great Britain is controlled by a combination of geological, geographical and 'political' factors. Ultimately, these will control the cost and hence the rate and extent of CBM exploration and production.
2. A complex history of deposition and basin inversion and uplift in Great Britain has resulted in the formation of a number of distinct structural settings hosting CBM target successions.
3. The degree of basin inversion and uplift during the Variscan orogeny and subsequent deformational events have further controlled the pattern of adsorbed methane volumes across Great Britain. Reactivation of basement lineaments during both extensional and compressional tectonic events has, in places such as the western Pennines, strongly compartmentalized prospective CBM target successions.
4. Identification of a consumer and an adequate infrastructure for development is of prime importance in initial CBM prospecting in any licence area. Also of major importance are planning and environmental issues which have greater implications at the well siting and completion stages of exploration.
5. The extent of former underground mining, depth of CBM targets successions, and, to a lesser extent, rank and measured methane content, are the principal geological factors which have so far controlled the initial choice of licence blocks.
6. Factors such as structural complexity, sedimentology, coal cleat orientation, *in situ* stress, hydrogeology and hydrology are likely to be considered in more detail at the well-siting stage. As a knowledge of British CBM increases, these will become more important in licence acquisition.
7. The current review of regulations relating to onshore licensing in Great Britain may help unify and clarify the position with regard to CBM development.

We are grateful to W. B. Ayers (Taurus Exploration, Inc.), A. Dickinson (EGSL), C. Thomas (NRA), D. W. Holliday, T. J. Charsley and N. J. P. Smith (all BGS) for their comments, criticisms and contributions which have helped to improve greatly the original manuscript. This paper is published with the permission of the Director of the British Geological Survey (NERC).

References

Ayers, W. B., Tisdale, R. M., Litzinger, L. A. & Steidl, P. F. 1993. Coalbed methane potential of Carboniferous strata in Great Britain. *In: Proceedings of the 1993 International Coalbed Methane Symposium, University of Alabama/Tuscaloosa, May 17–21*, 1–9.

Barker, C. E. & Goldstein, R. H. 1990. Fluid-inclusion technique for determining maximum temperature in calcite and its comparison to vitrinite reflectance. *Geology*, **18**, 1003–1006.

Barker-Read, G. R. 1984. *The gas dynamic behaviour of Coal Measures strata with particular reference to west Wales outburst-prone zones*. PhD Thesis, University College, Cardiff.

—— & Radchenko, S. A. 1988. Methane emission from coal and associated strata samples. *International Journal of Mining and Geological Engineering*, **7**, 101–126.

Besly, B. M. 1988. Palaeogeographic implications of late Westphalian to early Permian red-beds. *In:* Besly, B. M. & Kelling, G. (eds) *Sedimentation in a Synorogenic Basin Complex: the Upper Carboniferous of NW Europe*. Blackie, Glasgow, 200–221.

Bloxam, T. W. & Owen, T. R. 1985. Anthracitization of coals in the South Wales Coalfield. *International Journal of Coal Geology*, **4**, 299–307.

Bostick, N. H. 1979. Microscopic measurement of the level of catagenesis of solid organic matter in sedimentary rocks to aid exploration for petroleum and to determine former burial temperatures—a review. *In:* SCHOLLE, P. A. & SCHLUGER, P. R. (eds) *Aspects of Diagenesis.* Society of Economic Paleontologists and Mineralogists, Special Publication, **26**, 17–43.

BOTT, M. H. P. 1967. Geophysical investigations of the Northern Pennine basement rocks. *Proceedings of the Yorkshire Geological Society,* **33**, 1–20.

BRITISH GEOLOGICAL SURVEY 1987. *Hot Dry Rock Potential of the United Kingdom. Investigation of the Geothermal Potential of the UK.* British Geological Survey, Keyworth.

BROMILOW, J. G. 1959. The drainage and utilisation of Firedamp in Great Britain. *Colliery Guardian.* **199**, No. 5132, 61–68, 97–100.

CHADWICK, R. A., HOLLOWAY, S., HOLLIDAY, D. W. & HULBERT, A. G. 1993. The hydrocarbon potential of the Northumberland–Solway Basin. *In:* PARKER, J. R. (ed.) *Petroleum Geology of Northwest Europe: Proceedings of the 4th Conference.* Geological Society, London, 717–726.

CORNELIUS, C. T., HARTLEY, A., GAYER, R. & ROSS, C. 1993. Coal deposition and tectonic history of the South Wales Coalfield, UK: implication for coalbed methane resource development. *In: Proceedings of the 1993 International Coalbed Methane Symposium, University of Alabama/Tuscaloosa, May 17–21,* 161–172.

COWARD, M. P. & SMALLWOOD, S. 1984. An interpretation of the Variscan tectonics of SW Britain. *In:* HUTTON, D. H. W. & SANDERSON, D. J. (eds) *Variscan Tectonics of the North Atlantic Region.* Geological Society, London, Special Publication, **14**, 89–102.

CREANEY, S. 1980. Petrographic texture and vitrinite reflectance variation on the Alston Block, northeast England. *Proceedings of the Yorkshire Geological Society,* **42**, 553–580.

CREEDY, D. P. 1983. Seam gas-content data-base aids firedamp prediction. *The Mining Engineer,* **143**, August, 79–82.

—— 1985. *The origin and distribution of firedamp in some British coalfields.* PhD Thesis, University College, Cardiff.

—— 1986. Methods for the evaluation of seam gas content from measurements on coal samples. *Mining Science and Technology,* **3**, 141–160.

—— 1988. Geological controls on the formation and distribution of gas in British Coal Measures strata. *International Journal of Coal Geology,* **10**, 1–31.

—— 1991. An introduction to geological aspects of methane occurrence and control in British deep coal mines. *Quarterly Journal of Engineering Geology,* **24**, 209–220.

DAVIES, M. & BLOXAM, T. W. 1974. The geochemistry of some South Wales coals. *In:* OWEN, T. R. (ed.) *The Upper Palaeozoic and Post-Palaeozoic Rocks of Wales.* University of Wales Press, Cardiff, 225–261.

DAVIS, H. A., SIMPSON, T. E., LAWRENCE, A. W., MILLER, J. A. & LINZ, D. G. 1993. Coalbed methane produced water management strategies in the Black Warrior Basin of Alabama. *In: Proceedings of the 1993 International Coalbed Methane Symposium, University of Alabama/Tuscaloosa, May 17–21,* 317–338.

DIAMOND, W. P. & LEVINE, J. R. 1981. Direct method determination of the gas content of coal; procedures and results. *US Bureau of Mines Report of Investigations,* **8515**.

DOWNING, R. A., EDMUNDS, W. M. & GALES, I. N. 1987. Regional groundwater flow in sedimentary basins in the UK. *In:* GOFF, J. C. & WILLIAMS, B. P. J. (eds) *Fluid Flow in Sedimentary Basins and Aquifers.* Geological Society, London, Special Publication, **34**, 105–125.

DUNHAM, K. C. & POOLE, E. G. 1974. The Oxfordshire Coalfield. *Journal of Geological Society, London,* **130**, 387–391.

EVANS, C. J. 1987. *Crustal Stress in the United Kingdom. Investigation of the Continental Potential of the UK.* British Geological Survey, Keyworth.

—— & BRERETON, N. R. 1990. In situ crustal stress from borehole breakouts. *In:* HURST, A., LOVELL, M. A. & MORTON, A. C. (eds) *Geological Application of Wireline Logs.* Geological Society, London, Special Publication, **48**, 327–338.

——, KIMBELL, G. S. & ROLLIN, K. E. 1988. *Hot Dry Rock Potential in Urban Areas. Investigation of the Geothermal Potential of the UK.* British Geological Survey, Keyworth.

FOSTER, D., HOLLIDAY, D. W., JONES, D. M., OWENS, B. & WELSH, A. 1989. The concealed Upper Palaeozoic rocks of Berkshire and Oxfordshire. *Proceedings of the Geological Society,* **100**, 395–407.

FRANCIS, E. H. 1991. Carboniferous. *In:* CRAIG, C. Y. (ed.) *Geology of Scotland.* Geological Society, London, 347–392.

FRASER, A. J., NASH, D. F., STEELE, R. P. & EBDON, C. C. 1990. A regional assessment of the intra-Carboniferous play of northern England, *In:* BROOKS, J. (ed.) *Classic Petroleum Provinces.* Geological Society, London, Special Publication, **50**, 417–440.

GAYER, R., COLE, J., FRODSHAM, K., HARTLEY, A. J., HILLIER, M., MILIOROZOS, M. & WHITE, S. 1991. The role of fluids in the evolution of the South Wales Coalfield Foreland Basin. *Proceedings of the Ussher Society,* **7**, 380–384.

GLOVER, B. W., CORFIELD, S. M. & WATERS, C. N. 1990. Contrasting structural styles across the southern margin of the Pennine Basin, English Midlands. *Terra Abstracts,* **3**, 202.

——, HOLLOWAY, S. & YOUNG, S. R. 1993*a*. Geological controls on coalbed methane resources in Great Britain. *In: Proceedings of the 1993 International Coalbed Methane Symposium, University of Alabama/Tuscaloosa, May 17–21,* 741–746.

——, POWELL, J. H. & WATERS, C. N. 1993*b*. Etruria Formation (Westphalian C) palaeoenvironments and volcanicity on the southern margins of the Pennine Basin, South Staffordshire, England. *Journal of the Geological Society,* **150**, 737–750.

GREEN, P. F. 1986. On the thermo-tectonic evolution of Northern England: evidence from fission track analysis. *Geological Magazine*, **123**, 493–506.
—— 1989. Thermal and tectonic history of the East Midlands shelf (onshore UK) and surrounding regions assessed by apatite fission track analysis. *Journal of the Geological Society of London*, **146**, 755–733.
GUION, P. D. & FIELDING, C. R. 1988. Westphalian A and B sedimentation in the Pennine Basin, UK. *In:* BESLY, B. M. & KELLING, G. (eds) *Sedimentation in a Synorogenic Basin Complex: the Upper Carboniferous of NW Europe*. Blackie, Glasgow, 178–199.
HOLLIDAY, D. W. 1993. Mesozoic cover over northern England: interpretation of apatite fission track data. *Journal of the Geological Society of London*, **150**, 657–661.
INESON, J. 1967. *Groundwater Conditions in the Coal Measures of South Wales Coalfield.* Water Supply Paper, Institute of Geological Sciences, Hydrogeological Report No. 3.
JONES, J. 1989. The influence of contemporaneous tectonic activity on Westphalian sedimentation in the South Wales Coalfield. *In:* ARTHURTON, R. S., GUTTERIDGE, P. & NOLAN, S. C. (eds) *The Role of Tectonics in Devonian and Carboniferous Sedimentation in the British Isles*. Yorkshire Geological Society, Occasional Publication, **6**, 243–253.
—— 1991. A mountain model for the Variscan deformation of South Wales Coalfield. *Journal of the Geological Society of London*, **148**, 881–889.
KAISER, W. R. 1993. Abnormal pressure in coal basins of the Western United States. *In: Proceedings of the 1993 International Coalbed Methane Symposium, University of Alabama/Tuscaloosa, May 17–21*, 173–179.
——, AYERS, W. B., AMBROSE, W. A., LAUBACH, S. E., SCOTT, A. R. & TREMAIN, C. M. 1991. Geologic and hydrogeologic characterization of coalbed methane production, Fruitland Formation, San Juan Basin. *In:* AYERS, W. B., KAISER, W. R. *et al.* (eds) *Geologic and Hydrologic Controls on the Occurrence and Producibility of Coalbed Methane, Fruitland Formation, San Juan Basin*. The University of Texas at Austin Topical Report, Gas Research Institute, Contract No. 5087-214-1544 (GRI-91/00072). Gas Research Institute, Illinois, 97–119.
KELLAWAY, G. A. & WELSH, F. B. A. 1993. *The Geology of the Bristol District.* British Geological Survey, Memoir, 1:63 360 Geological Special Sheet (England and Wales). HMSO, London.
KELLEY, J. L. 1989. Coalbed methane: from nuisance to new source. *Gas Research Institute Digest*, **12**, 3–13.
KELLING, G. 1974. Upper Carboniferous sedimentation in South Wales. *In:* OWEN, T. R. (ed.) *The Upper Palaeozoic and Post-Palaeozoic Rocks of Wales*. University of Wales Press, Cardiff, 185–224.
—— 1988. Silesian sedimentation and tectonics in the South Wales Basin: a brief review. *In:* BESLY, B. M. & KELLING, G. (eds) *Sedimentation in a Synorogenic Basin Complex: the Upper Carboniferous of NW Europe*. Blackie, Glasgow, 38–42.
KNOTT, D. 1993. U.K. missing out on coalbed methane. 1993. *Oil & Gas Journal*, **July 12**, 34.
KUUSKRAA, V. A., BOYER, C. M. & KELEFANT, J. A. 1992. Hunt for quality basins goes abroad. *Oil & Gas Journal*, **October 5**, 49–54.
LEEDER, M. R. 1982. Upper Palaeozoic basins of the British Isles—Caledonide inheritance versus Hercynian plate margin processes. *Journal of the Geological Society of London*, **139**, 479–491.
—— 1987. Tectonic and Palaeogeographic models for Lower Carboniferous Europe. *In:* MILLER, J., ADAMS, A. E. & WRIGHT, V. P. (eds) *European Dinantian Environments*. Wiley, Chichester, 1–20.
—— 1988. Recent developments in Carboniferous geology: a critical review with implications for the British Isles and N.W. Europe. *Proceedings of the Geologists' Association*, **99**, 73–100.
—— & HARDMAN, M. 1990. Carboniferous geology of the Southern North Sea Basin and controls on hydrocarbon prospectivity. *In:* HARDMAN, R. F. P. & BROOKS, J. (eds) *Tectonic Events Responsible for Britain's Oil and Gas Reserves*. Geological Society, London, Special Publication, **55**, 87–105.
—— & MCMAHON, A. H. 1988. Upper Carboniferous (Silesian) basin subsidence in northern Britain. *In:* BESLY, B. M. & KELLING, G. (eds) *Sedimentation in a Synorogenic Basin Complex: the Upper Carboniferous of NW Europe*. Blackie, Glasgow, 43–52.
LEWIN, J. L., SIRIWARDANE, H. J. & AMERI, A. 1993. New perspectives on the indeterminacy of coalbed methane ownership. *In: Proceedings of the 1993 International Coalbed Methane Symposium, University of Alabama/Tuscaloosa, May 17–21*, 305–316.
LEWIS, C. L. E., GREEN, P. F., CARTER, A. & HURFORD, A. J. 1992. Elevated K/T palaeotemperatures throughout Northwest England: three kilometres of Tertiary erosion? *Earth and Planetary Science Letters*, **112**, 131–145.
MITCHELL, C. 1990. Coal-bed methane in the UK. *Energy Policy*, **November**, 1990.
MOORHOUSE, J. S. 1992. Mines gas fired power plants. *Mining Technology*, **April**, 99–101.
NATIONAL COAL BOARD 1979. *The Coalfields of Great Britain—Location of Main Classes of Coal.* Hobart House, London.
ORTIZ, I., WELLER, T. F., ANTHONY, R. V., FRANK, J., LINZ, D. & NAKLES, D. 1993. Disposal of produced waters: underground injection in the Black Warrior Basin. *In: Proceedings of the 1993 International Coalbed Methane Symposium, University of Alabama/Tuscaloosa, May 17–21*, 339–352.
OWEN, T. R. & WEAVER, J. D. 1983. The structure of the South Wales Coalfield and its margins. *In:* HANCOCK, P. L. (ed.) *The Variscan Fold Belt in the British Isles*. Hilger, Bristol, 47–87.
PATCHING, T. H. 1965. Variations in permeability of coal. *In: Proceedings of the Rock Mechanics Symposium*, Mines Branch, Department of Mines and Technical Surveys, Ottawa, Canada, 185–199.

POOLE, E. G. 1988. The concealed coalfield of Canonbie: comment. *Scottish Journal of Geology*, **24**, 305–306.

PRICE, L. C. 1983. Geologic time as a parameter of organic metamorphism and vitrinite reflectance as an absolute palaegeothermometer. *Journal of Petroleum Geology*, **6**, 5–38.

RAMSBOTTOM, W. H. C., CALVER, M. A., EAGAR, R. M. C., HODSON, F., HOLLIDAY, D. W., STUBBLEFIELD, C. J. & WILSON, R. B. 1978. *A Correlation of Silesian Rocks in the British Isles*. Geological Society, London, Special Report, **10**.

READ, W. A. 1988. Controls on Silesian sedimentation in the Midland Valley of Scotland. *In:* BESLY, B. M. & KELLING, G. (eds) *Sedimentation in a Synorogenic Basin Complex: the Upper Carboniferous of NW Europe*. Blackie, Glasgow, 178–199.

ROBINSON, N. & GRAYSON, R. 1990. Natural methane seepages in the Lancashire Coalfield. *Land & Minerals Surveying*, **8**, 333–340.

SCOTT, J. & COLTER, V. S. 1987. Geological aspects of current onshore Great Britain exploration plays. *In:* BROOKS, J. & GLENNIE, K. W. (eds) *Petroleum Geology of North West Europe*. Graham & Trotman, London, 95–107.

SHEPHARD-THORN, E. R. 1988. *Geology of the Country Around Ramsgate and Dover*. British Geological Survey, Memoir, Sheets 274 and 279 (England and Wales). HMSO, London.

SMITH, N. J. P. 1993. The case for exploration of deep plays in the Variscan fold belt and its foreland. *In:* PARKER, J. R. (ed.) *Petroleum Geology of Northwest Europe: Proceedings of the 4th Conference*. Geological Society, London, 667–675.

SPARKS, D. P., LAMBERT, S. W. & McLENDON, T. H. 1993. Coalbed gas well flow performance controls, Cedar Cove Area, Warrior Basin, USA. *In: Proceedings of the 1993 International Coalbed Methane Symposium, University of Alabama/Tuscaloosa, May 17–21*, 529–548.

STACH, E., MACKOWSKI, M. Th., TEICHMÜLLER, M., TAYLOR, G. H., CHANDRA, D., TEICHMÜLLER, R. 1982. *Textbook of Coal Petrology*, 3rd Edn. Gebruder Borntraeger, Berlin.

STEIDL, P. F. 1993. Evaluation of induced fractures intercepted by mining. *In: Proceedings of the 1993 International Coalbed Methane Symposium, University of Alabama/Tuscaloosa, May 17–21*, 675–686.

SUGGATE, R. P. 1976. Coal ranks and geological history of the Nottinghamshire–Yorkshire Coalfield. *Mercian Geologist*, **6**, 1–24.

THOMAS, L. P. 1974. The Westphalian (Coal Measures) in South Wales. *In:* OWEN, T. R. (ed.) *The Upper Palaeozoic and Post-Palaeozoic Rocks of Wales*. University of Wales Press, Cardiff, 133–160.

TONKS, L. H., JONES, R. C. B., LLOYD, W. & SHERLOCK, R. L. 1928. *Geology of Manchester and the South-east Lancashire Coalfield*. Geological Survey, England and Wales, Memoir.

TROTTER, F. M. 1948. The devolatilization of coal seams in South Wales. *Quarterly Journal of the Geological Society of London*, **104**, 387–437.

TURTON, F. B. 1981. Colliery explosions and fires; their influence upon legislation and mining practice. *The Mining Engineer*, **141**, 157–164.

ULERY, J. P. 1988. *Geologic Factors Influencing the Gas Content of Coalbeds in Southwestern Pennsylvania*. United States Department of the Interior, Bureau of Mines, Report of Investigations, **9195**.

WATERS, C. N., GLOVER, B. W. & POWELL, J. H. 1994. Structural synthesis of the S. Staffordshire, UK: implications for the Variscan evolution of the Pennine Basin. *Journal of the Geological Society, London*, **151**, 697–713.

WHITE, S. 1991. Palaeo-geothermal profiling across the South Wales Coalfield. *Proceedings of the Ussher Society*, **7**, 368–374.

WILLS, L. J. 1956. *Concealed Coalfields*. Blackie, Glasgow.

Pre-sedimentary palaeo-relief and compaction: controls on peat deposition and clastic sedimentation in the Radnice Member, Kladno Basin, Bohemia

STANISLAV OPLUŠTIL [1] & PETR VÍZDAL [2]

[1] *Department of Mineral Deposits, Faculty of Science, Charles University, Albertov 6, 128 43 Prague 2, Czech Republic*
[2] *Gekon Ltd, Slavojova 12, 120 00 Prague 2, Czech Republic*

Abstract: Pre-sedimentary palaeo-topography was a major control on facies distribution during the deposition of the basal unit (the Radnice Member) in the Kladno Basin. The undisturbed accumulation of peat took place in calm valleys protected by surrounding ridges from the effect of fluvial dynamics, whereas in open depressions only thin or split coals occurred. Facies distribution and the thickness of deposits depend on the palaeo-relief which influenced compaction rates and peat bog development without a major role of local syn-sedimentary tectonism within the study area. Different positions of both the maximum thicknesses of the clastic interseam intervals and lines of splitting of the coals have been documented. The thickest coals (up to 7 m) originated near the base of the Carboniferous deposits in the axes of depressions, either locked or protected by the adjacent ridges. They are underlain mostly by fine-grained sediments, responsible for an even compaction rate compensated by peat accumulation. Stratigraphically higher coals have maximum thicknesses of only 2.5 m at the depression margins or in shallow depressions. Towards the axes of the main depressions they are either split or thinner due to the rapid compaction rate of a thick sequence of fine-grained sediments and organic matter. Moreover, the drainage pattern changed as a result of the burial of relief. Owing to compaction, however, the river systems occupied routes along the palaeo-valley axes. No mineable coals occur in the area underlain by a thick sequence of coarse-grained sediments as a result of weak compaction.

The deposition of basal Upper Carboniferous sediments in central and western Bohemia was controlled by the pre-sedimentary relief and compaction. Their influence on deposition has been demonstrated by many workers (Klener 1973, 1975, 1982; Havlena & Pešek 1980). Some of these results, confirming the dependence of the quality and distribution of coals on the depocentre palaeo-topography, have been successfully applied to the exploration and evaluation of coal-bearing deposits outside the mining area (Spudil, unpublished data 1992).

An attempt has been made to evaluate the influence of pre-sedimentary relief on facies distribution to identify the sedimentological controls (compaction and syn-sedimentary tectonism) of peat formation in the southern part of the Kladno Basin, central Bohemia (Fig. 1).

Stratigraphy and geological setting

The Kladno Basin belongs to a WSW–ENE elongated complex of late Carboniferous basins in central and western Bohemia. The Carboniferous sediments unconformably overlie the late Proterozoic basement, composed mainly of folded shales and occasionally volcanic rocks and cherts.

The basinal deposits are divided into four lithostratigraphic formations (Fig. 2) and two of them are further divided into members. The deposition of the oldest unit, the Radnice Member, began around the Westphalian B/C boundary in a tectonically established depocentre with a NNE–SSW trending axis. In the Kladno Basin there are as many as five mineable coals within the Radnice Member. All except the lowest seam are exploited commercially (Fig. 2). There are also three volcanic layers (tuffs and tuffaceous rocks) between the coals and tonsteins within most of the coals. Structural influences caused the bottom of the depocentre to form a system of ridges and valleys with maximum elevation differences of up to 150 m. The relief was established mainly during the Viséan (Holub *et al.* 1991) by an intensive denudation of the Variscan mountain ridge and influenced deposition in the lowermost part of the following unit (the Nýřany Member). Moreover, after burial and compaction the

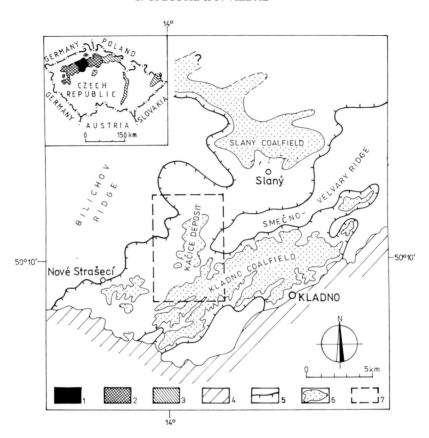

Fig. 1. General view of the distribution of Upper Carboniferous deposits in the Czech Republic and location of the study area within the southern part of the Kladno Basin, Central Bohemia, after Klener (1982). 1, Kladno Basin; 2, Carboniferous deposits of western and central Bohemia; 3, Carboniferous deposits elsewhere in the Czech Republic; 4, outcrop of the Upper Proterozoic basement; 5, distribution of the Radnice Member; 6, major coalfields; 7, study area.

effect of the pre-sedimentary relief was felt almost up to the Týnec Formation (Spudil, unpublished data 1992).

Methods of study

Data used in this study were derived from 60 boreholes drilled from the surface to the basement and from about 400 subsurface boreholes drilled through part of the Radnice Member (Fig. 3). Some data were also derived from galleries.

The construction of the pre-sedimentary relief is based on the occurrence of a widespread isochronous horizon, e.g. a tuff layer or coal seam. The method uses sections where the interval between the basement and the isochronous horizon is preserved without tectonic reduction. Isopach maps of this interval express a negative topography relative to the original relief. A real relief was obtained by subtracting the interval thickness from a selected constant.

The base of the upper Radnice Seam can be considered as a suitable horizon in the southern part of the study area. In the northern part, where the Upper Radnice coal is absent, the base of the 'Green tuff' marker bed (Fig. 2) was used. Unification of both parts in a simple model was possible in the area of overlap of the two horizons by comparison of contour lines based on these horizons (Fig. 4).

The use of this method eliminates the influence of both the post-sedimentary tectonism and secondary dips of strata (usually between 5° and 8° to the NE in the study area). However, the influence of compaction, syn-sedimentary movements (not identified in the study area) and erosion of the basement elevations before burial by the sediments are not included.

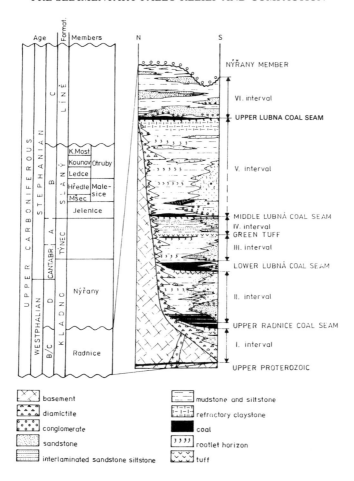

Fig. 2. Stratigraphy of the Kladno Basin and an idealized cross-section through the Radnice Member within the study area.

The commonly used decompaction methods are not applicable in the study area because of the absence of data on sediment porosity. In contrast, due to the partial compaction of the basal deposits during sedimentation of the relevant interval, a restricted influence of compaction is assumed. Some compaction models show a rapid decrease in porosity in the first few metres of burial. The model of Teodorovich & Chernov (1968) suggests a decrease in the porosity of clay sediments of about 20–25% at a depth of 8–10 m.

Erosion of palaeo-relief during the sedimentation of the Radnice Member was estimated from the rates of both denudation and deposition. The average intensity of erosion in the Bohemian Massif during the late Carboniferous was calculated by Kukal (1984) at $30 \, \text{cm}/10^3 \, \text{a}$. The depositional rate of clastic sediments ranges from 1 m in five to ten years (as documented by rooted upright fossil trees) to 1 m in 350 to 900 years (seasonal sedimentation; Skoček 1968). As a result of considerable local deviations of both denudation and deposition, the estimates of erosion of the highest elevations (about 150 m) vary greatly between 10 and 30 m.

Pre-sedimentary relief in the southern part of the Kladno Basin

The Kladno depression, elongated WSW–ENE, and an embayment to the NNE, called the Kačice depression, are the principal negative morphological elements of the study area and its surroundings (Figs 1 and 5). There are two main elevations: the Bilíchov ridge in the west and the Smečno-Velvary ridge in the east. The former is almost isometric, delineating the distribution of

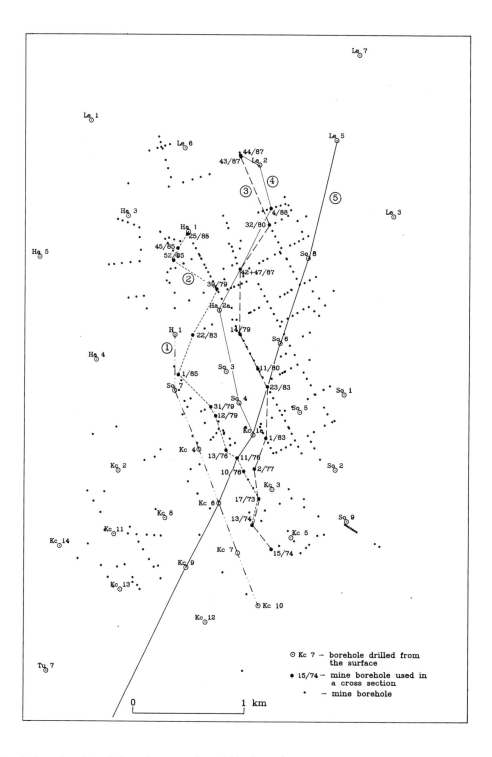

Fig. 3. Location of boreholes and cross-sections within the study area.

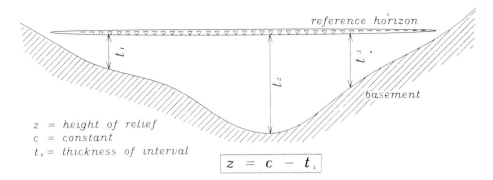

z = height of relief
c = constant
t_x = thickness of interval

$$z = c - t_x$$

Fig. 4. Reconstruction of palaeotopography using the interval thickness (t_x) between the reference horizon and the basement. The real relief is composed of the values (z) obtained by subtracting the interval thickness (t_x) from a constant (c).

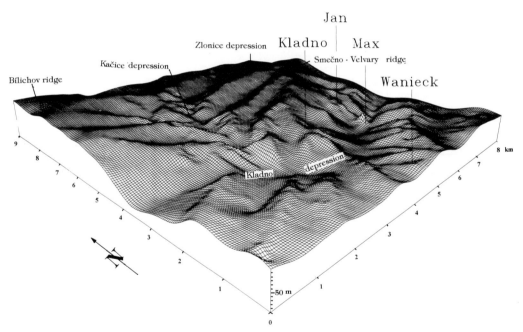

Fig. 5. Block diagram of the pre-sedimentary relief in the study area and its surroundings. The names of the most significant morphological elements and coal mines are given. Vertical exaggeration × 5.

the Radnice Member towards the northwest, and the latter, elongated in a WSW–ENE direction, divides the Kladno depression in the south from the Zlonice depression in the north. The connection between them through the Kačice depression has been documented. Adjacent protrusions of both ridges could have different trends; however, the SW–NE predominates.

Facies analysis and the depositional environment of the Radnice Member

The Radnice Member has been divided into six sub-isochronous intervals defined by coals or tuff layers. For each interval, isopachs and contour lines of percentage conglomerates and medium- to coarse-grained sandstones were constructed. Deposits of high viscosity flows

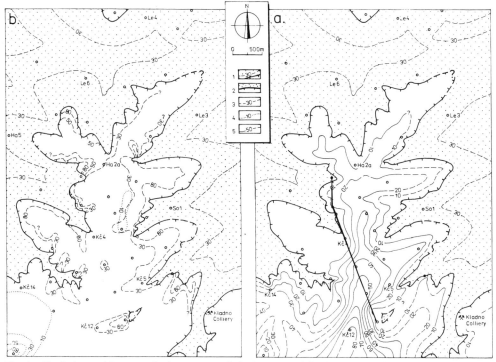

Fig. 6. Clastic interval between the basement and the Upper Radnice Seam: (**a**) thickness; (**b**) percentage of coarse-grained sediments. 1, Sedimentary area at the beginning of the interval of deposition; 2, sedimentary area at the end of the interval of deposition; 3, isopachs; 4, isolines of percentage of conglomerate and medium- to coarse-grained sandstones; 5, isolines of percentage of diamictites.

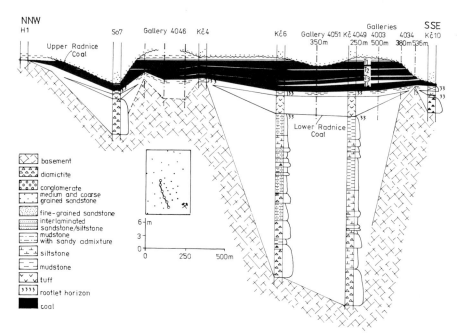

Fig. 7. North–south cross-section across the study area showing the interval between the basement and the Upper Radnice Seam. Distribution of tonsteins (T_1–T_4) within the coal indicates both post-sedimentary erosion and gradual transgression of peat bog. For location, see Figs 3 (cross-section 1) and 6.

(mud flows and debris flows composed of unsorted local material) were counted separately.

Considering that the density of boreholes is high, this method of analysis allows the determination of the influence of relief on facies distribution and sedimentary body geometry. It also enables an examination of the influence of both the compaction and syn-sedimentary tectonics on peat deposition.

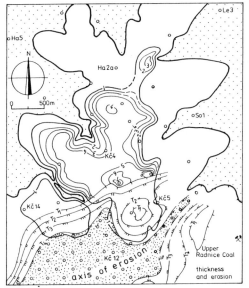

Fig. 8. Thickness and erosion (below the levels of the tonsteins T_1–T_4) of the Upper Radnice Seam.

Interval between the basement and the Upper Radnice Seam

Description. As seen from Fig. 6, the thickness of this interval varies between 0 and 80 m. The maximum thickness corresponds to the axis of the Kačice depression.

The whole interval is dominated by non-laminated, mainly massive, mudstones or siltstones, locally with a sandy admixture (Fig. 7). They are grey in colour, often mottled red and green. No fossil remains occur within these sediments. Only a rooted horizon and a thin coal (Lower Radnice Seam) occur in the upper part of the interval. The horizon is overlain by up to 4 m of a thick volcanic layer (Whetstone Horizon).

Along the margins of the depressions there are diamictites, usually found as fan-like or tongue-like bodies. The diamictites are unsorted, composed mainly of angular to subrounded clasts of local material: quartz, cherts, rhyolite and Upper Proterozoic shales up to 15 cm in size. They exhibit both matrix-supported and clast-supported textures. Within the interval, diamictites are up to 25 m thick, intercalated with the above-mentioned sediments towards the axis of depression. The interval is overlain by the Upper Radnice Seam with a maximum thickness of 7 m (Fig. 8). The thickness of coal suddenly decreases on the palaeo-slopes. Therefore the entire area occupied by the Upper Radnice Seam does not significantly exceed the area of mineable thickness. Adjacent ridges divide the area of coal into several embayments. Sedimentary partings occur mainly along the palaeo-slopes.

Depositional environments. Lacustrine environments dominated during the first interval with the deposition of mudstones and siltstones. The depth of the lake reached several metres, as documented by the absence of any plant remains. During this interval, debris flows and mud flows may have been generated on the palaeo-slopes of the Bílichov and Smečno-Velvary ridges, and they were transported towards the palaeo-valleys and were deposited as the gradient decreased. The lake gradually shallowed and locally changed into mires (Lower Radnice Seam) with volcanic activity before the end of the interval. The deposition of a thick tuff layer induced a further significant decrease in depth and increase in area of the lake, which changed into a peat bog. Peat deposition began in the southern part of the Kačice depression and gradually spread to the north. This can be deduced from the Upper Radnice Seam tonsteins gradually approaching the base of this coal (Fig. 7). An extensive peat accumulation was disturbed along the slopes of palaeo-relief by mud flows and debris flows. The compaction of fine-grained sediments in the Kačice and the Kladno depressions played an important part in long-term peat accumulation.

Interval between the Upper Radnice Seam and the Lower Lubná Seam

Description. As seen in Fig. 9, the maximum thickness (up to 20 m) is partly related to both the axis of the Kačice depression and the increased percentage of medium-grained sandstones to conglomerates. In the south the percentage isolines of conglomerates to medium-grained sandstones run WSW–ENE.

The Upper Radnice Coal is directly overlain

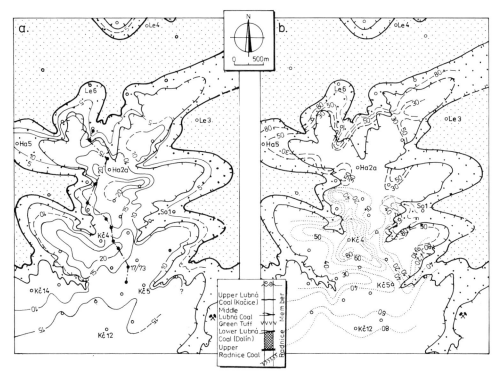

Fig. 9. Clastic interval between the Upper Radnice Seam and the Lower Lubná Seam. (**a**) Isopachs of the interval in metres; (**b**) percentage of coarse-grained sediments. For key, see Fig. 6.

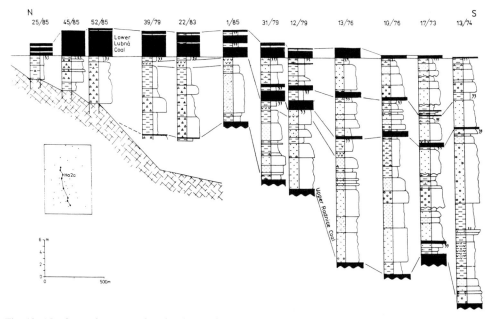

Fig. 10. North–south cross-section showing the interval between the Upper Radnice Seam and the Lower Lubná Seam and its parting. See Fig. 7 for key and Figs 3 (cross-section 2) and 9 for location.

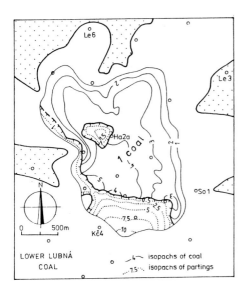

Fig. 11. Thickness and splitting of the Lower Lubná Seam.

by mudstones with identifiable plant remains or by cross-laminated fine-grained sandstones. In the upward direction the northern and central part of the Kačice depression is dominated by siltstones and mudstones with several rootlet horizons containing siderite spherulites (approximately 1 mm in size). In the southern part of the study area several upward-fining sequences with predominantly conglomerates and sandstones (5–11 m thick) are developed through the whole section of the interval (Fig. 10). These deposits infill the WSW–ENE trending erosional channels in the Upper Radnice Seam, reaching below the tonstein T_4 (Fig. 8), and are generally fining to the north. There are some diamictites, mainly along the northern and eastern margins of the Kačice depression, up to 6 m thick. The interval is topped by a discontinuous tuff layer due to unfavourable facies conditions.

The interval is overlain by the Lower Lubná Seam, developed in mineable thickness and quality only in the central and northern parts of the Kačice depression (Fig. 11). Its contour is

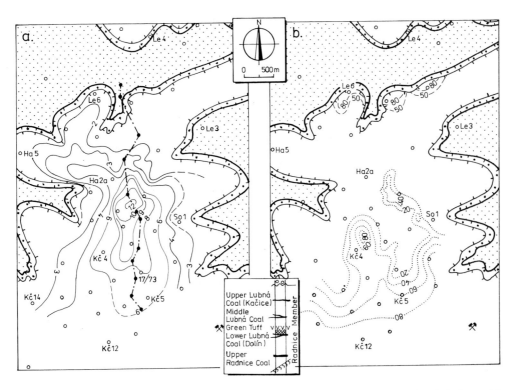

Fig. 12. Clastic interval between the Lower Lubná Seam and the Green Tuff. (a) Isopachs of the interval in metres; (b) percentage of coarse-grained sediments. For key, see Fig. 6.

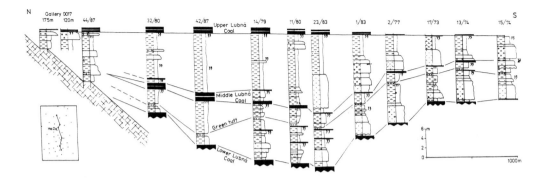

Fig. 13. Detailed north–south cross-section showing the intervals between the Lower Lubná Seam–Green Tuff–Middle Lubná Seam–Upper Lubná Seam. See Fig. 7 for key and Figs 3 (cross-section 3), 12, 14 and 16 for location.

simple, without narrow embayments. The coal is split in the south; its parting gradually reaches up to 10 m in thickness with increasing contents of coarse-grained sediments towards the south.

Depositional environments. Owing to the increase in the sedimentary area during this interval, the southern part of the Kačice depression lost the protection of the palaeotopography against the channel, which ran through the Kladno depression, as seen from the post-sedimentary erosion of the Upper Radnice Seam. Its central and northern parts, however, were protected by the pre-sedimentary relief. This is reflected in the deposition of fine-grained sediment in the lake produced by subsidence. The floodplain, where the deposition was controlled mainly by compaction, was gradually established here.

The northern and eastern margins of the floodplain were influenced by high viscosity flows, whereas in the south the fluvial influence is evident (crevasse splays). Rootlet horizons indicate that the floodplain lake shallows were colonized by plants, but due to a high compaction rate the peat deposition did not take place. The rate of subsidence decreased at the end of the interval and the floodplain lake changed into the peat bog of the Lower Lubná Seam. In the northern part, even compaction of the underlying fine-grained and organic deposits took place and thick coal formed. However, the compaction rate increased to the south, as did the thickness of lacustrine sediments of the interval below the Upper Radnice Seam. The accumulation of coarse-grained deposits over fine-grained deposits led to rapid compaction and seam splitting.

Interval between the Lower Lubná Seam and the Green Tuff

Description. The maximum thickness of this interval follows the trend of the axis of the Kačice depression in its central part (Fig. 12) and is represented by the maximum thickness of both the Lower Lubná Seam and the previous interval. The interval is thinner either at the depression margins or in the south of the study area, underlain here by a thick coarse-grained sequence unfavourable for significant compaction during this interval.

The central and northern parts of the Kačice depression are dominated by fine-grained sediments of the same character as in the previous interval. The area with the largest amount of coarse-grained sediments (around 80%) is located in the south and has a WSW–ENE trend. Sediments exhibit an upward fining structure with a predominance of coarse-grained members (Fig. 13). There are two tongue-like bodies which wedge out laterally in the central part of the depression, orientated perpendicular to the previous trend. They both fine upwards and laterally. The interval is topped by the widespread Green Tuff layer.

Depositional environments. The central and northern parts of the Kačice depression were protected by the elevation from the fluvial zone and the floodplain lake was created by subsidence. It induced drowning of the Lower Lubná Seam swamp. Local shallows originating during the interval were colonized by plants (rootlet horizon or thin coals). High viscosity flows influenced only the northern margin of the depression. The area of maximum thickness of

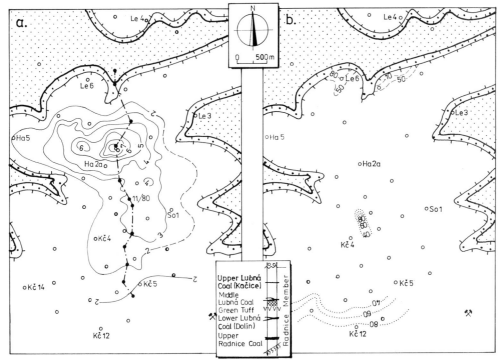

Fig. 14. Clastic interval between the Green Tuff and the Middle Lubná Seam. (a) Isopachs of the interval in metres; (b) percentage of coarse-grained sediments. For key, see Fig. 6.

the interval is superimposed on the area of maximum thickness of the underlying Lower Lubná seam, which in its unconsolidated stage had a high compactional potential.

Interval between the Green Tuff and the Middle Lubná Seam

Description. The area of maximum thickness of the interval is located in the northern part of the Kačice depression and is aligned perpendicular to the axis of the depression (Fig. 13). The minimum thickness in the south is induced by a low compaction rate due to the predominance of coarse-grained deposits below.

Within the whole area of the Kačice depression fine-grained sediments predominate (Fig. 13), with the volcanic admixture decreasing upward (Opluštil 1991). The rocks are horizontally laminated. Slump structures and local erosion surfaces are common. Except for a rootlet horizon directly underlying the Middle Lubná Seam, there are no identifiable plant remains within the interval. In the central and southern part of the Kačice depression this sequence shows a slight coarsening upward from mudstones to sandy siltstones. Deposits of the fluvial zone located only along the southern margin of the study area exhibit the same upward-fining structure as in the previous interval. Diamictites were found only along the northern margin of the depression.

The Middle Lubná seam occurs above this interval. It reaches its maximum thickness (around 2 m) on the northeast margin of the depression, where it is underlain by a ridge. In the centre of the Kačice depression the coal is split (Fig. 15) by mudstones into two benches (high compaction rate).

Depositional environments. The connection between the Kačice and the Zlonice depressions (outside the study area to the northeast) probably existed during this interval; however, no river system occupied this route.

At the beginning of the interval, a shallow floodplain lake covered almost the entire Kačice depression as a result of compaction. The area of the lake was gradually reduced; the small-scale upward coarsening unit probably records its infilling by small deltas (Scott 1978). After shallowing at the end of the interval the lake changed into a swamp. Owing to high compaction rates in the axis of the depression, occupied by a thick sequence of fine-grained and organic-rich sediments, the unsplit coal is developed only

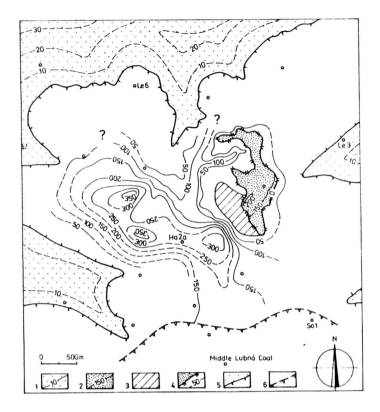

Fig. 15. Splitting of the Middle Lubná Seam. 1, Uncovered pre-sedimentary relief; 2, thickness of seam in centimetres; 3, parting with high content of coal admixture (30–60%); 4, line of seam splitting and thickness of parting (less than 30% of coal admixture); 5, border of a sedimentary area during the deposition of the Middle Lubná Seam; and 6, area of post-sedimentary erosion of the upper bench of the Middle Lubná Seam.

at the northeast margin of the depression (Fig. 15). The absence of a mineable coal in the south is explained by a low compaction rate of coarse-grained sediments (Fig. 14).

Interval between the Middle Lubná Seam and the Upper Lubná Seam

Description. The maximum thickness partly coincides with the axis of the Kačice depression (Fig. 16). Except for depression margins, the minimum thickness is located in the southeastern part of the study area, where it coincides with the maximum percentage of coarse-grained sediments.

In the central and northern part of the depression non-laminated mudstones with spherulites of siderite (1 mm in size) and roots dominate throughout the interval. They can be intercalated by up to 2 mm thick beds of fine-grained sandstones. In the southern part of the depression the upward-fining sequence (conglomerates of coarse sandstones to mudstones) occurs. It has a WSW–ENE trend, perpendicular to the trough axis. At the end of the interval refractory claystones were deposited over the whole area. The interval is overlain by the Upper Lubná Seam, developed over almost the whole Kačice depression. Coal 1 m thick occurs only in the northern and northeastern part of the depression, where no very thick fine-grained sequences underlie the mire. Owing to either low (in the south) or high subsidence rates (in the west and the centre of the depression) there is only a thin coal present.

Depositional environments. Despite the significant burial of pre-sedimentary relief during the interval, no fluvial deposits have been documented in the Kačice depression (Fig. 16), as seen from predominance of fine-grained deposits. Only the southern part was influenced by the fluvial zone occupying the Kladno depression (60–70% of coarse-grained sediments). The

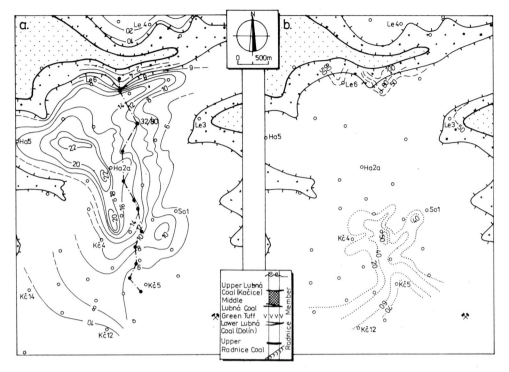

Fig. 16. Clastic interval between the the Middle Lubná and the Upper Lubná Seams. (a) Isopachs of the interval in metres; (b) percentage of coarse-grained sediments. For key, see Fig. 6.

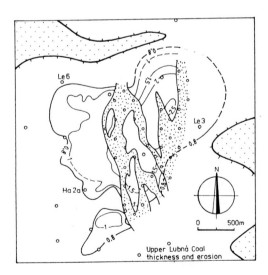

Fig. 17. Thickness and erosion (below the level of the tonstein in the upper part of the coal) of the Upper Lubná Seam.

central and northern parts of the Kačice depression contain floodplain deposits with shallow lakes produced by subsidence. The lakes and the areas between the lakes were colonized by plants (fossil soils, rootlet horizons). A low energy environment prevailed at the end of the interval during the deposition of refractory claystones within the Kačice depression. Only high viscosity flows occasionally disturbed the northern margin of the depression.

The Upper Lubná Seam swamp gradually spread over the Kačice depression, but favourable compaction rates prolonged peat deposition only in its shallow, northeastern part.

Interval between the Upper Lubná Seam and the top of the Radnice Member

Description. The isopachs of this interval show a SSW–NNE trend, conforming approximately with both the axis of the Kačice depression (Fig. 18) and erosional channels in the Upper Lubná Seam (Fig. 17). The maximum thickness (up to 50 m) is located in the northeast of the study area. The percentage of coarse-grained sedi-

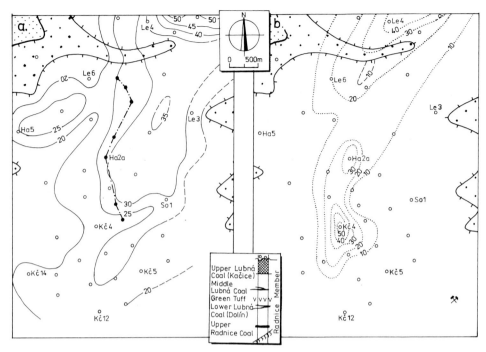

Fig. 18. Clastic interval between the Upper Lubná Seam and top of the Radnice Member. (**a**) Isopachs of the interval in metres; (**b**) percentage of coarse-grained sediments. For key, see Fig. 6.

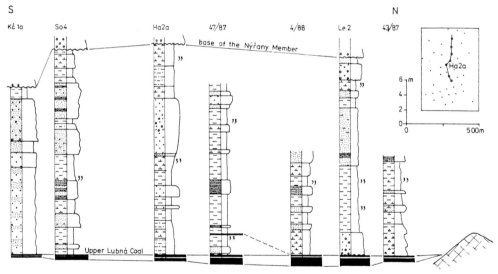

Fig. 19. Cross-section showing the interval between the Upper Lubná Seam and top of the Radnice Member. See Fig. 7 for key and Figs 3 (cross-section 4) and 18 for location.

ments rarely exceeds 50% within the study area. Where it does not exceed 20%, the mudstones and siltstones are occasionally interbedded with fine-grained sandstones or interlaminated sandstones and siltstones about 0.6 to 1 m thick. Rootlet horizons commonly occur and thin coals are locally present. In the areas with a higher percentage of coarse-grained sediments, the upward coarsening thick sandstones to conglomerates are overlain by thin layers of fine-

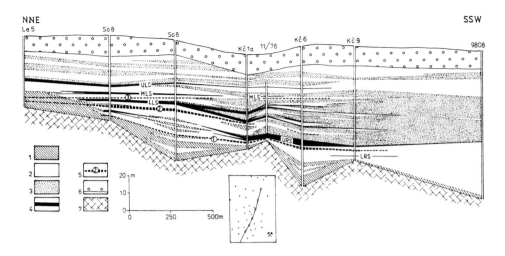

Fig. 20. Cross-section of the Radnice Member across the study area (based on Spudil *et al.* 1980, unpublished data). For location see Fig. 3 (cross-section 5). 1, Diamictite; 2, claystone to siltstone; 3, sandstone to conglomerate; 4, coal; 5, tuff layer (numbers in circle: 1, Whetstone Horizon; 2, Middle Horizon; 3, Green Tuff); 6, base of the Nýřany Member; 7, late Proterozoic basement; LRS, Lower Radnice Seam; URS, Upper Radnice Seam; LLS, Lower Lubná Seam; MLS, Middle Lubná Seam; and ULS, Upper Lubná Seam.

grained sediments (Fig. 19). Diamictites are very rare (found in one borehole).

Depositional environments. A significant change in drainage took place after the drowning of the Upper Lubná Seam swamp as a result of subsidence. The axis of the Kačice depression was reoccupied by a river system (erosion of the Upper Lubná Seam). Lakes produced by compaction were laid down on the adjacent floodplain. No significant coals occur within this interval, probably as a result of low compaction rates. The decreasing basement subsidence of the depocentre within the Kladno Basin and infilling of the depressions with sediments resulted in the end of deposition of the Radnice Member.

Summary and conclusions

The results of analyses of both interseam clastic intervals and coals show a close relationship between the palaeo-topography and facies distribution (Fig. 20). This resulted in differential compaction of the sediments, which is believed to have controlled the spatial distribution of the peat and clastic deposits as documented by the connection between the area of maximum thicknesses of the interseam clastic intervals or coal seams and the compaction potential of their underlying sediments. No syn-sedimentary fault movements have been documented in the study area.

Depositional environments changed during the life of the basin as a result of a gradual decrease in the influence of the palaeo-relief (Fig. 21). As a result of the pre-sedimentary relief at the beginning of the deposition (first interval, Fig. 21a), no river system occupied either the Kladno or the Kačice depression. Rivers occupied the Kladno depression after partial burial (second interval) of the palaeo-relief (erosion of the Upper Radnice Seam). It influenced the southern part of the Kačice depression adjacent to the previous depression. The floodplain existed from the second interval in the central and northern parts of this depression, protected by the surrounding ridges from fluvial dynamics. Owing to the facies distribution, differential compaction took place. In the higher intervals, areas with maximum thicknesses of the clastic intervals moved to the north as a result of the increased compaction potential of the sediments. Extensive burial of the pre-sedimentary relief resulted in the occupation of the Kačice depression by the river system (erosion of the Upper Lubná Seam) at the end of the Radnice Member deposition. The end of the deposition is related to a reduction in the basement subsidence and the infill of morphological depressions.

The influence of palaeo-topography on coal development is documented by both the protec-

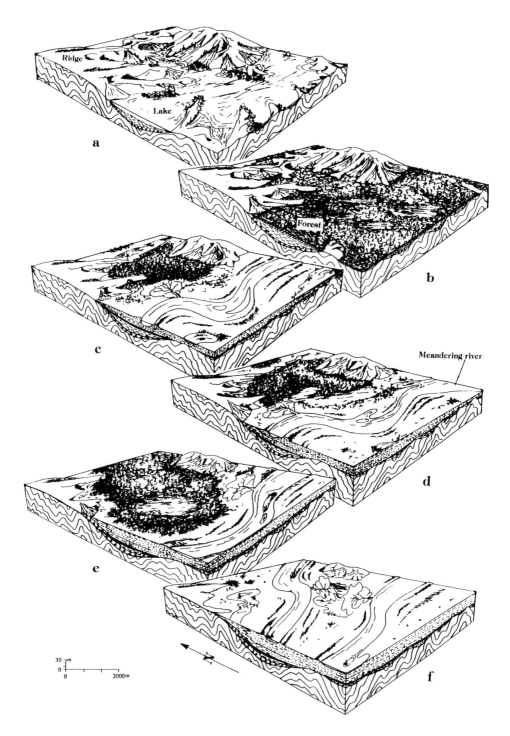

Fig. 21. Sedimentary environments and the basement configuration in the study area during the deposition of the Radnice Member. (**a**) Before deposition of the Upper Radnice Seam; (**b**) during deposition of the Upper Radnice Seam; (**c**) deposition of the Lower Lubná Seam; (**d**) during deposition of the Middle Lubná Seam; (**e**) during deposition of the Upper Lubná Seam; and (**f**) after deposition of the Upper Lubná Seam.

tion of an area with peat deposition against fluvial dynamics and the delimitation of the peat bogs. The thickest seams and the split coals are underlain by a thick sequence of fine-grained organic sediments. No mineable coal occurs in the area underlain by a thick sequence of coarse-grained deposits (Fig. 20). In the upward direction the conditions favourable for peat deposition gradually moved to the north into the protected parts of the Kačice depression. The stratigraphically lowermost Upper Radnice Seam reaches up to 7 m thick above the axis of the Kačice depression (compaction of lithologically homogeneous fine-grained sediments). Its mineable area is divided into several embayments. The overlying Lower Lubná Seam is up to 5 m thick above the axis of the central part of the Kačice depression. To the south the coal is split as a result of the rapid compaction of fine-grained sediments and peat induced by the loading of coarse-grained deposits during the previous interval. The stratigraphically higher Middle Lubná and Upper Lubná Seams attain maximum thicknesses of 2 and 2.5 m, respectively, only at the depression margins or in the shallow depressions. As a result of the intensive compaction of a thick sequence of fine-grained sediments in the axis of the Kačice depression, only thin or split coals occur. The low compaction potential of coarse-grained deposits in the south allowed only a thin accumulation of organic matter.

The authors acknowledge the management of the Kladno Mine, especially the geologists J. Kružík and J. Slavík, for their support of the study. Z. Čermáková and M. Malý are thanked for the preparation of diagrams for this paper.

References

HAVLENA, V. & PEŠEK, J. 1980 Permocarboniferous basins of the Czech Massif: stratigraphy, palaeogeography and the basic structural frame and principal structural division. *Sborník Příroda*, **34**, 1–144 (in Czech with English summary).

HOLUB, V., PEŠEK, J. & SKOČEK, V. 1991. Morphology and character of a basement of the Carboniferous basins and their source area of the Central Bohemian area and the Bohemian Massif between the Upper Devonian and the Westphalian B. *In:* OPLUŠTIL, S., PEŠEK, J. & VÍZDAL, P. (eds) *Proceedings of VIth Coal Conference*. Faculty of Science, Prague, 35–43 (in Czech).

KLENER, J. 1973. *Palaeogeography of the Radnice Member in the relationship on the palaeorelief of the basement in the Kladno-Rakovník basin*. MS thesis. Geofond, Prague [in Czech].

—— 1975. The Paleorelief of the basement of the Upper Carboniferous in the Kladno-Rakovník basin. *Bulletin de la Société Belge de Géologie*, **84**, 51–55.

—— 1982. Coal potential of the Radnice Member (Westphalian B–C) in the Kladno Basin (ČSSR). *Výzkum. práce Ústř. úst. geol.*, **30**, 1–35 [in Czech].

KUKAL, Z. 1984. Granitoids plutons were the main source of feldspar of Permo-Carboniferous sediments. *Časopis pro mineralogii a geologii*, **29**, 193–196 [in Czech].

OPLUŠTIL, S. 1991. Upper volcanic layer in the NW part of the Kladno Depression (Upper Carboniferous, Kladno Basin). *Věstník Českého Geologického Ústavu*, **66**, 379–385.

SCOTT, A. C. 1978. Sedimentological and ecological control of Westphalian B plant assemblages from West Yorkshire. *Proceedings of the Yorkshire Geological Society*, **41**, 461–508.

SKOČEK, V. 1968. Upper Carboniferous varvites in coal basins of Central Bohemia. *Věstník Ústředního Ústavu Geologického*, **43**, 113–121 [in Czech].

TEODOROVICH, G. I. & CHERNOV, A. A. 1968. Character of changes with depth in productive deposits of Apsheron oil-gas-bearing region. *Soviet Geology*, **4**, 83–93.

Sequence stratigraphy and lithofacies geometry in an early Namurian coal-bearing succession in central Scotland

W. A. READ

Department of Geology, Leicester University, Leicester LE1 7RH, UK

Abstract: High frequency glacial–eustatic allocycles have been identified in the upper part of the early Pendleian Limestone Coal Formation in the north-central Midland Valley of Scotland. Repeated eustatic oscillations caused extensive lateral migrations of a wide spectrum of depositional environments, ranging from quasi-marine to fluvial. This throws doubt on the assumptions that underlie some existing depositional models of similar late Carboniferous cyclic coal-bearing successions.

The study interval can be divided into three facies associations. The distal association was dominantly controlled by glacial–eustatic oscillations, which resulted in a remarkably regular 'layer cake' succession with laterally persistent coals. In the more sandy and variable transitional facies association, local episodic fluvial processes tended to obscure the underlying allocycles. Here coals may split and also amalgamate to form thick composite seams. The highly variable proximal association was dominantly controlled by irregular autocyclic fluvial sedimentary processes which largely obscured the background changes in relative sea level. Sequence stratigraphy is easy to apply to the distal facies association, harder to apply to the transitional association and very difficult to apply to the variable proximal association.

Many, if not most, sedimentological studies of late Carboniferous coal-bearing successions are still tacitly based on depositional models such as those originally proposed by Ferm (1970) and Horne *et al.* (1978) for the Pennsylvanian and by Fielding (1984) for the Westphalian. Although these models do acknowledge periodic widespread marine incursions, they nevertheless tend to imply that large areas stayed within the same broad palaeoenvironmental zones for relatively long periods. Such models may need to be modified in the light of the evidence of glacial–eustatic sea-level oscillations during the Silesian. Veevers & Powell (1987) have documented evidence of continental-scale ice-caps in Australia from the Namurian onwards and computer simulations by Crowley & Baum (1991, 1992) have suggested that major Silesian climatic variations acting on the Gondwanaland ice-cap may have produced glacial–eustatic oscillations in sea level with amplitudes of the order of 50 m. The Silesian seems to have been a period of 'icehouse' global climate, resembling the Pleistocene, during which a series of orbital parameters within the Milankovitch spectrum produced variations in insolation which, in turn, forced variations in global climate. These superimposed parameters must ultimately have given rise to a complex eustatic curve with a series of superimposed glacial–eustatic allocycles, as suggested by the simulated Late Carboniferous eustatic curve of Maynard & Leeder (1992, fig. 8). The author (Read 1994a, 1994b) has suggested that some of these allocycles had a relatively high frequency within the Milankovitch spectrum.

Such rapidly repeated eustatic oscillations must have had a profound effect on coastal and deltaic depositional environments. Furthermore, they must inevitably have affected contemporaneous fluvial environments far upstream as river channels adjusted their profiles to the changes in base level (see Schumm 1977). Thus during the Silesian we might reasonably expect rapidly repeated laterally extensive migrations of broad palaeoenvironments on scales of both time and distance that were not envisaged when most of the existing depositional models were formulated.

Detailed information from the Scottish Limestone Coal Formation (formerly Group) in the north-central part of the Midland Valley of Scotland has enabled the construction of a more realistic model (Read 1994a), which incorporates the effects of high frequency glacial–eustatic oscillations. One aim of this paper is to refine and amplify this new model, particularly with regard to lithofacies associations and their variations in both space and time. Another aim is to assess whether and to what extent we can validly apply the Exxon Production Research (EPR) Company's concepts of sequence stratigraphy (see Posamentier *et al.* 1988; Posamentier & Vail 1988; Mitchum & Van Wagoner 1991; Posamentier & Weimer 1993) to these lithofacies associations in particular and to late Carboniferous coal-bearing cyclical successions in gen-

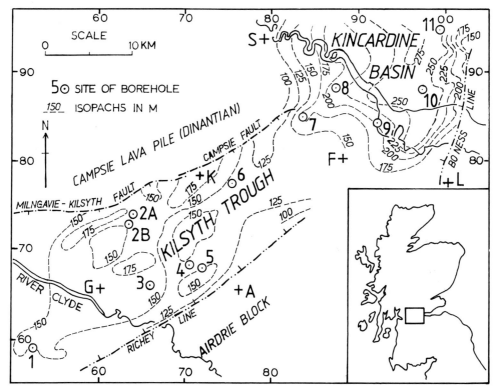

Fig. 1. Sketch map of the area chosen for study within the Glasgow–Stirling–Linlithgow region, showing the principal structural elements (after Read 1988), isopachs of the study interval in metres and sites of boreholes shown in Figs 2, 3 and 5. The names of these boreholes are as follows: (1) Darnley No. 1 (1936); (2A) Torrance No. 3 (1950–1951); (2B) Torrance No. 2 (1953–1954); (3) Queenslie No. 1 (1949); (4) Gartcosh No. 1 (1966); (5) Glenboig No. 1 (1967); (6) Dullatur No. 2 (1932); (7) Torwood (1960–1961); (8) Doll Mill (1965); (9) Orchardhead (1956); (10) Rigghead (1953–1956); and (11) Solsgirth (1941–1961). Numbers along the margins of the map refer to 100 km squares NS and NT of the British National Grid. Abbreviations: A, Airdrie; F, Falkirk; G, Glasgow; K, Kilsyth; L, Linlithgow; and S, Stirling.

eral. With its emphasis on a dominantly eustatic control, EPR sequence stratigraphy should be readily applicable to any 'coal measures' palaeoenvironments that were strongly influenced by glacial–eustatic oscillations and consequent base-level changes.

Area and succession studied

The Scottish Limestone Coal Formation is of early Pendleian (E_1) age. The stratigraphic interval chosen for detailed study lies in the upper part of that formation, between the major marine transgressions of the Black Metals and the Index Limestone (Read 1988, his fig. 16.2), in the Glasgow–Stirling–Linlithgow area of the Midland Valley (Fig. 1). Information from within this interval provides a suitable factual basis for the formulation of new depositional models and the assessment of EPR sequence stratigraphy because the lithofacies geometry can be determined in considerable detail from the records of hundreds of closely spaced fully cored boreholes. A high proportion of the cores has been examined by geologists. Additional information has been provided by mine plans of extensive multi-seam coal workings.

The upper part of the Limestone Coal Formation, which has been closely studied for more than 30 years (see Read & Forsyth 1989, 1991, and references cited therein), is generally similar in its lithofacies to the Westphalian A Coal Measures of Scotland and northern England, but was subject to stronger marine influences. It embraces a wide spectrum of palaeoenvironments, ranging from quasi-marine to fluvial, and sedimentation was subject to a variety of controls, including allocycles considered to be induced by glacial–eustatic oscillations, tectonic movements and local, largely

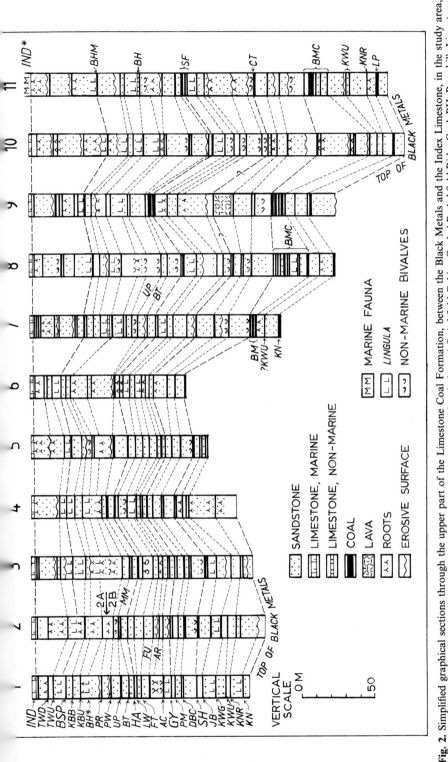

Fig. 2. Simplified graphical sections through the upper part of the Limestone Coal Formation, between the Black Metals and the Index Limestone, in the study area, showing coals and other marker horizons. For sites of boreholes, see Fig. 1. Abbreviations: AC, Ashfield Coking Coal; AR, Ashfield Rider Coal; BH, Berryhills Limestone (non-marine) and Coal; BHM, Blairhall Main Coal (= BSP); BM, Bannockburn Main Coal (= KWG + JB + SH); BMC, Bannockburn Main Complex; BSP, Bo'ness Splint Coal; BT, Batchie Coal; CT, Comrie Two-foot Coal (= GY); DBC = Dumbreck Cloven Coal; FT = Fourteen-inch Under Coal; FU = Fourteen-inch Under Coal; GY = Greenyards Coal; HA = Hartley Coal; IND = Index Limestone; JB = Jubilee Coal; KBB = Kilsyth No. 1 Blackband Coal; KBU = Kilsyth No. 1 Blackband Under Coal; KN = Knott Coal; KNR = Knott Rider Coal; KWG = Knightswood Gas Coal; KWU = Knightswood Under Coal; LP = Lochgelly Parrot Coal (= KN); LW = Lower Wee Coal; MM = Meiklehill Main Coal (= BT + UP); PM = Possil Main Coal; PR = Possil Rider Coal; PW = Possil Wee Coal; SF = Seven-foot Coal (= HA + BT); SH = Shale Coal; TWD = Twechar Dirty Coal; TWU = Twechar Under Coal; and UP = Upper Possil Coal. Denotes stratigraphic horizon equivalent to named marker.

fluvial, episodic sedimentary processes (Read 1994a).

The area selected for detailed study is shown in Fig. 1, which also shows isopachs of the study interval. Structurally, it comprises the graben-like Kilsyth Trough in the west and the asymmetrical Kincardine Basin in the east. This basin, which was the structural precursor to the end-Carboniferous Clackmannan Syncline, was the most rapidly subsiding Namurian basin in Scotland (Read 1988). The basinal areas were hemmed in by the low subsidence areas described by Read (1988) and shown and named on Fig. 1.

Siliciclastic sediments entered the Kincardine Basin, principally from the northwest and northeast, passed westwards from the basin into the Kilsyth Trough and then were transported west-south-west down the trough (Read 1988, his fig. 16.8). Smaller amounts of sandy sediment also entered the Kilsyth Trough from the north.

Figure 2 shows skeleton graphical sections from 11 representative geologist-examined cored boreholes which cut through all or most of the study interval. The borehole sites are given in Fig. 1. These skeletal sections show the 'upstream' transition from the remarkably regular 'layer cake' successions in the west of the Kilsyth Trough, where marine influences were strongest, to the laterally variable fluvially influenced successions on the eastern flank of the Kincardine Basin. About 25 coal horizons which, together with their underlying palaeosols (seat earths) and overlying claystones, have been traced over the greater part of the study area provide a firm basis for stratigraphic correlation.

To avoid the unnecessary repetition of previously published material, the following account deliberately omits some of the detailed evidence for correlations, the identification of palaeoenvironments and the identification of controls on deposition within the study interval. This detailed evidence may be found in papers by Forsyth & Read (1962), Read & Forsyth (1989, 1991) and Read (1994a). The overall palaeogeographical, palaeoclimatic and structural setting has been described by Read (1988).

Allocycles, autocycles and lithofacies associations

Allocycles

Two, or possibly more, orders of allocycle may be detected in the Scottish Pendleian succession (Read 1994a, b). 'Low frequency' or 'long' allocycles were responsible for a series of widespread marine transgressions, including those marked by the Black Metals and the Index Limestone, which form the lower and upper limits of the study interval. These major transgressions had a recurrence interval of about 1 Ma (Read 1994b), so the 'low frequency' allocyclicity is broadly comparable with the third-order cyclicity of EPR sequence stratigraphy (Posamentier & Weimer 1993). Of greater relevance to the present study are the 'high frequency' or 'short' allocycles which are probably equivalent to EPR fifth-order cyclicity (Read 1994b). These controlled the formation of the 25 or so laterally persistent coal horizons within the study interval (Fig. 2; Forsyth & Read 1962). An additional order of allocycle with a frequency between those of the low and high frequency allocycles, possibly equivalent to EPR fourth-order cyclicity, may also be present. This intermediate frequency allocyclicity is difficult to detect in the study interval or within the rest of the Limestone Coal Formation, although it is more obvious in the succeeding Scottish Upper Limestone Formation. Thus it will be described elsewhere.

Autocycles

Allocyclical control is strongly dominant in the western part of the study area. Under allocyclical control the number of cycles, as indicated by horizons of vegetation colonization, tends to remain fairly constant, despite any variations in net subsidence (Weedon & Read in press). Although this is true in the western part of the Kilsyth Trough (Figs 1 and 2), there is a definite tendency for the number of vegetation horizons to increase further east. Finally, in some parts of the succession in the Kincardine Basin, the number of vegetation horizons tends to increase proportionally with net subsidence.

Such a relationship has also been found in autocyclically deposited successions later in the Scottish Namurian in which cyclicity was controlled by purely local, episodic, sedimentary processes (as described later in the Summary of controls section). This autocyclicity, which in the study interval seems to be superimposed on the background allocyclicity, is more obvious in the Kincardine Basin, where fluvial influences were stronger, than in the Kilsyth Trough (Read and Forsyth 1989, their fig. 5). Autocycles in the Limestone Coal Formation tend to be thinner and less regularly spaced than allocycles. They also have a shorter recurrence interval.

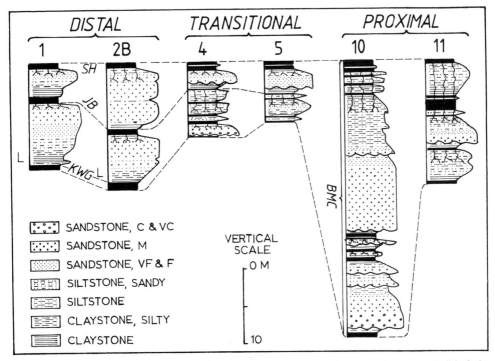

Fig. 3. Graphical sections, all through exactly the same stratigraphic interval, showing the changes in lithofacies as this interval is traced eastwards ('upstream') from the distal facies association, through the transitional facies association, into the proximal facies association. For borehole sites, numbers and names, see Fig. 1. For lithological ornaments and symbols not shown in key, see Fig. 2. Abbreviations: C, coarse-grained; F, fine-grained; M, medium-grained; VC, very coarse-grained; VF, very fine-grained. For other abbreviations, see Fig. 2.

Lithofacies associations

In earlier publications (Read & Forsyth 1989, 1991; Read 1994a) the upper part of the Limestone Coal Formation was divided into two lithofacies associations, termed the distal and the proximal facies associations, although it was fully realized that these were essentially end-members of a continuum. The variations in lithofacies can now be traced in greater detail by recognizing three associations, here named the distal, transitional and proximal facies associations. Characteristic sections of each are illustrated in Fig. 3. The same lithological members occur in all three associations, which, however, differ in their geometry and lateral variability, and the relative thicknesses of individual members.

Distal facies association. The remarkable 'layer cake' successions in the western part of the Kilsyth Trough largely belong to the distal facies association and are dominated by evenly spaced high frequency allocycles. Almost all the lithological members, apart from channel sandstones, are laterally persistent. Most individual cycles (which have conventionally been measured from the top of one horizon of vegetation colonization to the top of the next above) are less than 5 m thick, post-compaction. As cycle thicknesses remain fairly constant, correlation lines linking equivalent coal horizons are approximately parallel (Fig. 2).

Coals seams, which are seldom more than 1 m thick and contain few siliciclastic partings, have sharply defined tops. They are usually immediately overlain by thin black carbonaceous claystones, but locally a few centimetres of silty or even sandy strata may intervene. These silty or sandy beds may contain broken fragments of the quasi-marine brachiopod *Lingula*. The thin black claystones grade upwards into dark grey, less carbonaceous claystones, containing either *Lingula* or the non-marine bivalves *Curvirimula* or *Naiadites*. In the western part of the Kilsyth Trough more than a third of the allocycles

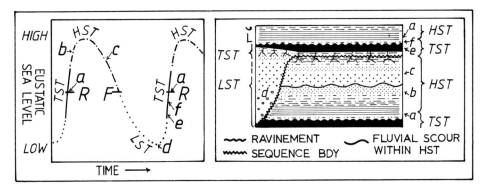

Fig. 4. Hypothetical eustatic curve showing inflexion points and systems tracts in a high frequency type 1 sequence plus part of the following sequence, together with an equivalent schematic lithological succession from within the allocyclic distal facies association. Lower case letters a to f indicate points on the eustatic curve and the corresponding positions in the lithological succession. For explanation of lithological ornaments and symbols, see Figs 2 and 3. R and F denote the inflexion points in the eustatic curve that mark the maximum rates of eustatic rise and eustatic fall, respectively. Other abbreviations: HST, highstand systems tract; LST, lowstand systems tract; and TST, transgressive systems tract.

contain *Lingula*, albeit not necessarily in every borehole section. Furthermore, in the Pendleian these non-marine bivalve genera are associated with the marine end of a spectrum of environments that range from marine to brackish (Eagar *in* Read & Forsyth 1989). Thus the fauna indicates persistent marine influences.

The claystones become more silty upwards and pass by alternation into micaceous siltstones. These in turn pass by alternation upwards into whitish grey ripple-laminated or cross-stratified sheet sandstones which are generally fine- or very fine-grained. Both the siltstones and the overlying sandstones are commonly intensively bioturbated and they probably represent bay-fill and distributary mouth bar deposits. Within some of the sheet sandstones an upward-coarsening finer grained lower portion is succeeded by an upward-fining upper portion. The base of the upper portion is commonly medium-grained and there are occasionally traces of a scoured surface between the lower and upper portions, indicating a two-storey sheet sandstone profile. In a number of allocycles, erosive-based upward-fining channel-fill sandstones cut right through the sheet sandstone member and down to, or even through, the underlying coal. Almost all of the sheet sandstones grade upwards into silty or clayey palaeosols (seat earths), which underlie the succeeding coal. Most of these represent immature gleysols that formed under waterlogged conditions, but locally remnants are preserved of thicker brownish palaeosols with mature leached profiles that must have been formed during prolonged periods of lowered water-table.

Read & Forsyth (1991) considered that the distal association was dominantly deltaic, but that the upper storeys of sheet sandstones, together with the erosive-based channel sandstones, were fluvial. They thought that these channels were incised during periods of lowered sea level. A schematic section through an allocycle in the distal association is illustrated in Fig. 4.

Transitional facies association. In the transitional facies association the proportion of sandstone increases and the allocycles become harder to delimit and correlate compared with the distal facies association. Most of the individual horizons of vegetation colonization may still be traced over wide areas, but correlation lines linking these commonly diverge or converge sharply (Fig. 2).

Coals tend to be thicker and seam splits and amalgamations become fairly common. *Lingula* bands occur less commonly in the claystone roofs, which themselves become more silty. Sheet sandstones are thicker and coarser grained. Many show a well marked two-storey profile in which an upper fluvial storey scours down into an underlying deltaic storey (Read & Forsyth 1989, their fig. 3). Some individual sandstone sheets have erosive bases. Erosive-

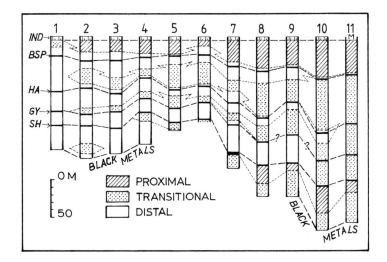

Fig. 5. Same borehole sections as illustrated in Fig. 2, but showing the different facies associations present within each section. Note how the wedges of the transitional facies and proximal facies associations thicken eastwards to occupy the whole study interval and taper out westwards into the allocyclic layer cake successions of the distal facies association. For borehole sites, numbers and names, see Fig. 1; for abbreviations, see Fig. 2.

based sandstone-filled channels are more common and more deeply incised than in the distal facies association, but few show multi-storey profiles. They tend to form fairly straight channels which may be as much as 12 m deep and 1.5 km wide. Palaeosols are generally thicker. Most are still immature gleysols, but relict mature soil profiles seem to be more common.

Compared with the distal association, marine and deltaic influences were weaker and fluvial influences stronger, with episodic fluvial processes tending to interfere with and confuse the effects of the background rises and falls in relative sea level.

Proximal facies association. The proximal association is by far the most variable of the three. Here the dominant controls were local and episodic (see later) and it becomes very difficult to trace either the background allocycles inferred from laterally equivalent intervals in the distal facies association, or even individual stratigraphic horizons.

Coals tend to be thick, particularly in the Bannockburn Main Complex (Figs 2 and 3), where composite seams must represent successions of raised mires. However, they are impersistent and laterally variable and are commonly split by siliciclastic partings (Read 1961) which thicken abruptly. Erosive-based channel sandstones are abundant. Many of these are thick and one, admittedly composite, example reaches 42 m. Most are medium-grained, but many are coarse- and very coarse-grained or even pebbly, and multi-storey profiles are common. Many channel-fills include immature arkoses with sporadic concentrations of pink garnets that suggest a source area in the southern Highlands of Scotland. Some finer grained sheet-like sandstones are also present, but these tend to be thin and laterally impersistent. Palaeosols are locally thick, but most are immature and impersistent. Claystones are mostly thin, impersistent and barren. In the proximal association local episodic fluvial sedimentary processes were strongly dominant and marine influences generally imperceptible, although rare records of *Lingula* show that the sea could occasionally penetrate into this dominantly fluvial environment.

Later variations and lithofacies geometry

Lateral facies variations are illustrated in Fig. 5, which shows the 11 borehole sections of Fig. 2 in terms of the three facies associations. Section 1, west-southwest of Glasgow in the western part of the Kilsyth Trough, lies in a distal situation open to marine influences from the west, but remote from fluvial influences. Thus apart from a narrow band of transitional facies associations at the very top, the whole study interval is composed of the distal facies association.

As the succession is traced eastwards and 'upstream' along the Kilsyth Trough, a wedge of the proximal association appears at the very top and three irregular wedges of the transitional facies association appear lower in the succession. These wedges tend to thicken eastwards and occupy a greater proportion of the succession in the Kincardine Basin, where a second wedge of the proximal facies association appears below the horizon of the top of the Shale Coal (Fig. 2) in the lower part of study interval. Finally, in section 11, in the northeast of the Kincardine Basin, the whole interval is composed of the transitional and proximal facies associations.

The overall facies changes are accompanied by corresponding changes in coal seam geometry, which may be traced in Fig. 2. Such changes greatly affected commercial exploitation. Coals in the distal association are relatively thin, but persistent, and wash-outs, seam splits and amalgamations are rare. Within the eastward-thickening wedges of the transitional facies, coals are generally thicker, but may amalgamate to form composite seams such as the Meikelhill Main seen in sections 2–6. Such amalgamations tend to be related to the appearance of thick bodies of relatively uncompacted sand immediately below the combined seams. As the wedges of the transitional facies association are traced eastwards, seam splits and wash-outs become increasingly more common.

The most striking example of seam amalgamation is seen on the southwest flank of the Kincardine Basin, where the Knightswood Gas, Jubilee and Shale coals unite to form the thick composite Bannockburn Main Coal (Fig. 2, section 7). This seam is also associated with thick sandstones immediately below. Figure 3 shows some of the lateral facies changes within this particular stratigraphic interval in greater detail. Here, the regular allocycles and three separate Knightswood Gas, Jubilee and Shale coals within the distal facies successions in sections 1 and 2B pass eastwards into the more variable transitional facies succession in sections 4 and 5, where the coals are split by thick silty and sandy partings and erosive-based sandstones appear.

Within the Kincardine Basin, east of where the three original coals combine to form the composite Bannockburn Main Seam, the whole of this part of the succession passes into the highly variable Bannockburn Main Complex. This has been classed with the proximal facies association and is seen in sections 10 and 11. Here, thick but laterally impersistent coals are split by numerous silty and sandy partings. Thick peats forming ombrotrophic raised mires seem to have lain closely adjacent to active fluvial channels transporting sand (Read 1961). Because of repeated seam splits, amalgamations and wash-outs it has now become almost impossible to determine which of the coals in the complex relate to the three separate seams present in the distal association.

The two wedges of proximal facies association shown in Fig. 5 may represent rather different palaeoenvironments. Whereas thick, laterally impersistent coals, including composite seams, characterize the Bannockburn Main Complex in the lower wedge, coals are thinner and more widely spaced in the upper wedge, in which channel sandstones and thick palaeosols tend to be more prominent. However, both wedges have similar, highly irregular, grain size profiles and both show abrupt lateral changes in lithology.

The four westward-tapering wedges of transitional facies association, followed to the east by the proximal facies association in the bottom and top wedge, represent periods of increased sand input into the basinal areas, possibly accompanied by the regrading of stream profiles. During such periods the overall gradient between the source area and the shoreline seems to have increased, because the lateral passage from the distal association, through the transitional association and finally into the proximal association took place over shorter distances than usual. During the deposition of other intervals, such as that between the claystone roof of the Shale Coal and the base of the Greenyards Coal, the distal facies association extended over most of the study area and the lithological succession was much more uniform.

Summary of controls

The controls on sedimentation within the upper part of the Limestone Coal Formation have already been analysed in considerable detail by Read & Forsyth (1989, 1991) and Read (1994a). Thus, to avoid repetition, this section attempts only to summarize the results of these earlier papers. These have concluded that the principal controls were as follows: (a) eustatic changes in sea level; (b) rapid tectonic subsidence of the Kilsyth Trough and Kincardine Basin; (c) episodic tectonic and/or isostatic uplift of parts of the Highland source area (which only became important later in the Namurian); and (d) local episodic dominantly fluvial sedimentary processes.

The low-frequency (third-order) allocyclicity which was responsible for the widespread

Pendleian marine transgressions, such as those of the Black Metals and the Index Limestone, is likely to have been eustatic, because these transgressions can be traced not only over the greater part of the Midland Valley but also into England and, in some examples, far beyond. Thus, for example, ammonids (goniatites) which are said by Ramsbottom (1977) to be characteristic of the Index Limestone horizon have been traced as far as the Donetz Basin (Aisenverg et al. 1979).

Spectral analysis (see Weedon 1991) of grain size data from within the study interval strongly suggests that the high frequency allocyclicity that characterized the distal facies association was regular in both space and time (Weedon & Read in press). This in turn indicates orbital forcing. As continental-scale ice sheets are known to have existed in Gondwanaland during the Namurian, the underlying controlling mechanism was probably glacial–eustatic. The high frequency allocycles are also strongly asymmetrical (Read 1994a), further supporting a glacial–eustatic control (see later). Read (1994b) estimated the frequency of the 'short' allocycles to be about 38–43 ka, which is markedly higher than the dominant frequencies (95 and 123 ka) of the 'short' eccentricity orbital parameter (Berger 1988, 1989). Thus the high frequency allocycles are comparable with the fifth-order cycles of EPR sequence stratigraphy (Posamentier & Weimer 1993), which can only be preserved in areas of rapid subsidence and rapid sedimentation (Mitchum & Van Wagoner 1991).

Both tectonic subsidence and sedimentation are known to have been rapid in the Kilsyth Trough and the Kincardine Basin, particularly the latter. A post-decompactional thickness of 1100–1200 m of sediment was deposited in this basin during the Pendleian and Arnsbergian, whose combined duration is now thought to have been about 10 Ma (Claoué-Long, pers. comm. 1993). This would imply an overall rate of subsidence in excess of $0.11 \, m \, ka^{-1}$ and decompaction would obviously increase this figure. Thus the study area must have provided optimum conditions for the formation and preservation of high frequency allocycles. Although basin subsidence, which was probably largely attributable to thermal sag (Read 1988), was fairly rapid, there is no evidence that it took place episodically in a series of tectonic 'jerks'. Episodic tectonic uplift, followed by increased erosion, took place in parts of the Highland source area later in the Namurian (Read 1989) and a similar mechanism may possibly have influenced the deposition of one or more of the four wedges of transitional and proximal facies association shown in Fig. 5.

The local episodic sedimentary processes which were active during the deposition of the dominantly fluvial proximal facies association are thought to have included channel avulsion, meander migration and cut-off and crevassing during floods. Some thin local sheet-like sand bodies may have been produced by the switching of minor freshwater deltas which prograded into flood basin lakes (see Read 1994a). Beerbower (1964) classed such episodic processes as 'autocyclic' The quantitative analysis of Read & Dean (1982) of dominantly fluvial strata which were deposited in the Kincardine Basin later in the Namurian showed that such episodic fluvial processes had produced an autocyclic succession in which the total number of cycles was closely linearly correlated with net subsidence. In the present study interval these local autocyclic processes could have interrupted and effectively obscured the effects of the background oscillations in relative sea level, although the latter must inevitably have affected upstream areas. Marked contrasts in the short-term compaction potential of thick but impersistent peats and channel sandstones may also have helped to obscure the effects of changes in base level.

Exxon Production Research sequence stratigraphy: applications and limitations

Sequence stratigraphy was first applied to the upper part of the Limestone Coal Formation by Read & Forsyth (1991) and this application has been amended and refined by Read (1994a). In EPR sequence stratigraphy as originally described (Van Wagoner et al. 1988; Posamentier et al. 1988; Posamentier & Vail 1988) depositional sequences were considered to have been produced in open-coastal environments on passive margins by symmetrical, third order eustatic oscillations. These generally had a frequency between 0.5 and 3 Ma. A 'greenhouse' global climate lacking continental-scale ice-caps seems to have been tacitly assumed. The EPR sequence stratigraphy models have subsequently been updated and extended (e.g. Mitchum & Van Wagoner 1991; Posamentier & James 1993).

It is now realized that the formation of sequences is independent of both space and time (Posamentier et al. 1992: Posamentier & Weimer 1993) and depends solely on the rates of eustatic rise and fall in sea level relative to tectonic subsidence. Thus it is perfectly possible for sequences to develop in response to high frequency orbital forcing (Mitchum and Van Wagoner 1991). As the basic concepts of EPR sequence stratigraphy are now well known, the reader is referred to the papers quoted above for further details.

Figure 4 shows a schematic section through part of a distal facies association succession in relation to the systems tracts of a high frequency, type 1 sequence. Typically such sequences in this facies association cannot be divided into subsidiary parasequences. Because continental-scale ice-caps grow more slowly than they melt (Berger 1988), glacial–eustatic sea-level curves are asymmetrical, with rates of eustatic rise that are much more rapid than those of eustatic fall. However, Pleistocene examples (Williams 1988, figs 4 and 6) suggest that rates of fall may, nevertheless, be sufficiently rapid to outpace tectonic subsidence, even in rapidly subsiding basins. This certainly seems to have been the situation in the upper part of the Limestone Coal Formation, where channels were incised during glacial lowstands, producing erosive type 1 sequence boundaries. Thus the high frequency allocycles of the distal facies association can readily be described in terms of EPR sequences by choosing the erosive base of the appropriate incised lowstand channel as the arbitrary base of the cycle, instead of (conventionally) the top of the coal (Fig. 4).

System tracts and the distal facies association

Read (1994a) has described how each lithological member of the distal facies association can be allocated to an appropriate systems tract. This has facilitated the construction of a new depositional model, which has been illustrated by a series of palaeogeographical and palaeoenvironmental sketch maps showing the study area at successive stages of a high frequency, glacial–eustatic cycle (see Read 1994a, Fig. 4). Figure 4 attempts briefly to summarize this information.

Maximum flooding is thought to have occurred at the R inflexion point, when the eustatic sea level was rising most rapidly. At this stage *Lingula* colonized the area (Fig. 4, stage a). Shortly after this point the transgressive systems tract (TST) gave place to the early highstand systems tract (HST) as the rate of eustatic rise decelerated. Brackish water clays started to grade upwards into silts and then upward-coarsening sand sheets as highstand delta lobes filled the Kincardine Basin before starting to prograde westwards along the Kilsyth Trough (Fig. 4, stage b). Later in the HST when the eustatic rise had peaked, rivers succeeded deltas during 'normal' progradation and scoured down into the earlier deltaic sands, producing the scoured surface below the upper storey of the two-storey sheet sands (Fig. 4, stage c). Sandy fluvial deposition probably stopped rather abruptly at about the F inflexion point, which marks the maximum rate of eustatic fall.

After this point, the surface of the HST was exposed and subaerially weathered and eroded. Fluvial channels, which became graded to glacial lowstand sea levels, were also deeply incised. Sediments were transported through these channels towards the distant continental slope, bypassing the present study area. The erosive surface that bounds the incised channels, plus the eroded top of the previous HST, together constitute the type 1 sequence boundary. The lowest and coarsest parts of the sandy fills of the incised channels may belong to the lowstand systems tract (LST) (Fig. 4, stage d), whereas the higher parts of these fills probably belong to the early TST. The erosive type 1 boundary surface was progressively flooded during the deposition of the TST when eustatic sea level started to rise from its lowstand position. Relicts of mature palaeosols which had formed subaerially on this surface have occasionally been preserved but these, together with large areas of the sequence boundary, were generally obscured by later root bioturbation and by overprinting by gleysols, which formed as the water-table rose. (Thus, using the earlier nomenclature, each cyclothem or allocycle includes a prolonged hiatus at about the seat earth horizon, but this break has usually been obscured by root reworking and pedogenesis.)

Peat then accumulated very rapidly to form extensive ombrotrophic raised mires. For a time peat growth was able to keep pace with the accelerating rate of sea-level rise plus tectonic subsidence, but eventually saline water invaded the mires and killed the lycopsid flora. The sharp surface at the top of the coal is equivalent to a transgressive surface of marine erosion or 'ravinement' and the thin impersistent sandy or silty beds containing *Lingula* fragments which locally overlie the coal reflect the reworking of earlier sandy fluvio-deltaic deposits (Fig. 4, stage f).

The upper part of the Limestone Coal Formation is definitely dominated volumetrically by TSTs and HSTs, because LSTs are represented only by the fills of fairly narrow incised channels. This situation is in sharp contrast with that in the Namurian Millstone Grit in the southern Pennines, which tends to be dominated volumetrically by LSTs (Read 1991).

This difference can be explained by the proximal position of the study area relative to its closely adjacent principal source area in the Highland massif. Both the Kincardine Basin and Kilsyth Trough were also hemmed in during the early Namurian by areas of low subsidence, which funnelled and hence accelerated the westward progadation of HST fluvio-deltaic systems. During glacial lowstands the sea probably lay far to the west, possibly in northwest Ireland. The proximal enclosed palaeogeographical setting for the study interval contrasts with the more 'open coast' Namurian settings of the southern Pennine basins. The latter also lay further away from their source area.

Limitations of Exxon Production Research sequence stratigraphy

Although it is comparatively easy to apply sequence stratigraphy to the distal facies association, where deposition was dominantly controlled by regular glacial–eustatic oscillations, the EPR concepts are more difficult to apply to the transitional facies association, in which the underlying fluvial 'signal' was progressively subject to interference from episodic 'noise' effects. Schumm (1977) has shown how fluvial systems are inherently 'noisy' because they exist in a state of dynamically unstable equilibrium, which is subject to the influence of intrinsic 'geomorphic thresholds'. If these thresholds are exceeded, this affects the stability of the system and causes abrupt changes, without any alterations in external influences.

These changes include not only the episodic autocyclic processes within the depositional basin (e.g. channel migration, meander cut-off and lake–delta switching), but also processes further upstream, such as changes in sediment storage within the valley, slope failure and the onset of gullying. Short-term climatic changes, which affect vegetation cover within the catchment area, will also greatly influence the amount and grade of the sediment transported by rivers into the depositional basin.

According to Posamentier & James (1993), the effects of variations in stream discharge and sediment flux, together with those of tectonic activity, tend to become progressively more important upstream in drainage systems. Thus they would be likely to influence deposition in the proximal facies association to a greater extent than in the other two facies associations. In the complex laterally variable successions of the dominantly fluvial proximal facies association, the episodic 'noise' effects listed above seem to have largely obscured the background 'signal' produced by relative sea-level changes, making it very difficult to apply EPR sequence stratigraphy.

Schumm (1993) has suggested that the upstream effects of sea-level changes are likely to be only moderate and relatively short-lived. Thus relative falls in sea level are unlikely to lead to the rejuvenation of whole drainage systems, because unconfined streams will adjust by changing their sinuosity, channel geometry and bed roughness. Judging from the relatively high proportion of sand which entered the basinal areas during the study interval, the river valley deposits upstream from these basins must have been predominantly sandy, so that stream channels were probably unconfined. Thus stream adjustments may help to explain the limited extent to which the base-level effects of eustatic oscillations seem to have been transmitted upstream into the dominantly fluvial palaeoenvironments of the proximal facies association.

The problems that have been encountered in the present study when trying to apply EPR sequence stratigraphy to successions deposited in more proximal settings highlight the danger of attempting to 'force-fit' all European Silesian cyclical successions into such a rigid framework. This is particularly true of any succession in which it is suspected that the effects of any underlying eustatic oscillations may have been significantly modified and obscured by episodic allocyclic fluvial processes, or by tectonic activity.

Conclusions

Regularly spaced high frequency allocycles, which were most probably caused by glacial–eustatic oscillations, have been identified in the upper part of the Limestone Coal Formation in the north-central part of the Midland Valley. Here a combination of rapid subsidence and rapid sedimentation provided optimum conditions for the formation and preservation of such allocycles, whose frequency (broadly equivalent to that of fifth-order EPR cyclicity) was markedly less than the dominant frequencies (95 and 123 ka) of the 'short' eccentricity orbital parameter. The rapidly repeated sea-level oscillations resulted in extensive migrations of a wide spectrum of depositional environments, which ranged from quasi-marine to fluvial. This phenomenon throws doubt on the assumptions that underlie some existing depositional models of late Carboniferous cyclical coal-bearing successions.

Records from numerous closely-spaced boreholes and from underground mines have allowed the lithofacies geometry of the study interval to be determined in considerable detail and enabled the succession to be divided into three facies associations. The distal facies association was dominantly controlled by eustatic oscillations, which resulted in a remarkably regular 'layer-cake' succession containing fairly thin but laterally persistent coals. In the transitional facies association, episodic fluvial processes started to interfere with and obscure the effects of eustasy. Here the succession becomes more sandy and more variable. Coals are generally thicker but may be split, or else amalgamate to form thick composite seams. The more highly variable proximal facies association was dominantly controlled by irregular autocyclic fluvial processes which largely obscured the effects of the background changes in relative sea level, although the latter must inevitably have affected sedimentation in such upstream areas. In this facies association, coals tend to be thick but impersistent, and the original peats lay close to active fluvial channels transporting sand.

Four westward-tapering tongues of the transitional facies, sometimes followed by the proximal facies, reflect periods of increased sand input into the depositional basin and an increase in overall stream gradient. These tongues are fairly regularly spaced and they may possibly reflect EPR fourth-order cyclicity. Between these tongues, the distal facies association extended over most of the study area.

Exxon Production Research sequence stratigraphy may readily be applied to the distal facies association. Here the asymmetrical high frequency allocycles, which include channel sandstones, are equivalent to high frequency, type 1 sequences. However, the EPR concepts are harder to apply to the transitional facies association and become very difficult to apply to the proximal facies association. This demonstrates the danger of attempting to 'force-fit' all late Carboniferous cyclic coal-bearing successions into the rigid framework of EPR sequence stratigraphy.

The non-confidential borehole records held by the BGS in Edinburgh have provided most of the information used in this study and the BGS staff in Edinburgh, particularly R. Gillanders, are thanked for their kind assistance and cooperation. The author thanks S. Flint and an unknown referee for their constructive comments, which have even led him to modify some of his original views. He also thanks I. Chisholm, D. Holliday and J. Rippon for reading an earlier draft of this paper and suggesting improvements.

References

AISENVERG, D. E., BRAZHNIKOVA, N. E., VASSILYUK, N. P., RETLINGER, E. A., FOMINA, E.V. & EINOR, O. L. 1979. The Serpukhovian Stage of the Lower Carboniferous of the USSR. *In:* WAGNER, R. H., HIGGINS, A. C. & MEYEN, S. V. (eds) *The Carboniferous of the U.S.S.R.* Yorkshire Geological Society, Occasional Publication, **4**, 109–124.

BEERBOWER, J. R. 1964. Cyclothems and cyclic depositional mechanisms in alluvial plain sedimentation. *Kansas Geological Survey Bulletin*, **169**, 31–42.

BERGER, A. 1988. Milankovitch theory and climate. *Reviews of Geophysics*, **26**, 624–657.

—— 1989. The spectral characteristics of pre-Quaternary climatic records, an example of the relationship between the astronomical theory and geo-sciences. *In:* BERGER, A., SCHNEIDER, S. & DUPLESSY, J.Cl. (eds) *Climate and the Geo-Sciences, a Challenge for Sciences and Society in the 21st Century.* Kluwer, Dordrecht, 47–76.

CROWLEY, T. J. & BAUM, S. K. 1991. Estimating Carboniferous sea-level fluctuations from Gondwanan ice extent. *Geology*, **19**, 975–977.

—— & —— 1992. Modeling late Paleozoic glaciation. *Geology*, **20**, 507–510.

FERM, J. C. 1970. Allegheny deltaic deposits. *In:* MORGAN, J. P. (ed.) *Deltaic Sedimentation, Modern and Ancient.* Society of Economic Paleontologists and Mineralogists, Special Publication, **15**, 246–255.

FIELDING, C. R. 1984. A coal depositional model for the Durham Coal Measures of N.E. England. *Journal of the Geological Society*, **141**, 919–931.

FORSYTH, I. H. 1979. *The* Lingula *Bands in the Upper Part of the Limestone Coal Group (E_1 Stage of the Namurian) in the Glasgow District.* Report of the Institute of Geological Sciences, **79/16**.

—— & READ, W. A. 1962. The correlation of the Limestone Coal Group above the Kilsyth Coking Coal in the Glasgow–Stirling region. *Bulletin of the Geological Survey of Great Britain*, **19**, 29–52.

HORNE, J. C., FERM, J. C., CARUCCIO, F. T. & BAGANZ, B. P. 1978. Depositional models in coal exploration and mine planning in the Appalachian region. *Bulletin of the American Association of Petroleum Geologists*, **62**, 2379–2411.

MAYNARD, J. R. & LEEDER, M. R. 1992. On the periodicity and magnitude of Late Carboniferous glacio-eustatic sea-level changes. *Journal of the Geological Society*, **149**, 303–311.

MITCHUM, R. M. & VAN WAGONER, J. C. 1991. High-frequency sequences and their stacking patterns: sequence-stratigraphic evidence of high-frequency eustatic cycles. *Sedimentary Geology*, **70**, 131–160.

POSAMENTIER, H. W. & JAMES, D. P. 1993. An overview of sequence-stratigraphic concepts: uses and abuses. *In:* POSAMENTIER, H. W., SUMMERHAYES, C. P., HAQ, B. U. & ALLEN, G. P. (eds) *Sequence Stratigraphy and Facies Associations.* International Association of Sedimentologists, Special Publication, **18**, 3–18

—— & VAIL, P. R. 1988. Eustatic controls on clastic

deposition II—sequence and tract models. *In:* WILGUS, C. K., HASTINGS, B. S., ROSS, C. A., POSAMENTIER, H., VAN WAGONER, J. & KENDAL, C. G. St. C. (eds) *Sea-level Changes: an Integrated Approach.* Society of Economic Paleontologists and Mineralogists, Special Publication, **42**, 125–154.

—— & WEIMER, P. 1993. Silicilcastic sequence stratigraphy and petroleum geology—where to from here? *Bulletin of the American Association of Petroleum Geologists,* **77**, 731–742.

——, ALLEN, G. P. & JAMES, D. P. 1992. High resolution sequence stratigraphy—the East Coulee Delta, Alberta. *Journal of Sedimentary Petrography,* **62**, 310–317.

——, JERVEY, M. T. & VAIL, P. R. 1988. Eustatic controls on clastic deposition I—conceptual framework. *In:* WILGUS, C. K., HASTINGS, B. S., ROSS, C. A., POSAMENTIER, H., VAN WAGONER, J. & KENDAL, C. G. St. C. (eds) *Sea-level Changes: an Integrated Approach.* Society of Economic Paleontologists and Mineralogists, Special Publication, **42**, 109–124.

RAMSBOTTOM, W. H. C. 1977. Correlation of the Scottish Upper Limestone Group (Namurian) with that of the North of England. *Scottish Journal of Geology,* **13**, 327–330.

READ, W. A. (1961) Aberrant cyclic sedimentation in the Limestone Coal Group of the Stirling Coalfield. *Transactions of the Edinburgh Geological Society,* **18**, 271–292.

—— 1988. Controls on Silesian sedimentation in the Midland Valley of Scotland. *In:* BESLY, B. K. & KELLING, G. (eds) *Sedimentation in a Synorogenic Basin Complex. The Upper Carboniferous of NW Europe.* Blackie, Glasgow, 222–241.

—— 1989. The interplay of sedimentation, volcanicity and tectonics in the Passage Group (Arnsbergian, E_2 to Westphalian A) in the Midland Valley of Scotland. *In:* ARTHURTON, R. S., GUTTERIDGE, P. & NOLAN, S. C. (eds) *The Role of Tectonics in Devonian and Carboniferous Sedimentation in the British Isles.* Yorkshire Geological Society, Occasional Publication, **6**, 143–152.

—— 1991. The Millstone Grit of the southern Pennines viewed in the light of eustatically controlled sequence stratigraphy. *Geological Journal,* **26**, 157–165.

—— 1994a. High frequency, glacial-eustatic sequences in early Namurian coal-bearing fluviodeltaic deposits, central Scotland. *In:* DE BOER, P. L. & SMITH, D. G. (eds) *Orbital Forcing and Cyclic Sequences.* International Association of Sedimentologists, Special Publication, **19**, 413–428.

—— 1994b. The frequencies of Scottish Pendleian allocycles. *Scottish Journal of Geology,* **30**, 91–94.

—— & DEAN, J. M. 1982. Quantitative relationships between numbers of fluvial cycles, bulk lithological composition and net subsidence in a Scottish Namurian basin. *Sedimentology,* **29**, 181–200.

—— & FORSYTH, I. H. 1989. Allocycles and autocycles in the upper part of the Limestone Coal Group (Pendleian E_1) in the Glasgow–Stirling region of the Midland Valley of Scotland. *Geological Journal,* **24**, 121–137.

—— & —— 1991. Allocycles in the upper part of the Limestone Coal Group (Pendleian, E_1) of the Glasgow–Stirling Region viewed in the light of sequence stratigraphy. *Geological Journal,* **26**, 85–89.

SCHUMM, S. A. 1977. *The Fluvial System.* Wiley, New York.

—— 1993. River response to baselevel change: implications for sequence stratigraphy. *Journal of Geology,* **101**, 279–294.

VAN WAGONER, J. C., POSAMENTIER, H. W., MITCHUM, R. M., VAIL, P. R., SARG, J. F., LOUTIT, T. S. & HARDENBOL, J. 1988. An overview of the fundamentals of sequence stratigraphy and key definitions. *In:* WILGUS, C. K., HASTINGS, B. S., ROSS, C. A., POSAMENTIER, H., VAN WAGONER, J. & KENDAL, C. G. St. C. (eds) *Sea-level Changes: an Integrated Approach.* Society of Economic Paleontologists and Mineralogists, Special publication, **42**, 39–45.

VEEVERS, J. J. & POWELL, C. McA. 1987. Late Paleozoic glacial episodes in Gondwanaland reflected in transgressive–regressive depositional sequences in Euramerica. *Geological Society of America Bulletin,* **98**, 475–487.

WEEDON, G. P. 1991. The spectral analysis of stratigraphic time series. *In:* EINSELE, G., RICKEN, W. & SEILACHER, A. (eds) *Cycles and Events in Stratigraphy.* Springer, Berlin, 840–854.

—— & READ, W. A. in press. Orbital-climatic forcing of Namurian cyclic sedimentation from spectral analysis of the Limestone Coal formation, Central Scotland. *In:* HOUSE, M. (ed.) *Orbital forcing timescales and cyclostratigraphy.* Geological Society, London, Special Publication.

WILLIAMS, D. F. 1988. Evidence for and against sea-level changes from the stable isotopic record of the Cenozoic. *In:* WILGUS, C. K., HASTINGS, B. S., ROSS, C. A., POSAMENTIER, H., VAN WAGONER, J. & KENDAL, C. G. St. C. (eds) *Sea-level Changes: an Integrated Approach.* Society of Economic Paleontologists and Mineralogists, Special Publication, **42**, 31–36.

Unusual enrichment of U, Mo and V in an Upper Cretaceous coal seam, Hungary

OTTO TOMSCHEY

Laboratory for Geochemical Research, Hungarian Academy of Sciences, Budapest, Hungary

Abstract: The Late Cretaceous coal basin of Ajka lies in the Bakony Mountains, Transdanubia, Hungary, above Triassic–Jurassic dolomites–limestones and/or Upper Cretaceous calcareous marly sediments. The concentrations of U, Mo and V are unusually high (U = 80 ppm, Mo = 138 ppm and V = 600 ppm on average in ash). Analyses were carried out to determine the trace metal concentrations in the various solid phases within the coal rock and the results compared with the U, Mo and V contents of fly ash from the mine-mouth thermal power plant.

The Ajka coalfield has supplied the thermal power plant at Ajka for more than 40 years. The coalfield lies in Central Transdanubia, Hungary, in the western part of the Bakony Mountains (Fig. 1). The significance of these coal reserves was recognized in the 19th century and the first geological description of the area was given by Szabó (1871). The stratigraphic and lithological problems of the coal sequences in the Bakony Mountains (Transdanubia, Hungary), were later discussed by Vadász (1960). More recent results have been obtained in the fields of stratigraphy, palynology and palaeogeography (Haas *et al.* (1986) and by Góczán *et al.* (1986).

The unusual behaviour of trace elements in the coals of this region was first reported by Szádeczky-Kardoss and Földvári (1955). In the 1980s, as a result of investigations of the trace element geochemistry of the coal seams, new data were collected by Tomschey (1988) and these were supplemented by data on the prospective coalfield lying in the continuation of the recent mining activity (Tomschey 1990).

The results obtained so far in the recently mined and planned mining areas are reported, with special emphasis on the unusually high amounts of U, Mo and V; the values obtained for ash samples are compared with those of fly ash produced by the thermal power plant.

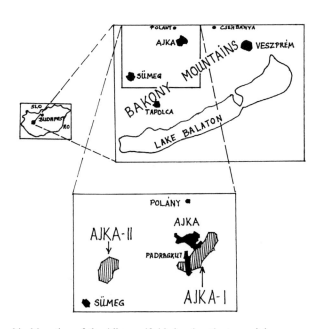

Fig. 1. Geographical location of the Ajka coalfield showing the two mining areas.

From Whateley, M. K. G. & Spears, D. A. (eds), 1995, *European Coal Geology*,
Geological Society Special Publication No. 82, pp. 299–305

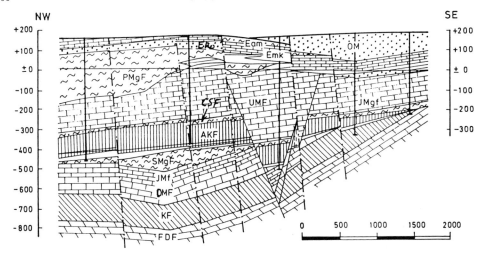

Fig. 2. Geological profile (NW–SE) across the Ajka-II coalfield (after Pera *et al.* 1987). Legend: OM, Oligocene–Miocene rocks; Eam, Eocene clay–marl to marl; Emk, Eocene limestones; Eko, Eocene conglomerate; PMgF, Polány Marl Formation; JMgF, Jákó Marl Formation; AKF, Ajka Coal Formation; CSF, Csehbánya Formation; SMgF, Sümeg Marl Formation; JMf, Jurassic Limestone Formation; DMF, Dachstein Limestone Formation; KF, Kössen Formation; FDF, Main Dolomite Formation; and UMF, Ugod Limestone Formation.

Geological setting

The Ajka coalfield lies in central Transdanubia (Hungary) in the Transdanubian Mid-Mountains (Fig. 1), in a relatively stable tectonic environment. The basement formations consist of older Mesozoic strata; in the southeastern part Triassic calcareous–dolomitic formations are found, and towards the northwestern part Jurassic formations are found. The Triassic and Jurassic sedimentary rocks are overlain by Cretaceous marls that thicken northeastwards. This marl formation represents an impermeable layer between the older Triassic–Jurassic formations and the coal-bearing sequence. The eroded surface of the marl sequence is overlain by the Senonian coal-bearing strata: the Ajka Coal Formation. The lower part of the formation consists of clayey sediments and these are overlain by the coal sequence above which clays, marls and clayey sands alternate in a thickness of 100–150 m. An overall geological profile of the NW–SE direction of the Ajka prospective coal basin is presented in Fig. 2 (after Pera *et al.* 1987).

The profile of the coal seam, the different types of sediment and the preliminary results of trace element distributions in the Ajka area have been described by Tomschey (1989).

Materials and methods

A total of 274 samples from the Ajka mine and from boreholes and 30 samples from the fly ash produced by the thermal power plant at the mouth of the mine were investigated. The in-mine and borehole samples represent the so-called lower seam of the coal sequence (Fig. 3).

The ash contents of the samples were determined by combustion at 1000°C for one hour. Data for the representative samples, i.e. the positions of in-mine samples (Fig. 3) within the lower coal seam (samples Nos. 1–4) and for the borehole samples (abbreviated names of boreholes, depth intervals, samples Nos. 5–12; see Fig. 4) and the corresponding ash contents are shown in Table 1.

The three elements (U, Mo and V) were determined from solution after digesting the fly ash with $LiBO_2$: Mo and V were determined by atomic absorption spectrometry (Perkin-Elmer 5000) and U by photometry (Pye Unicam 1800 SP).

Results and discussion

The concentrations of U, Mo and V in representative samples are given in Table 2. For the in-mine samples the values refer to 5 cm thick samples from the coal, whereas in borehole samples the values relate to the average sample in the depth interval given in Table 1. The high concentrations in samples Nos. 1 and 4 are worthy of mention, although only these two samples displayed such extreme values among the 274 samples studied; if we disregard these two then the maxima and minima are 260 and 10 ppm for U, 473 and 10 ppm for Mo and 1520

ENRICHMENT OF U, Mo AND V IN A COAL SEAM

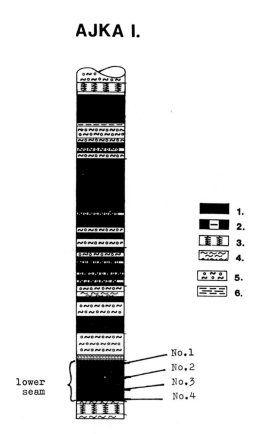

Fig. 3. Profile of the Ajka coal seam showing the lower seam and the position of in-mine samples. Legend: 1, brown coal; 2, clayey coal; 3, coaly clay; 4, marl; 5, molluscan marl; and 6, clay.

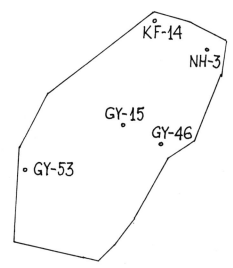

Fig. 4. Position of the boreholes in the planned mining area (abbreviations correspond to those in the tables).

The Clarke number of Mo is 1.5 ppm in the Earth's crust (Taylor 1964). In vitrain ash Otte (1953) found 0.6% Mo (an extremely high value), but usually the Mo concentrations in coal ash vary between several ppm to several tens of ppm. Mo averages of 30 to 40 ppm were reported Leutwein & Rösler (1956) from German coals of early Carboniferous to early Permian age (ash values). Foscolos et al. (1989) described high Mo values (565 and 712 ppm) from lignites in northern Greece (ash values). In the Ajka coals the average Mo concentration is 137 ppm in ash; the extreme value is 473 ppm (disregarding the in-mine samples Nos. 1 and 4).

Table 1. *Samples, positions in seam and ash contents*

	Ash content (%)
Mine samples (lower seam)	
1 1.9 m above floor	8.3
2 1.3 m above floor	25.7
3 0.6 m above floor	51.0
4 0.1 m above floor	22.6
Core samples (depth below surface in m)	
5 NH-3 797.0–798.3	22.0
6 KF-14 752.4–753.4	8.9
7 KF-14 757.2–757.7	21.1
8 KF-14 760.3–760.9	14.9
9 KF-14 637.0–637.9	24.6
10 GY-15 558.6–558.9	21.8
11 GY-46 540.0–540.7	16.3
12 GY-53 686.3–686.6	21.0

and 80 ppm for V (see also Table 6). Figure 5 gives histograms of the concentrations of these elements in the coal ash.

The ash content varies within a wide range (Table 1), so the ash values were converted into whole coal data (Table 3). The extreme values became closer and the values have a more homogeneous distribution.

Uranium is known to occur commonly in coals. In Hungary, the U contents of Eocene brown coals have been reported by Földväri (1952) and Szalay (1954, 1957). In the broader environment, the U concentrations of coals have been reported from the Kladno area, Czech Republic (Bouska 1981), as in the range 2–90 ppm (in ash). The average U content of the coals at Ajka is 23 ppm and the average U concentration of the fly ash is 80 ppm.

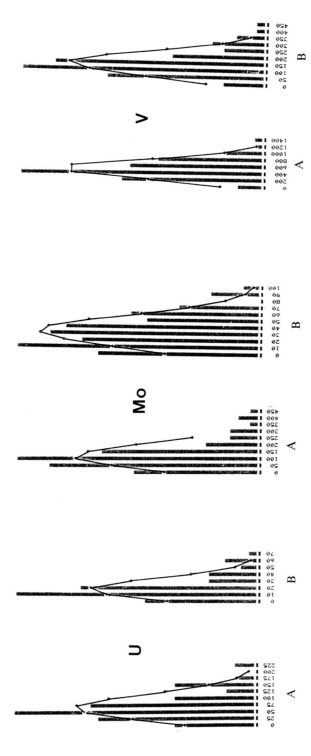

Fig. 5. Histograms of frequency distributions of the U, Mo and V concentrations in the ash (A) and in the coal (B).

In coals V is also a common trace element and, similar to other trace elements, very different V concentrations are reported from the different coal regions of the world (Bouska 1981). Rankama & Sahama (1950) have reported an Argentinian coal with maximum contents of vanadium pentoxide of 21.4% in ash (unfortunately, no data are reported on the ash content). This is an extreme value, of course, but V concentrations in coal ash show averages of 500–1000 ppm in extreme cases; usually the V content is 200–300 ppm or less. In the Ajka coalfield the average V content of the ash is round 600 ppm and is 167 ppm in coal.

Table 2. *Trace element contents of ash burnt at 1000°C for one hour*

Sample No.	U (ppm)	Mo (ppm)	V (ppm)
1	1300	2800	6850
2	260	400	1380
3	250	210	370
4	310	1560	3950
5	110	140	810
6	230	180	1190
7	170	300	1210
8	160	230	770
9	150	320	640
10	160	470	930
11	150	270	850
12	100	440	700

Table 3. *Trace element contents of whole coal recalculated from ash data*

Sample No.	U (ppm)	Mo (ppm)	V (ppm)
1	108	232	569
2	67	103	355
3	129	108	191
4	70	353	893
5	24	31	178
6	20	16	106
7	36	63	255
8	24	34	115
9	37	79	157
10	35	102	203
11	24	44	139
12	21	92	147

The relatively high concentrations of these elements in this coal suggested an investigation of the form of chemical bond of U, Mo and V in the coal. Different views have been reported on the possible trace element-bearing phases. It has been traditionally presumed that U is bound by organic matter. According to Golovko (1960), Mo may be absorbed onto clays or accumulated by sorption on organic matter in the early phase of coal-forming processes and subsequently concentrated in sulphides. Uzunov (1980) reported that in Bulgarian coals the major proportion of V is bound by vitrite.

Shimko & Kuznetzov (1978) developed a method to determine the host phases of trace elements in organic matter-rich sediments. This method, consisting of seven steps with increasingly strong chemical agents, was successfully applied to Slovakian lignites by Tomschey et al. (1986). When investigating the Ajka coal, the procedure was reduced to three steps as suggested by Tomschey (1991): (1) the coal was treated with sodium acetate at pH 7 to remove the elements present in a water-soluble state, in weak and strong surface sorption; (2), the remaining material was treated with a 1:1 mixture of sodium acetate and acetic acid at pH 3 to remove the elements bound by carbonates; and (3) the remaining material was treated with 30% hydrogen peroxide to remove the elements bound by the organic matter and/or by sulphides (unfortunately, this method is unsuitable for differentiating between the chemical bonds of organic matter and sulphides). Material left after this three-step treatment was considered as an insoluble residue.

Table 4 gives information about the form of chemical bond of U, Mo and V in the Ajka coal samples 6 and 7. In Table 4 the letters A, B and C correspond to steps 1, 2 and 3 described above and the values are given as percentages. The element-bearing phases are heterogeneous, e.g. 57% of the total V content is bound to organic matter and/or sulphides in sample 6, but only 27% in sample 7. Taking into account the common geochemical behaviour of V, it is probably bound by organic matter, not by sulphides (pyrite and marcasite in the Ajka coals).

Table 4. *Percentage distribution of U, Mo and V in the various phases of the coal*

Sample No.	U (%)			Mo (%)			V (%)		
	A	B	C	A	B	C	A	B	C
6	6	6	12	—	—	13	—	—	57
7	4	7	5	6	4	17	—	4	27

A: Element in water-soluble state and/or bound adsorptively (solvent, Na acetate; pH 7.0).
B: Element bound by carbonates (solvent, 1:1 mixture of Na acetate and acetic acid, pH 3.0).
C: Element bound by organic matter and/or by sulphides (solvent, 30% hydrogen peroxide).

U and Mo display a similar behaviour, the only difference is that several per cent of the total amounts of these elements are bound to easily soluble phases; their major proportions are bound by the inorganic ash-forming phases.

It is important to know the element-bearing phases as the form of chemical bond in the coal will determine that in the fly ash. Elements in an easily soluble form or bound to the organic matter and/or sulphides will probably occur in the ash as oxides. In the Ajka coals the concentrations of these elements means that it is necessary to consider the ash as a source of these elements. The form of chemical bond in the ash may affect the technology to be used.

To illustrate the fact that in the coals and ashes of the Ajka region U, Mo and V display an unusual enrichment, Table 5 gives the values from Taylor (1964) for the average of the Earth's crust. These values correspond fairly well to those reported earlier by Vinogradov (1962) and by Turekian & Wedepohl (1961) for average marly sediments. The enrichment factor of Mo and, to a lesser extent, those of U and V verify the unusual abundance of these elements in this coal seam. A comparison was made with Austrian coals of a similar age and formation (Sachsenhofer & Tomschey 1992). These investigations supported the preliminary results that these Upper Cretaceous coals in the Transdanubian Mid-Mountains are a unique sequence with high U, Mo and V concentrations.

Table 5. *Crustal abundances of Taylor (1964), average whole coal rock and ash concentrations at Ajka, and the related enrichment factors*

Element	Taylor (1964) (ppm)	Ajka whole coal rock (ppm)	Enrichment factor	Ajka fly ash (ppm)	Enrichment factor
U	2.7	23	8.5	80	29.6
Mo	1.5	37	24.7	138	92.0
V	135	167	1.2	597	4.4

Table 6 compares the average U, Mo and V concentrations of the ash values of borehole coal samples with those of ash samples produced by the thermal power plant. The values are similar, so the power plant ash also provides a representative view of trace element concentrations on a scale of several tonnes.

Table 6. *Comparison of average minimum and maximum concentrations of elements in Ajka coalfield ash and Ajka thermal power plant ash*

	U (ppm)	Mo (ppm)	V (ppm)
Ajka coalfield			
Miniumum	10	10	80
Maximum	230	473	1520
Average (n = 174)	80	138	597
Ajka power plant			
Minimum	40	80	430
Maximum	170	125	725
Average (n = 30)	106	92	585

Conclusions

Based on the results obtained from the geochemistry of U, Mo and V in the Ajka region, it can be stated that in the so-called lower seam: the enrichment of Mo and U in the ash exceeds by about two orders of magnitude the average sedimentary rock concentrations; the enrichment of V is less, but is five times higher than the sedimentary rock average; compared with the concentrations observed in the neighbouring coals of similar age and formation, the enrichment of these elements is seen to be unique in the Eastern Alps–Carpathian region; based on the comparison of these values with those of the ash of the mine-mouth thermal power plant, it is seen that these high concentrations are valid throughout the whole lower seam; and, finally, with the aid of technology still to be developed, the ash could be considered as a secondary raw material for these elements.

References

BOUSKA, V. 1981. *Geochemistry of Coal*. Academia, Prague, 284pp.

FOSCOLOS, A. E., GOODARZI, F., KOUKOUZAS, C. N., & HATZIYANNIS, G. 1989. Reconnaissance study of mineral matter and trace elements in Greek lignites. *Chemical Geology*, **76**, 107–130.

FÖLDVÁRI, A. 1951. The geochemistry of radioactive substances in Mecsek Mountains. *Acta Geologica Hungarica*, **1**, 37–48.

GÓCZÁN, F., SIEGEL-FARKAS, Á., MÓRA-CZABALAY, L., RIMANÓCZY, Á., VICZIÁN, I., RÁKOSI, L., CSALAGOVITS, I. & PARTÉNYI, Z. 1986. Ajka Coal formation: biostratigraphy and geohistory. *Acta Geologica Hungarica*, **29**, 221–231.

GOLOVKO, V. A. 1960. Distribution of trace elements in the coal-bearing strata of the central region. *Doklady AN SSSR*, **132**, 911–914 [in Russian].

HAAS, J., JOCHA-EDELÉNYI, E. CSÁSZÁR, G. & PARTÉNYI, Z. 1986. *Genetic Circumstances of the*

Senonian Coal Measures of the Bakony Mountains. Hungarian Geological Survey, Annual Report 1984, 343–354 [in Hungarian].

LEUTWIN, F. & RÖSLER, H. J. 1956. Geochemische Untersuchungen an palaeozoischen und mezozoischen Kohlen Mittel- und Ostdeutschlands. *Freibergische Forschungshefte,* **C 19**, 1–196.

OTTE, M. U. 1953. Spurenelemente in einigen dutschen Steinkohlen. *Chemie der Erde,* **16**, 239–294.

PERA, F. et al. 1987. *The Projected Ajka-II Mine.* Veszprém Coal Mines, 1–32 [in Hungarian].

RANKAMA, K. & SAHAMA, T. G. 1950. *Geochemistry.* University of Chicago Press, Chicago, 912pp.

SACHSENHOFER, R. F., TOMSCHEY, O. 1992. Gosautype coals of Austria and Hungary—a preliminary geochemical comparison. *Acta Geologica Hungarica,* **35**, 49–57.

SHIMKO, G. A. & KUZNETZOV, V. A. 1978. Analytical methods of rocks and water during geochemical exploration. *Geokhimija i geofizka AN BelSSR* [in Russian].

SZABÓ, J. 1871. The Ajka coal field in the Bakony Mountains. *Földtani Közlöny* [in Hungarian].

SZÁDECZKY-KARDOSS, E. & FÖLDVÁRI-VOGL, M. 1955. Geochemical investigations on Hungarian coals. *Földtani Közlöny* **85**, 7–43 [in Hungarian].

SZALAY, S. 1954. The enrichment of uranium in some brown coals in Hungary. *Acta Geologica Hungarica,* **2**, 299–310.

—— 1957. The role of humus in the geochemical enrichment of U in coal and other bioliths. *Acta Physica Hungarica,* **8**, 25–35.

TAYLOR, S. R. 1964. Abundance of chemical elements in the continental crust: a new table. *Geochimica Cosmochimica Acta,* **28**, 1273–1278.

TOMSCHEY, O. 1988. *Trace Metals and Radioactive Elements in the Coals of the Ajka-II Coal Field and Comparison with Ajka-I.* Research Report, Budapest, Laboratory for Geochemical Research, 25 pp (manuscript) [in Hungarian].

—— 1989. *Comprehensive Evaluation of the Trace Element Geochemical Investigations on the Coals of the Ajka-II Region.* Research Report, Budapest, Laboratory for Geochemical Research, 71pp (manuscript) [in Hungarian].

—— 1990. Trace elements in the Ajka-II Upper Cretaceous coal basin, Transdanubia, Hungary. *Acta Geologica Hungarica,* **33**, 121–135.

—— 1991. Distribution of trace elements in coal and their host phases in a Lower Eocene coal seam of Hungary. *Bulletin de la Société Géologique de France,* **162**, 267–270.

——, HARMAN, M. & BLASKO, D. 1986. Trace element distribution in the Pukanec lignite deposit. *Geologicky Zbornik,* **37**, 137–146.

TUREKIAN, K. K. & WEDEPOHL, K. H. 1961. Distribution of the elements in some major units of the Earth's crust. *Bulletin of the Geological Society of America,* **72**, 175–191.

UZUNOV, J. 1980. Geochemical nature of petrographic components of coal and ways of vanadium concentration in them. *Geologica Balcanica,* **10**, 57–74.

VADÁSZ, E. 1960. *Geology of Hungary.* Akadémiai Kiadó, Budapest, 646 pp [in Hungarian].

VINOGRADOV, A. P. 1962. Average occurrences of chemical elements in the main magmatic rock formations of the Earth's crust. *Geokhimija,* 555–572 [in Russian].

Origin and distribution of sulphur in the Neogene Beypazari Lignite Basin, Central Anatolia, Turkey

M. K. G. WHATELEY[1] & E. TUNCALI[2]

[1] *Department of Geology, University of Leicester, Leicester LE1 7RH, UK*
[2] *Directorate of Mineral Research and Exploration, Ankara, Turkey*

Abstract: During the Miocene a number of fault-bounded basins developed in Central Anatolia, Turkey. These basins were filled initially with coarse clastic material. Upward fining of the clastics during basin-fill, with an increase in the amount of clay and carbon content, led to the development of relatively shallow limnic basins in which extensive peat deposits accumulated. One such basin at Çayirhan, near Beyparazi, contains thick laterally extensive lignite seams. These lignites are characterized by their high sulphur content (up to 8.2% on an air-dried basis). It is suggested that hydrothermal processes are responsible for the increased sulphur contents of the Çayirhan lignites, resulting from sulphate and sulphide precipitation. The mineral matter contains ubiquitous zeolites. The presence of heulandite in the first seam and analcime in the second seam may be a result of depth–temperature control on the distribution of zeolites in the lignite; but these differences are more probably the result of variations in the chemistry of the circulating fluid. Study of the sulphur content reveals three types of sulphur distribution, namely, vertical variation within individual seams, variation between seams and lateral variation across the basin. The first two are related to variations in the chemistry of the mineral matter in the lignite, and the last is probably related to structural/topographic control of the mire at the time of formation.

The large Neogene Beyparazi Basin lies about 100 km northwest of Ankara in Central Anatolia. The basin is filled with mainly lacustrine and volcano-sedimentary rocks and the sequence contains economic resources of lignite, bituminous shale and trona.

The Beyparazi Basin (Fig. 1) was first investigated by Kalafatçioglu & Uysalli (1964), who studied the stratigraphy, sedimentology and tectonics of the basin and the adjacent area. This was followed with studies by Altinli (1977), Saner (1979) and Tunc (1980), who further

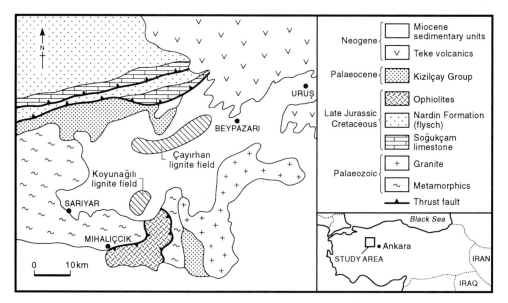

Fig. 1. Location of the Beypazari Basin showing the position of the Çayirhan lignite field and the main rock units in the region. Modified after Yagmurlu et al. (1988).

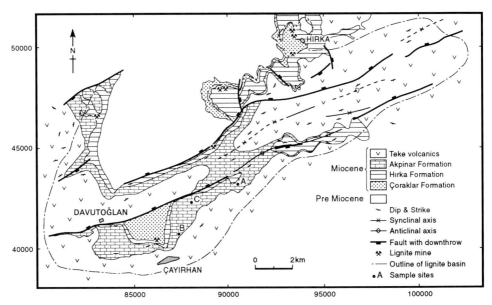

Fig. 2. Simplified geological map of the Çayirhan lignite field. Modified after Gökmen et al. (1993). A, B and C mark the location of the underground sites at which the upper lignite seam was sampled.

refined the stratigraphic, sedimentological and tectonic models of the basin. The lignite in the basin underwent initial prospecting between 1939 and 1954 (Gökmen et al. 1993) and a 1:25 000 scale geological map was produced in 1963 (Gökmen 1965). Mining of the lignite started in 1966. Extensive drilling took place in 1976, and Narin (1980) and Siyako (1984) used the results to describe the geological aspects of the lignite.

The results showed that the Çayirhan and Koyunağili lignite fields in the Beyparazi Basin (Fig. 1) contain more than 400 Mt of lignite resources, enough to support the two 150 MW thermal power stations sited on the Çayirhan lignite field. To alleviate the problem of high sulphur content, both are equipped with desulphurization plants. The lignite is extracted by underground methods, including both mechanized longwall and modified manual sections.

In general, the inorganic constituents of lignite can contribute to numerous technological problems such as the abrasion of mining equipment, boiler fouling and slagging and environmental pollution. Sulphur is one of the main constituents that contributes to the sulphur dioxide emissions and resulting acid rain problems. Generally, lignite with less than 0.6% sulphur on an air-dried basis would meet environmental regulations in the USA (Casagrande 1987). Exploration projects therefore tend to look for lignite with the lowest sulphur content.

The aim of this paper is to review the sedimentology of the basin-fill, including the lignite, and to present the preliminary results of the investigation into the chemistry of the lignite and the origin and distribution of the sulphur in the Çayirhan lignite.

Geological setting

The Çayirhan lignite field extends NE–SW along the northern edge of the Beyparazi Basin (Fig. 1). The southern and northern margins of the lignite-bearing strata appear to be defined by NE–SW trending faults in places (Fig. 2). In the Çayirhan lignite field the strata are folded along a NE–SW trending, asymmetrical, antiformal axis, with the southern limb steeply dipping to the south (Fig. 2). Yagmurlu et al. (1987) suggest that the basin developed as a half-graben formed under an extensional tectonic regime, with the northern margin acting as the active downthrown side.

For convenience, the rocks in the area are divided into pre-Neogene and Neogene age groups (Fig. 3) The pre-Neogene rocks consist mainly of Palaeozoic metamorphic schists which were intruded by granite, Jurassic and lower Cretaceous limestones and ophiolites and upper Cretaceous and Palaeocene clastic sediments. The Jurassic and Cretaceous units have been

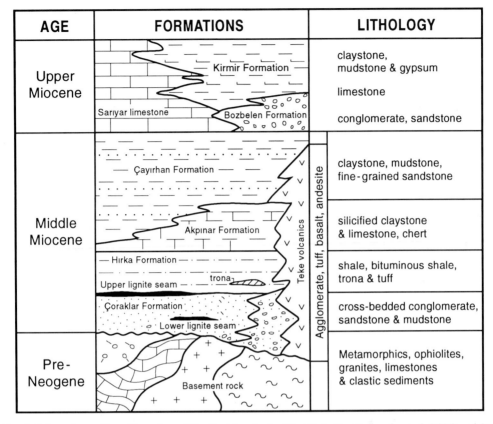

Fig. 3. Schematic stratigraphic section of the Çayirhan basin. Modified after Yagmurlu et al. (1988) and Inci (1991).

thrust southward over the Palaeozoic and Palaeocene rocks (Altinli 1977; Yagmurlu et al. 1987).

The Neogene sequence (currently thought to be of middle and upper Miocene age in the study area) consists mainly of clastic sediments, carbonate rocks, volcanic rocks and evaporites, which show widespread lateral and vertical lithofacies variations. The sediments are divided into eight formations (Fig. 3).

The Çoraklar Formation rests unconformably on the pre-Neogene rocks and consists of conglomerates, sandstones, siltstones and two separate lignite zones (Fig. 3). The lower lignite seam was deposited in the lower part of the Çoraklar Formation, whereas the thicker, economically important, upper lignite seam was deposited at the top of the formation.

The conglomerates are yellow-green, poorly packed and sorted and show large-scale trough and planar cross-bedding. There is a general fining upwards. The clasts mainly consist of metamorphic, volcanic and limestone pebbles, cobbles and boulders. They are interpreted by Yagmurlu et al. (1988) as channel-fill conglomerates. Inci (1991) interprets these conglomerates as being deposited in proximal alluvial fans, which interfinger with the Teke volcanic rocks (Fig. 3).

The sandstones above the conglomerates are laterally discontinuous, but are stacked vertically and are 0.75–4.50 m thick. They are composed of epiclastic grains of similar origin to those in the conglomerate. The grains are moderately well sorted and are angular to subrounded. There is a vertical change from planar to trough cross-bedding followed by parallel laminations. The tops of the sandstones show bioturbation. Palaeocurrent directions are towards the south (Yagmurlu et al. 1988). The sedimentological data have been interpreted by Yagmurlu et al. (1988) to represent high to mid-flow regimes of meandering stream systems, although Inci (1991) interprets them as being deposited in a braided river environment.

The mudstones towards the top of the

Table 1. *Average proximate analyses, sulphur contents and calorific values of the upper lignite seam in the Çayirhan lignite field, illustraing lateral variation in quality between the eastern and western areas, and vertical variation between the first and second seams throughout the basin. DMMF, Dry mineral matter-free*

	Proximal analysis (as-received)			
	Western area	Eastern area	First seam	Second seam
Moisture content (%)	21.71	26.44	24.69	23.99
Ash content (%)	34.35	25.36	28.46	30.86
Volatile content (%)	25.67	25.92	26.18	24.23
Fixed carbon (%)	21.42 (DMMF 48.75)	23.50 (DMMF 48.76)	18.10	18.91
Total sulphur content (%)	4.04	2.74	3.59	3.24
Calorific value (kcal kg^{-1})	2557	2839	2682	2686
$R_{(max)}$ (%)			0.37	0.35

formation are generally light green with occasional silty and sandy lenses with some volcanic material. The amount of carbonaceous material increased upwards until the economically important lignite seam was deposited at the top of the formation in a lacustrine environment (Inci 1991).

There is a very sharp contact between the lignite seam and the overlying Hirka Formation. The Hirka Formation consists mainly of alternating shales, mudstones, trona, bituminous shales, tuffs and silicified limestones. These sediments have been interpreted as being deposited in a lacustrine environment (Yagmurlu *et al.* 1988; Inci 1991). The sharp change in sediment type and sedimentation style has been put down to a change in climate, from a wet (for lignite) to a hotter and drier (formation of trona) climate. Alkaline springs originating in contemporaneous Teke volcanics are believed to have contributed to the trona brine in the ephemeral lake (Inci 1991). The climate appears to have become increasingly arid until evaporites, including gypsum, were deposited in the Kimir Formation (Fig. 3).

Çayirhan lignites

The lower lignite seam is areally restricted and laterally discontinuous, has numerous seam partings and varies from 1.0 to 10.95 m thick. It was only discovered in 1982 during the development of an underground adit. It is a low quality seam with high ash (52%) and sulphur (3%) contents. This seam is not exploited at present.

There are about 150 m of parting sediments between the lower seam and the upper seam. These sediments fine upwards from sandy braided river channel-fill facies above the lower seam, through interbedded sandstones, siltstones and claystones to shales and carbonaceous shales deposited in a lacustrine environment, below the upper seam.

The upper lignite seam is laterally extensive and varies from 1.0 to 4.9 m thick, but averages about 3.0 m thick. A 1 m thick parting composed of siltstone with chert nodules splits the upper seam into two lignite beds, referred to colloquially as the first and second seams. These seams show lateral variations in quality between the northeastern and southwestern parts of the basin (Table 1) as well as vertical variations (Fig. 4).

The upper seam

The upper seam was intersected in 139 boreholes and a sample of each of the first and second seams was analysed for total sulphur in 130 of these boreholes.

The macroscopic characteristics of the first and second seams, which make up the upper seam (Fig. 4), were described in detail at three places in the underground workings. The lignite is brownish to black in colour. The first seam is mainly bright with some dull bands. The second seam is generally dull with bright bands. Reflectance measurements (R_{max}) have an average of 0.359, which puts them into the lignite rank category.

Lignite at the three sites was sampled at 10 cm intervals, giving a total of 87 samples. For each sample the following were produced: petrographic analysis, palynological description, proximate analysis, total sulphur content and calorific value. Ash oxide analysis was also carried out on each sample.

Lignite petrography

The petrographic composition of each sample was determined and the variation in the three main maceral groups (huminite, liptinite and inertinite) was plotted adjacent to a vertical

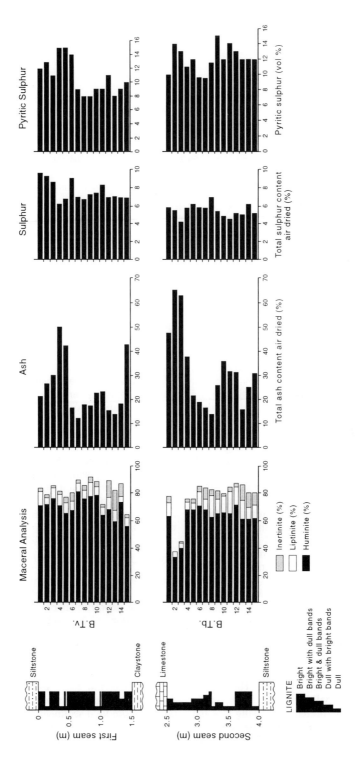

Fig. 4. Vertical distribution of macerals and air-dried ash and sulphur contents of the first and second seam as seen at representative site B in the Çayirhan lignite

Table 2. *Average maceral content (vol.%) of the 87 samples from the first and second seams of the upper lignite from the three underground sample sites, A, B and C, in the Çayirhan lignite field, Beyparzari*

	First seam		Second seam	
	Average	Range	Average	Range
Textinite	3	0–15	3	0–17
Texto-ulminite	11	1–26	9	0–33
Eu-ulminite	8	0–20	7	0–22
Detrohuminite	3	0–17	6	0–23
Attrinite	2	0–7	2	0–11
Densinite	8	0–22	5	0–23
Gelinite	34	12–81	32	11–58
Corpohuminite	1	0–6	1	0–7
Total huminite	**71**	**56–86**	**63**	**33–83**
Sporinite	5	0–16	7	0–14
Cutinite	<1	0–3	<1	0–4
Resinite	<1	0–4	<1	0–3
Total liptinite	**6**	**0–16**	**8**	**0–15**
Micrinite	<1	0–4	<1	0–2
Macrinite	<1	0–4	1	0–7
Fusinite	4	0–12	3	0–8
Semifusinite	1	0–8	<1	0–3
Sclerotinite	<1	0–2	<1	0–4
Total inertinite	**6**	**0–23**	**4**	**0–19**

profile of the macroscopic constituents of the seam (Fig. 4).

The total huminite content for the first seam varies between 56 and 86 vol.% and for the second seam between 33 and 83 vol.% (Table 2). The main macerals in the huminite group are gelinite, densinite, texto-ulminite and eu-ulminite. Liptinite contents are low to moderate (1–16 vol.%). The predominant maceral is sporinite, with cutinite and resinite accounting for up to 4 vol.%. Macerals of the inertinite group form up to 23 vol.%. Significant contributions come from fusinite (12 vol.%) and semifusinite (8 vol.%). Other inertinite macerals such as micrinite and macrinite occur rarely. It can be seen that the amount of huminite decreases towards the top of the second seam, which corresponds to an increase in the mineral matter (ash) content. The huminite content is much less variable in the first seam. Pyrite (or marcasite) was seen in lignites as veinlets, disseminations and framboids.

Palynology

The spores and pollen types and their species were identified in each of the 87 lignite samples collected during this study (Table 3). The vertical distribution of the palynomorphs is shown in Fig. 5. They are interpreted as showing that the Çayirhan lignites are middle Miocene in age.

The palynological content of the samples shows that *Alnus*, *Quercus*, Cyrillaceae, Betulaceae, Fagaceae and Juglandaceae are dominant. The first seam appears to have higher percentages of Polypodiaceae, Taxodiceae, Pinaceae, Myricaceae and Ulmaceae. The *Myrica* and Cyrillaceae are interpreted as representing *in situ* vegetation contributing directly to the peat in a lake environment. The tree palynomorphs are thought to have originated from a high altitude (500–1000 m) forest vegetation which probably surrounded the lake and were carried into the lake by wind and streams. To support this range of vegetation the annual rainfall must have been high. There was probably a seasonal range in temperature, with summer temperatures being fairly high.

Lignite facies and depositional environment

Facies-critical macerals and petrographic indices derived from maceral analysis can be used to assess the depositional environment during the accumulation of the peat (Diessel 1992). The proportions of macerals in coal reflect the organic source materials contributing to the accumulation of peat and the conditions during accumulation, i.e. water-table height, pH, anoxic and oxic bacterial decay and mechanical breakdown of the organic matter related to transportation before final sedimentation.

In lignites, the facies-critical macerals are components such as humotelinite, which indi-

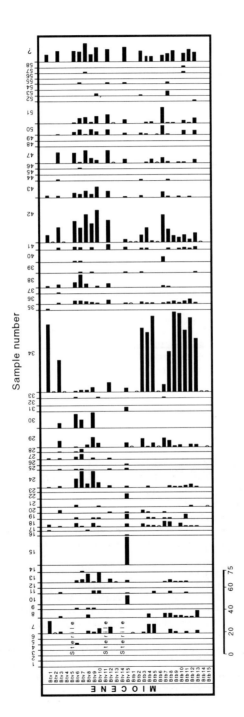

Fig. 5. Vertical distribution of palynomorphs at representative site B in the Çayirhan lignite field. For a complete description of the palynomorphs, see Table 3.

Table 3. *Description of the palynomorphs identified in the upper seam in the Çayirhan lignite field, Beyparzari. The numbers are cross-references to Fig. 5*

1 *Leiotriletes* (Naumova 1937) R. Potonie ve Kremp 1954.
2 *Leiotriletes adriennis* (Pot. ve Gell.) Krutzsch 1959
3 *Leiotriletes microadriennis* Krutzsch 1959
4 *Toroisporis* Krutzsch 1959
5 *Echinatisporis* Krutzsch 1959
6 *Baculatisporites primarius* (Wolff 1934) Thomson ve Pflug
7 *Laevigatosporites haardti* (R. Potonie ve Venitz) Th.ve Pf.
8 *Cycadopites* (Wodehouse)
9 *Monocolpopollenites tranquillus* (R. Potonie Th.ve Pf.
10 *Arecipites* (Wodehouse)
11 *Magnolipollis* Krutzsch
12 *Monogemmites* Krutzsch
13 *Inaperturopollenites* PFlug ve Thomson
14 *Inaperturopollenites dubius* (Pot.ve Ven) Th.ve Pf.
15 *Inaperturopollenites hiatus* (Pot.) Th.ve Pf.
17 *Sequoiapollenites* Thierg.
18 *Pityosporites microalatus* (Pot.) Th.ve Pf
20 *Pityosporite alatus* (Pot.) Th.ve Pf.
21 *Graminidites* Cookson
22 *Sparganiaceaeapollenites* Thiergart
23 *Cypera ceaepollis* Krutzsch
24 *Triatriopollenites* Pflug
25 *Triatriopollenites rurensis* Pf.ve Th
26 *Triatriopollenites bituitus* (Pot.) Th.ve Pf.
27 *Momipites punctatus* (Pot.) Nagy
28 *Triatriopollenites myricoides* (Kremp) Th.ve Pf.
29 *Triatriopollenites coryphaeus* (Pot.) Th.ve Pf.
30 *Triatriopollenites coryloies* Pf. in Th.ve Pf.
31 *Trivestibulopollenites betuloides* Pf. in Th.ve Pf.
32 *Caryapollenites* Raatz 1937 ve R. Potonie emend. Krutzsch
33 *Caryapollenites simplex* (Pot.) Raatz
34 *Polyvestibulopollenites verus* (Pot.) Th.ve. Pf.
35 Polyporopollenites Pf. in Th.ve Pf.
36 Polyporopollenites undulosus (Wolff) Th.ve Pf.
37 *Porocolpopollenites* Pf. in Th.ve Pf.
38 *Tricolporopollenites* Th.ve Pf.
39 *Tricolporopollenites henrici* (Pot.) Th.ve Pf.
40 *Tricolporopollenites asper* Th.ve Pf.
41 *Tricolporopollenites densus* Pf. in Th.ve Pf.
42 *Tricolporopollenites microhenrici* (Pot.) Th.ve Pf.
43 *Tricolporopollenites liblarensis* (Th.) Th.ve Pf.
44 *Tricolporopollenites retiformis* Th.ve Pf.
45 *Tricolporopollenites retimuratus* (Trevisan)
46 *Ephedripites* Bochovitina
47 *Tricolporopollenites* Pf.ve Th.
48 *Tricolporopollenites villensis* (Th.) Th.ve Pf.
49 *Tricolporopollenites pseudocingulum* (Pot.) Th.ve Pf.
50 *Tricolporopollenites cingulum* (Pot.) Th.ve Pf.
51 *Tricolporopollenites megaexactus* (Pot.) Th. ve Pf.
52 *Tricolporopollenites eupohrii* (Pot.) Th.ve Pf.
53 *Tricolporopollenites microreticulatus* Th.ve Pf.
54 *Ilexpollenites margaritatus* (Pot.) Thiergart ve Potonie
55 *Teracolporopollenites* Pf.ve Th.
56 *Periporopollenites multiporatus* Th.ve Pf.
57 *Ovoidites* Pot.
58 *Ovoidites parvus* (Cookson ve Dettmann)

cates an origin from wood-producing plants, and liptinite macerals such as sporinite and cutinite, which refer to specific original material, i.e. spores and leaves, respectively. Others such as fusinite are generally regarded as the burnt remains of plant material (charcoal) after a swamp fire (Jones *et al.* 1991). This must reflect a relatively low position of the water-table during peat accumulation. Subaquatic conditions are indicated by an association of humodetrinite, sporinite and clay minerals.

Diessel (1986) introduced two petrographic indices, namely the gelification index (GI) and the tissue preservation index (TPI). He compared sedimentologically well characterized strata to prevailing mire types such as dry forest swamp, wet forest, marsh, fen, etc. He was then able to use the GI and TPI values to help define depositional environments, e.g. back barrier, upper delta plain. This concept, originally developed on Permian coals from Australia, appears to be applicable to other coal-bearing sequences and geological ages (Yeo *et al.* 1988; Kalkreuth & Leckie 1989).

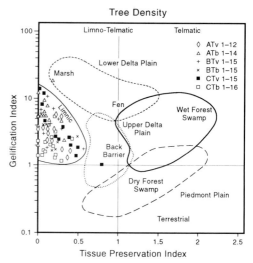

Fig. 6. Maceral-critical facies diagram and proposed depositional environment for the Çayirhan lignite. Modified after Diessel (1986).

In this study of the Çayirhan lignites slightly modified GI and TPI values were used to define mire types and the depositional environment from the petrographic data listed inTable 2. It is apparent that the most significant differences in facies-critical macerals occur within the humotelinite and humodetrinite groups. There appear to be insignificant changes to the liptinite and inertinite groups in the lignite-bearing succession.

The modified indices are:

$$GI = \frac{\text{All huminite (except texto-ulminite and detrohuminite)} + \text{macrinite}}{\text{Semifusinite} + \text{fusinite} + \text{inertodetrinite} + \text{texto-ulminite} + \text{detrohuminite}}$$

$$TI = \frac{\text{Humotelinite} + \text{corpohuminite } (in\ situ) + \text{fusinite} + \text{semifusinite}}{\text{Gelinite} + \text{macrinite} + \text{humodetrinite}}$$

When plotted (Fig. 6), all of the samples plot in a position interpreted as a limnic environment. Cohen et al. (1984) described periodic fires in the raised, forested areas of the Okefenokee Swamp producing charcoal which becomes incorporated into peat as fusinite. The drier forested areas develop between the open, waterlogged 'prairies' in which huminite-rich peats form. The presence of fusinite in the upper seam lignite indicates periodic desiccation in the mire caused by local lowering of the water-table, and possible forest fires.

Chemical characteristics of the Çayirhan lignite

The variable chemical characteristics of the Çayirhan lignite, which play an important part in their utilization for electric power generation, results from physicochemical processes which existed during the formation of the peat and its subsequent coalification. The lignite chemistry can also be used to interpret the environment of deposition of the lignite. A brief comparison of the available data for lignite quality and ash characterization is presented.

Lignite quality

The lateral variation in lignite quality is given in Table 1. Figure 4 illustrates how the air-dried ash and sulphur contents change upwards both in and between the first (BTv) and second (BTb) seams. Table 4 summarizes the statistics of the proximate analyses, sulphur contents and calorific values. The average ash content of the second seam is higher, but the same pattern of ash distribution emerges in both the second and first seams (Fig. 4). The basal sample is relatively high in ash, followed by alternating high and low ash content zones until near the top of each seam where there is a significant increase in ash content. This pattern suggests similar changes in depositional control during the deposition of the first and second seams.

The average total sulphur content of the first seam is higher than the second seam (Table 4 and Fig. 4).

The rank of the two seams is almost identical on a dry mineral matter-free (DMMF) fixed carbon basis (Table 1). The first seam has a slightly higher reflectance value (R_{max}) of 0.37% than the second seam, which has a value of 0.35%. The second seam has a higher ash content (Table 4) and has more dull coal in the section (Fig. 4), which may account for the lower reflectance values. Kavusan (1993) shows that the mean reflectance values of the upper seam increase towards major thrust faults. He suggests that the increase in rank results from accelerated coalification caused by tectonic factors.

The plot of the DMMF hydrogen to carbon and oxygen to carbon ratios (Fig. 7) indicates that the first seam has a higher carbon content than the second seam. The 'total' carbon thus derived does not include material derived from inorganic sources, as very few carbonate minerals were identified. The first seam has a higher hydrogen to oxygen ratio. Coalification increases from the upper right to the lower left of the graph. This suggests that the first seam has a slightly higher percentage of organic matter, but that it is of lower rank. This rank variation is extremely small, as are the results for reflectance measurements. They both confirm the lignite rank.

Mineral matter

The major mineral phases found in the as-received, raw lignites were identified using X-ray diffraction (XRD) and energy-dispersive X-ray (EDX) analyses. In the first seam, XRD traces (Fig. 8a) indicate the presence of heulandite [$(Ca,NA_2)(Al_2Si_7O_{18}) \cdot 6H_2O$], gypsum ($CaSO_4 \cdot 2H_2O$) and pyrite ($FeS_2$), with minor amounts of quartz, calcite and plagioclase feldspar. In the second seam (Fig. 8b) the major mineral phases include analcime [$Na(AlSi_2O_6) \cdot H_2O$], gypsum and pyrite, with minor amounts of quartz, hemihydrite, calcite and plagioclase feldspar.

Table 4. *Summary of the as-received proximate analyses, total sulphur content and calorific value of the first and second seams of the upper seam at site B in the Çayirhan lignite field, Beyparzari*

	Moisture (%)	Ash (%)	Volatile matter (%)	Fixed carbon (%)	Total sulphur (%)	Calorific value (kcal kg^{-1})
BTv (first seam)						
Average	27.59	19.18	26.83	26.40	4.75	3086
Minimum	18.51	9.59	19.20	14.71	4.12	1922
Maximum	31.60	42.90	29.30	32.05	6.13	3692
σ	3.61	10.02	2.65	5.16	0.72	578
σ^2	13.06	100.50	7.02	26.63	0.52	334 955
CV	0.13	0.52	0.10	0.20	0.15	0.19
BTb (second seam)						
Average	18.82	27.36	30.65	23.18	3.61	2748
Minimum	4.90	11.94	21.93	6.68	3.01	1202
Maximum	25.57	60.42	34.95	29.93	4.47	3559
σ	5.74	15.26	3.81	7.19	0.43	766
σ^2	32.91	232.97	14.51	51.67	0.19	588 265
CV	0.30	0.56	0.12	0.31	0.12	0.28

CV, Coefficient of variation.

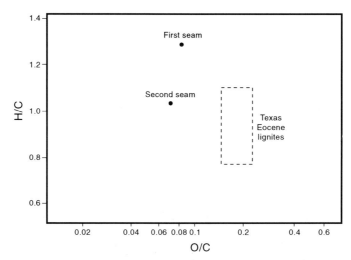

Fig. 7. Atomic H/C and O/C plot of the Çayirhan lignite. Modified after Tewalt (1987). The Texas lignites are plotted for comparative purposes.

Heulandite and analcime are zeolites which form as secondary minerals derived by hydrothermal alteration of the various aluminosilicates such as feldspars (usually in volcanic or igneous rocks) or as authigenic minerals (usually in sedimentary rocks). Blatt (1982) describes several types of alteration of volcaniclastic detrital fragments found in sedimentary rocks. One example of a situation where this alteration would occur is in sands deposited in basins adjacent to convergent plate margins, where there would be a high heat flow and access to saline water in rich sodium. A particularly common feature of this alteration is the crystallization of zeolites, which could be considered a product of either a high grade diagenesis or low grade metamorphism. They appear to form only where there is an absence of carbonate rocks.

Circulating saline low temperature groundwater in the closed Çayirhan basin probably derived the sodium, potassium and calcium from the overlying gypsiferous trona deposits. Westercamp (1981) recognized temperature-dependent zones of zeolites with heulandite forming at the lowest temperature (50–100°C), which altered to analcime at about 140°C. There may

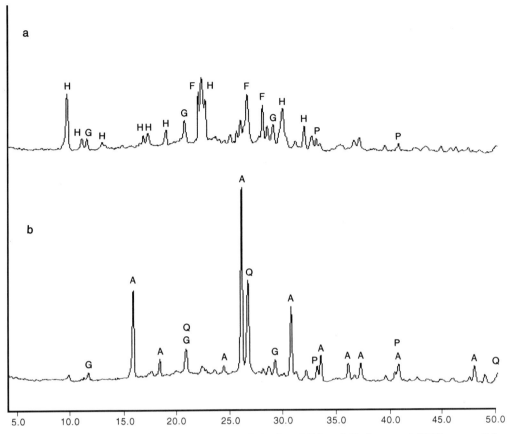

Fig. 8. Typical XRD patterns of raw lignite from (**a**) the first (BTv) and (**b**) the second (BTb) seams of the Çayirhan lignite, Beypazari Basin, Central Anatolia, Turkey. H, Heulandite; A, analcime; F, plagioclase feldspar; G, gypsum; P, Pyrite, and Q, quartz.

be a depth–temperature control on the distribution of zeolites in lignite, but these differences more probably relate to variations in the chemistry of the original volcaniclastic or clastic material associated with lignite or to variations in the chemistry of the circulating fluid.

During the initial circulation, the temperature of the hydrothermal fluid would have increased and under these conditions the hydrothermal solution has been shown to lose its magnesium and sulphate, but gain calcium, potassium, sodium and silica (Kristmannsdottir 1979). Sulphate is precipitated as $CaSO_4$ in joints and cleats. In a reducing environment, such as a lignite seam, sulphate would be reduced to pyrite. Kristmannsdottir (1979) has shown that hydrothermal solutions create a sequence of alteration minerals related to temperature and depth, with zeolites forming at temperatures < 100°C. It was also shown that rising hydrothermal solutions precipitated pyrite in veins over a wide (30–330°C) temperature range.

These processes would account for the increased sulphur contents of the Çayirhan lignites, resulting from increased sulphate and sulphide precipitation, and the ubiquitous presence of zeolites in the mineral matter. The presence of heulandite in the first seam and analcime in the second seam may be a result of depth–temperature control on the distribution of zeolites in the lignite but, as the seams are only 1 m apart, these differences are more probably the result of variations in the chemistry of the original volcaniclastic or detrital clastic material associated with the lignite or to variations in the chemistry of the circulating fluid.

Ash characterization

The lignite samples were reduced to ash at 750°C in accordance with ASTM procedures. The ash oxide analyses thus derived and the

Table 5. *Table of ash oxide analyses and correlation matrices of the 87 samples taken from three sites in the underground workings. The data are presented in terms of the first and second seams*

	SiO_2	Al_2O_3 $+ TiO_2$	MgO	K_2O	Fe_2O_3	CaO	Na_2O	SO_3
Ash oxide analysis of first seam only								
Mean	35.33	14.33	5.94	1.18	12.96	12.51	3.09	13.15
Minimum	18.48	3.63	1.60	0.40	9.57	5.55	0.90	4.53
Maximum	52.58	22.93	17.00	1.80	17.95	28.25	6.90	21.90
σ^2	189.23	31.98	25.36	0.24	8.23	65.25	3.11	51.26
σ	13.76	5.65	5.04	0.49	2.87	8.08	1.76	7.16
CV	0.39	0.39	0.85	0.41	0.22	0.65	0.57	0.54
SiO_2		0.675	−0.695	0.866	−0.166	−0.889	0.367	−0.983
$Al_2O_3 + TiO_2$			−0.766	0.606	−0.162	−0.822	0.375	−0.671
MgO				−0.869	−0.352	0.797	−0.426	0.637
K_2O					0.254	−0.810	0.234	−0.815
Fe_2O_3						−0.068	0.140	0.287
CaO							−0.545	0.836
Na_2O_3								−0.373
Ash oxide analysis of the second seam								
Mean	48.64	18.31	2.43	1.16	10.96	4.91	6.80	5.45
Minimum	43.20	14.98	1.40	0.80	9.97	1.20	3.00	0.87
Maximum	52.90	22.28	3.80	1.60	13.50	8.10	8.40	9.50
σ^2	14.77	6.42	0.99	0.08	1.36	6.73	3.95	9.94
σ	3.84	2.53	1.00	0.29	1.17	2.59	1.99	3.15
CV	0.08	0.14	0.41	0.25	0.11	0.53	0.29	0.58
SiO_2		0.413	−0.242	0.463	0.088	−0.632	−0.192	−0.834
$Al_2O_3 + TiO_2$			−0.881	−0.034	−0.080	−0.880	0.680	−0.802
MgO				−0.104	0.018	0.668	−0.572	0.636
K_2O					0.335	0.098	−0.698	−0.345
Fe_2O_3						−0.109	−0.396	−0.181
CaO							−0.554	0.871
Na_2O_3								−0.186

correlation coefficients are presented in Table 5. The silica values are lowest in the first seam, whereas the CaO and SO_3 values are highest. There is a strong positive correlation between SiO_2, Al_2O_3 and K_2O in the first seam. There is a strong correlation between CaO, MgO and SO_3 in both the first and second seams. There are strong negative correlations between MgO, CaO and SO_3, and SiO_2 and Al_2O_3 in both seams.

The high SiO_2 and Al_2O_3 contents of the second seam may be indicative of a detrital source of clastic sediment, such as water-borne sediment. Tewalt (1987) suggested that the high silica values in Texas Eocene lignites may be due to volcanic ash. The close association of volcanic rocks with the Çayirhan lignites (Fig. 3) and the presence of zeolites (often seen as the alteration product of volcaniclastic material) indicates a volcanic source of silica in the mineral matter. The lower amounts of SiO_2 and Al_2O_3 in the first seam indicate that detrital or volcanic sources of mineral matter were of lesser significance. The high SO_3 and CaO values in the first seam are derived from synthetic anhydrite. Most CaO will be derived from the zeolites. There appears to be very little calcite in the raw lignite. As the pyrite is oxidized during the high temperature ashing process, some of the sulphur is retained in the ash and combines with the calcium released from the zeolites to form anhydrite. The high correlation between these two oxides in both seams suggests that there is some sulphate present throughout the succession, but in greater amounts in the first seam.

Ash oxide concentrations affect the ash fusion temperatures, which in turn determine the viscosity of the ash and the formation of slag during combustion in the thermal power station (Tewalt 1987). Ash fusion temperatures can be lowered as a result of increased $SiO_2 : Al_2O_3$ ratios, increased Fe_2O_3 and SO_3 content and compounds formed from sodium, potassium, magnesium or sulphur. These factors should be

Table 6. *Average contents of the sulphur in the first and second seams in the Çayirhan lignite field, Beyparzari. All sulphur contents are reported as a percentage of the whole lignite on an as-received basis unless otherwise stated*

	Combustible sulphur (%)	Sulphur in ash (%)	Total sulphur (%)	Combustible sulphur/ total sulphur (%)	Total air dried sulphur (%)	Correlation coefficient combustible sulphur/ total sulphur
First seam						
No. of samples	102	102	125	102	129	0.92
Average	2.58	0.90	3.59	71.88	4.42	
Minimum	0.30	0.03	1.27	20.00	1.50	
Maximum	5.88	2.30	6.17	98.62	7.80	
Standard deviation	1.14	0.45	1.19	14.11	1.37	
Variance	1.30	0.20	1.41	199.09	1.88	
CV	0.44	0.49	0.33	0.20	0.31	
Second seam						
No. of samples	102	102	113	102	127	0.88
Average	2.45	0.78	3.24	75.06	3.87	
Minimum	0.68	0.09	0.00	39.58	1.76	
Maximum	4.61	1.94	5.90	98.00	5.91	
Standard deviation	0.79	0.39	0.91	11.27	0.90	
Variance	0.61	0.15	0.79	126.68	0.81	
CV	0.32	0.49	0.28	0.15	0.23	

taken into account when designing a power station boiler. The high sulphur content presents its own problems associated with the emission of SO_2.

Distribution of sulphur

The total sulphur (S_T) content appears to be uniformly high throughout both seams, although the first seam has a slightly higher average S_T (Table 6). There is a distinct increase in S_T towards the top of the first seam (Fig. 4) This coincides with an increase in the pyritic sulphur (S_{py}) content (vol.%) in the seam (Fig. 4). The second seam has a uniformly higher S_{py} content than the first seam.

The averages of the as-received sulphur contents derived from the borehole cores of the first and second seams have been plotted to show their lateral spatial distribution (Figs 9 and 10). These maps confirm the NE–SW differences (Table 1) and show that the highest sulphur contents are in the south and southwestern limb of the lignite field for both seams. The first seam (Table 6) has an overall higher average total sulphur content (3.59% on an as-received basis) than the second seam (3.24%). The correlation coefficients between the total sulphur and combustible sulphur (a term used here to describe the loss of sulphur during high temperature ashing) are high (Table 6). This suggests that there is a constant proportion of organic sulphur (S_o) and S_{py} in the lignite across the basin with the S_o and S_{py} contents increasing in the western limb of the basin.

Analyses of the combustible sulphur in the lignite and the sulphur retained in the ash (reported as a percentage of the lignite) (Table 6) show that the first seam has a slightly higher average combustible sulphur content (2.58%) than the second seam (2.45%). However, the average of the sulphur retained in the ash is lower in the second seam (0.78%) than the first seam (0.90%). During high temperature ashing the gypsum and S_{py} oxidize and combine with the calcium from the zeolites in the raw lignite to form anhydrite in the ash. This is reflected in the SO_3 content seen in the ash oxide analyses (Table 5). There is probably a lower gypsum and zeolite content in the second seam, but a higher S_{py} content (Fig. 4), hence the lower retained sulphur (anhydrite) in the ash. This means that the second seam is contributing a greater percentage of its sulphur content to atmospheric sulphur. This problem is compounded because at the boiler temperatures in the thermal power station (1100°C), anhydrite decomposes into CaO and SO_2 (Chinchon *et al.* 1991), adding to the 'combustible' sulphur.

The remaining (combustible) sulphur must be derived mainly from organic sulphur. The high combustible sulphur contents suggest that there

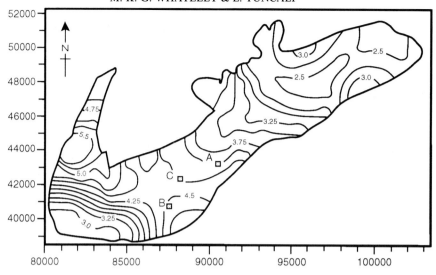

Fig. 9. Spatial distribution of the as-received total sulphur content (%) of the first seam in the Çayirhan lignite field. Data derived from 130 boreholes and the three underground sample sites, A, B and C (Fig. 2). The legend shows the contour intervals at 0.25% intervals. The outline of the lignite field was digitized from the geological map (Fig. 2). The coordinate system is metric and the eastings and northings are at 1000 m spacings. North is towards the top of the map.

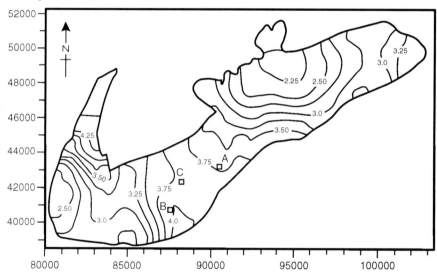

Fig. 10. Spatial distribution of the as-received total sulphur content (%) of the second seam in the Çayirhan lignite field. Data derived from 130 boreholes and the three underground sample sites, A, B and C (Fig. 2). The legend shows the contour intervals at 0.25% intervals. The outline of the lignite field was digitized from the geological map (Fig. 2). The coordinate system is metric and the eastings and northings are at 1000 m spacings. North is towards the top of the map.

must be a significant amount of organic sulphur in the lignite. The results of analyses for forms of sulphur are not yet available.

Origin of sulphur

Sulphur appears in three main forms in coal, namely pyritic, organic and sulphate sulphur. There are many possible origins of the sulphur in peat, such as marine roof rocks (Horne et al. 1978), marine influences during deposition (Casagrande et al. 1977), microbial action and changing pH conditions (Casagrande 1987). Casagrande (1987) believes that almost all the sulphur in coal can be accounted for during the peat-forming stage, where syngenetic sulphur is incorporated. Hydrogen sulphide, formed from the microbial reduction of sulphate, reacts with organic matter to form organic sulphur and reacts with ferrous iron (Fe^{2+}) to form pyrite. This reaction appears to proceed at a faster rate when the pH is higher as more ferrous iron is released. Therefore, those areas in a coal-forming environment influenced by a higher sulphate content will yield lignites with higher sulphur contents.

The contribution to the playa in the Hirka Formation by alkaline springs originating from the contemporaneous Teke volcanics (Inci 1991) suggests that similar (?hot) springs may have contributed to the mire water during peat formation. This may have led to increased sulphate contents, which resulted in lignite with a high sulphur content.

The strata immediately above the lignite in the Çayirhan basin contain trona and some carbonates (Fig. 3). At the time of peat formation, an increase in the pH of the peat water may have occurred as a result of a gradual increase in the build-up of carbonate in the peat water, especially before the deposition of the overlying Hirka Formation. This would increase the pH and result in an increase in syngenetic S_{py} and S_o contents.

There is no evidence for marine roof rocks in the Çayirhan basin, but there are higher than normal amounts of sulphate sulphur (gypsum) in the raw lignite, as seen on the XRD and EDX traces. The sedimentological model proposed by Yagmurlu et al. (1988) suggests that the lignite formed at the same time as, but spatially separate from, a gypsiferous, trona-rich playa lake. However, it is more probable, as Inci (1991) suggests, that the lignite formed in a lacustrine environment, and that the bituminous shales and trona of the Hirka Formation formed above the lignite (Fig. 3). Inci (1991) describes the formation of magnesium sulphate in desiccation cracks in the bituminous shales below the trona as immediate post-depositional, secondary mineralization.

It is therefore probable that the evaporite deposits in the Hirka Formation (Fig. 3) which overlie the lignite-bearing Çoraklar Formation were separated temporally from the lignite. Circulating saline low temperature hydrothermal solutions in the closed Çayirhan basin probably precipitated sulphate in joints and cleats in the lignite. In a reducing environment, such as the lignite seams, some of the sulphate may have been reduced to pyrite. This would explain the increased S_{py} contents in the lignite seams, particularly in the first seam (Fig. 4). Oxidation of the pyrite caused by the circulation of more recent oxidizing groundwater would add to the secondary gypsum.

Conclusions

The lignites in the Çayirhan basin formed in a limnic environment characterized by high rainfall and seasonally high temperatures. Periodic desiccation of the mire occurred locally as evidenced by the local abundance of fusinite. The upper lignite is separated into two seams, the first and second, split by a siltstone parting. Although both seams have relatively high ash contents, the first seam has a lower average ash content. The mineral matter appears to consist almost exclusively of zeolites, gypsum and pyrite. The heulandite in the first seam and analcime in the second seam are probably derived from the alteration of volcaniclastic mineral matter in the lignite. The apparent zonation may be a result of depth–temperature control on the distribution of zeolites in the lignite, but these differences are probably the result of variations in the chemistry of the original volcaniclastic or clastic material associated with the lignite.

These lignites are also characterized by their high sulphur content (up to 8.2% on an air-dried basis). The study of the mineral matter (ash oxides and trace element geochemistry) and the sulphur content reveals three types of sulphur distribution (Querol et al. 1992), namely, (1) vertical variation within individual seams (Fig. 4), (2) variation between the seams (Fig. 4) and (3) lateral variation across the basin (Figs 9 and 10).

The first two appear to be controlled by external factors. The central role of the sulphate and its impact on the eventual lignite sulphur content is well documented (Casagrande 1987). The sulphate may have been derived from the contemporaneous Teke volcanic rocks, resulting in the increased S_T contents. Increased volcanic activity during the formation of the first seam may be responsible for its higher S_T content.

In addition, the carbonate basement surrounding the basin and the carbonate deposition above the first seam would have contributed carbonate-rich water to the mire. The high carbonate content in the waters of the depositional environment may be responsible for the higher sulphur content in the first seam lignite, by allowing significant amounts of S_{py} and S_o to accumulate syngenetically.

Circulating saline low temperature hydrothermal fluids in the closed Çayirhan basin probably precipitated epigenetic sulphate in joints and cleats. In a reducing environment, such as the lignite seams, sulphate would be reduced to pyrite. This could also explain the increased S_{py} contents in the lignite seams. Oxidation of the pyrite caused by the circulation of subsequent oxidizing groundwater would form gypsum. The second seam has a lower S_T content but a higher S_{py} content. This may be a reflection of a lower pH in the initial peat water, but a subsequent increase in the epigenetic S_{py}. The overall higher S_T content of the first seam suggests that there is a higher S_o content in the lower half of the seam, but that epigenetic S_{py} forms a significant contribution to the S_T content of the top half of the seam.

Lateral variation, reflected in the increased organic and pyritic sulphur contents in the southwestern half of the basin for both the first and second seams, suggest tectonically controlled topography producing a deeper part of the basin. This would have allowed a build-up of syngenetic organic and pyritic sulphur in the permanently anoxic part of the basin where the water-table would not have periodically lowered sufficiently to allow the partial oxidation of sulphur.

We thank the Directorate of Mineral Research and Exploration for providing the data, S. Toprak for the petrographic analyses and N. Tulu and N. Gülgör for the palynological determinations. Our thanks also to M. J. Norry, J. P. Richards and D. A. Spears for their constructive criticism of earlier versions of this manuscript, to S. Button for producing the figures for this paper and to A. Smith for his help with the XRD traces.

References

ALTINLI, I. E. 1977. Geology of the eastern territory of Nallihan (Ankara Province). *Istanbul University Science Faculty Series*, **B.42** (1-2), 29–44.

BLATT, H. 1982. *Sedimentary Petrology*. Freeman, San Francisco, 564pp.

CASAGRANDE, D. 1987. Sulphur in peat and coal. *In:* SCOTT, A. C. (ed.) *Coal and Coal-bearing Strata: Recent Advances*. Geological Society, London, Special Publication, **32**, 87–105.

——, SIEFERT, L., BERSHINISKI, C. & SUTTON, N. 1977. Sulfur in peat forming systems of Okefenokee Swamp and Florida Everglades: origins of sulfur in coals. *Geochimica Cosmochimica Acta*, **41**, 161–167.

CHINCHON, J. S., QUEROL, X., FERNANDEZ-TURIEL, J. L. & LOPEZ-SOLER, A. 1991. Environmental impact of mineral transformations undergone during coal combustion. *Environmental, Geological and Water Science*, **18**, 11–15.

COHEN, A. D., ANDREJKO, M. J., SPACKMAN, W. & CORVINUS, D. 1984. Peat deposits of the Okefenokee Swamp. *In:* COHEN, A. D., CASAGRANDE, D. J., ANDREJKO, M. J. & BEST, G. R. (eds) *The Okefenokee Swamp: its Natural History, Geology and Geochemistry*. Wetlands Surveys, Los Alamos, 493–553.

DIESSEL, C. F. K. 1986. The correlation between coal facies and depositional environments. *In: Advanced Studies of Sydney Basin, University of Newcastle Symposium Proceedings*, 19–22.

—— 1992. *Coal-bearing Depositional Systems*. Springer-Verlag, Berlin, 721pp.

GÖKMEN, V. 1965. *Nallihan-Baypazari (Ankara) civarinda linyit ihtiva eden Neojen sahasinin jeolojisi hakkinda rapor*. Directorate of Mineral Research and Exploration Report 3802 [in Turkish].

——, MEMIKOGLU, O., DAGLI, M., ÖZ, D. & TUNCALI, E. 1993. *Türkiye Linyit Envanteri*. Directorate of Mineral Research and Exploration, 356pp.

HORNE, J. C., FERME, J. C., CARUCCIO, F. T. & BAGANZ, B. P. 1978. Depositional models in coal exploration and mine planning in Appalachian Region. *Bulletin of the American Association of Petroleum Geologists*, **62**, 2379–2411.

INCI, U. 1991. Miocene alluvial fan-alkaline playa lignite-trona bearing deposits from an inverted basin in Anatolia; sedimentology and tectonic controls on deposition. *Sedimentary Geology*, **71**, 73–97.

JONES, T. P., SCOTT, A. C. & COPE, M. 1991. Reflectance measurements and the temperature of formation of modern charcoals and implications for studies of fusain. *Bulletin de la Société Géologique de France*, **162**, 193–200.

KALAFATÇIOGLU, A. & UYSALLI, H. 1964. Geology of the Beypazari, Nallihan and Seben regions. *Bulletin of Mineral Resources and Exploration Institute of Turkey*, **62**, 1–12.

KALKREUTH, W. & LECKIE, D. A. 1989. Sedimentological and petrological characteristics of Cretaceous strandplain coals: a model for coal accumulation from the North American Western Interior Seaway. *In:* LYONS, P. C. & ALPERN, B. (eds) *Peat and Coal: Origin, Facies and Depositional Models. International Journal of Coal Geology*, **12**, 381–424.

KAVUSAN, G. 1993. Interrelations between tectonics and vitrinite reflectance of Beypazari Çayirhan lignites. *In: Turkish Geology Symposium, Ankara*, 357–363.

KRISTMANNSDOTTIR, H. 1979. Alteration of basaltic rocks by hydrothermal activity at 100–300°C. *In:* MORTLAND, A. & FARMER, B. (eds) *Developments in Sedimentology*, **27**, Elsevier, Amsterdam, 359–367.

NARIN, R. 1980. Beypazari, Beysehir lignite deposit in Central Anatolia, Turkey. *Bulletin of Mineral Resources and Exploration Institute of Turkey*, **17**, 21–50 [in Turkish with English abstract].

QUEROL, X., SALAS, R., PARDO, G. & ARDEVOL, L. 1992. Albian Coal-bearing Deposits of the Iberian Range in Northeastern Spain. Geological Society of America, Special Paper, **267**, 193–207.

SANER, S. 1979. Explanation of the development of the

western Pontid mountain and adjacent basins, based in plate tectonic theory, northwestern Turkey. *Bulletin of Mineral Resources and Exploration Institute of Turkey*, **93**, 1–20.

SIYAKO, F. 1984. *Beypazari (Ankara) Kömürlü Neojen Havzasi ve Çevresinin Jeolojisi*. Directorate of Mineral Research and Exploration, Special Report, 46pp [in Turkish].

TEWALT, S. J. 1987. Chemical characteristics of Gulf Coast lignites. *In:* FINKELMAN, R. B., CASAGRANDE, D. J. & BENSON, S. A. (eds) *Gulf Coast Lignite Geology*. Environmental and Coal Association, 201–210.

TUNC, M. 1980. *Davutoglan (Beypazari)-Seben (Bolu) arasinda kalan ve Aladag Çay boyunca olan bölgenin stratigrafisi*. PhD Thesis, University of Ankara.

WESTERCAMP, D. 1981. Distribution and volcanostructural control of zeolites and other amygdale minerals in the island of Martinique, FWI. *Journal of Volcanology and Geothermal Research*, **11**, 353–365.

YAGMURLU, F., HELVACI, C., INCI, U. & ONAL, M. 1987. Tectonic characteristics and structural evolution of Beypazari–Nallihan basin, Central Anatolia [abstract] *In: Melih Tokay Geology Symposium*. Middle East Technical University, 2–4.

——, HELVACI, C. & INCI, U. 1988. Depositional setting and geometric structure of the Beypazari lignite deposits, Central Anatolia. *International Journal of Coal Geology*, **10**, 337–360.

YEO, G., KALKREUTH, W., DOLBY, G. & WHITE, J. 1988. Preliminary report on petrographic, palynological and geochemical studies of coals from the Pictow coalfield, Nova Scotia. *Geological Survey of Canada, Current Research*, **88-1B**, 29–40.

Index

Page numbers in italics refer to Figures or Tables.

accommodation space 1–3
Aegiranum marine band 50–1, 61, 238
Ajka coalfield 300
  history of mining 299
  mine ash geochemistry
    methods of analysis 300
    results 300–3
    results discussed 303–4
Alaska 222
Alberta 222
analcime 315, 316, 317
Anatolia *see* Beypazari Lignite Basin
anthracitization 256
Anthracoceras 60
Anthracocerites 60
Anthracomya 89
Anthraconaia 61, 85
Anthracosia 61
antimony in coal *154*
Arenicolites 61, 62
arsenic in coal *151*, *154*
Asfordby coalfield *18*
Ashover anticline 118
Atchafalaya Basin 60
attrinite *312*
Australia 222
Ayrshire coalfield *18*

bassanite 185
bedding planes, effect on stability of 235–6
Belgium, energy imports *196*, *198*
  see also Campine Basin
Beypazari Lignite Basin
  Çayirhan lignites
    ash content 317–19
    depositional environment 312–15
    mineral content 315–17
    palynology 312
    petrography 310–12
    quality 315
    stratigraphy 310
    sulphur distribution 319–20
    sulphur origin 320–1
  geological setting 308–10
  history of research 307–8
bioturbation 62
bismuth in coal *151*
bituminous coal 55, *190*, *191*
bivalves 61, 62, 65, 85, 289
Blackhawk Formation 9
boghead coal 55
Bohemia *see* Kladno Basin
borehole seismic surveys
  methods 160, 163
  results 160–3, 163–4

  results discussed 164–7
boron in coal *151*
brachiopods 61, 85, 289
Breathitt Group 5, 8, 9
Bristol and Somerset coalfield *18*
brown coal *190*, *191*
  see also lignite
bunkers 24

cadmium in coal *154*
Calamites 59, 69
calcite *151*, 186
calorific value *175*, *257*
Cambriense marine band 51
Campine Basin 215–16
  depositional environment 216–20
    Westphalian A/B 220
    Westphalian C 220–6
  palaeogeography 227–30
  seismic studies 226–7
Canada 222
Cancellatum marine band 79
cank 32
cannel coal 55, 106
Canonbie coalfield *18*
carbon tax 200–1
Carbonicola 62, 85
Carbonita 62
Castlegate Sandstone 11
Çayirhan lignites
  ash content 317–19
  depositional environment 312–15
  mineral content 315–17
  palynology 312
  petrography 310–12
  quality 315
  stratigraphy 310
  sulphur distribution 319–20
  sulphur origin 320–1
Central Scottish coalfield *18*
channel bank collapse 120–1
channel facies *130*, 130
  major 130
    description 66–7
    effect on mine planning 32–3
    geometry *52*
    geotechnical properties *26*
    interpretation 67
    sedimentary discontinuities *124*, 125
  minor 130
    description 67–8
    effect on mine planning 33–4
    geometry *52*
    geotechnical properties *26*
    interpretation 67–8

sedimentary discontinuities *124*, 125
chlorite 187
chromium in coal *151*
clarain 55–6
clarodurain 55–6
Clay Cross (Vanderbeckei) marine band 49, 61, 238
clay minerals *151*
clay mylonite 235
cleat 236, 256
cleat fractures 102–3
coal, NCB characterization of 55
coalbed methane (CBM) in Britain
 degasification controls
  basin inversion 258
  faulting 258–9
  hydrogeology 259
  in situ stress 259
 estimated reserves 251, 259–60
 extraction economics 262
 formation, factors affecting
  cleat effects 256
  coalification effects 254–6
  composition 254
  magmatic effects 256
  pressure 256–8
  volume 254
 identifying prospects 261–2
 production controls 260–1
 provenance 253–4
 sources 251–3
coalification 254–6
cobalt in coal *151*, *154*
Cochlichnus 62
compaction 121–2, 126–7, 281–3
conchoidal fractures 103
copper in coal *151*, *154*
Çoraklar Formation 309
Cordaites 69
corpohuminite *312*
Cottingley Crow Coal 83
crassidurain *54*, 56
Crawshaw Sandstone *10*, 11
Cretaceous coals 299
crevasse splay facies 128, 129–30
 description 70
 effect on mine planning 34
 geometry *52*
 geotechnical properties *26*
 interpretation 70–1
 mapping 38–9
 sedimentary discontinuities *124*, 125
cross-hole seismic reflection survey
 methods 160
 results 160–3
 results discussed 164–7
crush 30
Cumberland Group 222
Cumbria coalfield *18*
Curvimula 289
cutinite *312*, 314
Czech Republic
 coal reserves 191–4
 history of coal consumption 189–90
 history of energy consumption 190–1

deep mining
 exploration stage 21–2
 forecasting conditions
  data evaluation 35
  drivage horizon selection 39–40
  face location 39
  facies distribution 36–9
  facies recognition 35–6
  seam correlation 35
  verification 40–1
 operational development 22–4
 use of facies analysis 24–34
delta facies *see* lacustrine delta facies
Denmark, energy imports *196*, *198*
densinite 178, *312*
Densosporites 56
detrohuminite 312
dickite 186, 187
dinting 30
dirt 55
dirt bands 122
discontinuities, sedimentary
 effect on face stability 122–5
 effect on groundwater 126
 orientation of 131
dolomite 186
drill rig penetrometer 99
 coal friability measurement 106–8
  methods 108–11
  results 111–13
Dunbarella 60
durain 55–6
Durham coalfield *18*
dust monitoring 181
 Ffos Las project
  methods 181–3
  results 183–7

East Fife coalfield *18*
East Midlands coalfields *8*
East Pennine coalfield
 cleat mineralization 256
 coal quality 213
 degasification 258
 geological continuity 213–14
 production figures 207–8
 recovery rates 211–13
 setting 208–10
  central zone 210
  eastern zone 210–11
  western zone 210
 stratigraphy *211*
easy slip thrusting (EST) 101–2
Edale Gulf *80*
energy policy, European Community 195–6
 competition 201
 energy imports *196*
 energy management 197–9
 environmental implications 200–1
 import dependency 196–7
 policy evolution 199–200
 role of coal 201–3
Etruria Formation 51

eu-ulminite 178, *312*
European Community energy policy 195–6
  competition 201
  energy imports *196*
  energy management 197–9
  environmental implications 200–1
  import dependency 196–7
  policy evolution 199–200
  role of coal 201–3
evaporite 310, 321
Exxon Production Research (EPR) sequence stratigraphy *see* sequence stratigraphy

Fabasporites 56
face location
  optimization of 39
  relation to channels *23*
facies analysis 18, 19–21, 51–2, 117
  Kladno Basin
    basement–U Radnice seam 273
    Green Tuff–M Lubná seam 277–8
    L Lubná seam–Green Tuff 276–7
    M Lubná seam–U Lubná seam 278–9
    U Lubná seam–top of member 279–81
    U Radnice seam–L Lubná seam 273–6
  Limestone Coal Formation
    distal 289–90
    proximal 291–2
    transitional 290–1
  Rough Rock Group 83–91
  use in face cutting *33*
  use in forecasting
    facies association effects 35–6
    facies distribution effects 36–9
  use in mine planning 24–34
faults
  controls on coalbed methane 258–9
  effect on stability of 236–8
feather fractures 104–5
feldspar 185
Ffos Las dust monitoring project
  methods 181–3
  results 183–7
Ffyndaff coalfield
  site geology 241–2
  tectonic setting 238–41
  thrust structural analysis
    methods 242–4
    results 244–7
    results discussed 247–8
firedamp explosions 251
flexural slip 235
flooding surfaces 9
floor lift 30
fluvial environment analysis, Campine Basin 220–6
Fort Union Formation 222
fractures in coal
  effects 104–5
  petrography relations 105–6
  processes 101–2
  site study
    methods 108–11
    results 111–13
    styles 102–3
France, energy imports *196*, *198*
friability
  causes 99
  methods of measurement 106–8
  S Wales coalfield study 99–101
    fracture effects 104–5
    fracture patterns and styles 102–3
    fracture processes 101–2
    fracture/petrography relations 105–6
    methods 108–11
    results 111–13
    significance of results 113–14
fusain 55–6
fusinite *151*, *312*, 314

Gainsborough Trough *80*
gannister 26, 30, 58
gas content map *252*
  see also coalbed methane
Gastrioceras 85
Geisina 62
gelifaction index (GI) 314, 315
gelinite *312*
geochemistry
  Çayirhan lignites 315–19
  mine ash
    methods of measurement 300
    results 300–1, *302*, *303*
    results discussed 303–4
  tonsteins 141–3
  see also trace elements
geotechnical properties 233–5
  effect on mining of *26*
  Ffyndaff case study
    methods 243–4
    results 244–7
    results discussed 247–8
    site geology 241–2
    tectonic setting 238–41
  Pennine coal 24–5
  structures and stability
    bedding planes 235
    faults 236–8
    joints 236
Germany, energy imports *196*, *198*
glacial–eustatic oscillations 285
goaf 24
gob 24
goniatites 61, 85
Great Britain coalfield map *18*
Greece
  coals of Tertiary age
    calorific value *175*
    depositional environment *174*
    geological setting 171–3
    production 179–80
    quality 176–8
    reserves 178–9
  energy budget 171
  energy imports *196*, *198*
gypsum 185, 310, 315, 321
Gyrochorte 62

halite 185–6
Haslingden Flags 83, 85, 86
hemihydrite 315
heulandite 315, 316, 317
highstand systems tracts 9, 294, 295
    Rough Rock Group 93
Hirka Formation *309*, 310, 321
histosol 56
hole to surface seismic reflection survey
    methods 163
    results 163–4
    results discussed 164–7
humic coal 55, 106
huminite 178, *311*, *312*, 312
humodetrinite 314
humotelinite 312–14
Hungary *see* Ajka field
hydrogeology, controls on coalbed methane of 259

illite 186–7
incised valleys 5–7, *11*
India 222
inertinite 178, *311*, *312*
    friability 105–6
initial flooding surface 93
interfluve sequence boundaries 7–8
Ireland, energy imports *196*, *198*
iron carbonate (siderite) 59, 62
isopach maps, use of 36–7, 129
Italy, energy imports *196*, *198*

joints, effect on stability of 236

Kaiparowits Plateau 5
kaolinite 137, 186
Kent coalfield *18*, 258
kettlebottoms 34
kettles 34
Kilsyth Trough/Kincardine Basin
    Limestone Coal Formation study 286–8
        allocycles 288
        autocycles 288
        lithofacies associations 289–91
        lithofacies geometry 291–2
        sequence stratigraphy 293–5
Kimir Formation *309*, 310
Kincardine Basin *see* Kilsyth Trough/Kincardine Basin
Kladno Basin
    facies analysis
        basement–U Radnice seam 273
        Green Tuff–M Lubná seam 277–8
        L Lubná seam–Green Tuff 276–7
        M Lubná seam–U Lubná seam 278–9
        U Lubná seam–top of member 279–81
        U Radnice seam–L Lubná seam 273–6
    geological setting 267–8
    pre-sedimentary relief study
        methods 268–9
        results 269–71
    topography/facies relations 281–3

lacustrine delta facies 129
    description 63–5
    effect on mine planning 32
    geometry *52*
    geotechnical properties *26*
    interpretation 62–5
    sedimentary discontinuities *124*, 125
lacustrine facies
    description 62
    effect on mine planning 31
    geometry *52*
    geotechnical properties *26*
    interpretation 65–6
lacustrine sedimentation 273, 276, 277, 279, 281
Laevigatosporites minor 56
Lancashire coalfield *18*
laser ablation microprobe–inductively coupled plasma mass spectrometry (LAMP–ICP–MS)
    method 148–9
    quantification techniques 151–3
    results 149–51
lead in coal *154*
leaves defined 55
Lepidodendron 62
lignites
    Çayirhan, Turkey
        ash content 317–19
        depositional environment 312–15
        mineral content 315–17
        palynology 312
        petrography 310–12
        quality 315
        stratigraphy 310
        sulphur distribution 319–20
        sulphur origin 320–1
    Greece
        distribution 171, 173
        production 179–80
Limestone Coal Formation
    allocycles 288
    autocycles 288
    lithofacies
        associations
            distal 289–90
            proximal 291
            transitional 290–1
        geometry 291–2
    regional setting 286–8
    sequence stratigraphy
        application 293–4
        limitations of method 295
        systems tracts 294–5
Lingula 61, 81, 85, 289
liptinite *151*, *311*, *312*
lithotype classification 55
Lothian coalfield *18*
lowstand systems tracts 11, 294, 295
    Rough Rock Group 91–3
Luxembourg, energy imports *196*, *198*
Lycospora pusilla 56

macrinite *312*
magma intrusion, effect on coal of 256

INDEX

marcasite *151, 154*
marine bands 9–11, 50–1, 81
marine facies
   description 61
   effect on mine planning 31
   geometry *52*
   geotechnical properties *26*
   interpretation 61–2
mathematical modelling *see* modelling
maximum flooding surface 9, 93
metasomatism, effect on coal of 256
methane *see* coalbed methane (CBM)
micrinite *312*
Midland Valley coalfield 258
Milankovitch spectrum 285
mineralization, effect on coal of 256
mineralogy
   airborne dust 185–7
   tonsteins 139–41
mire facies
   components *4*
   description 55–6
   effect on mine planning 25–30
   geometry *52*
   interpretation 56–8
   role in coal formation 3
modelling coal formation
   depositional environment 1
   stratigraphy 3–4
molybdenum in coal *154*, 300–4
montmorillonite 187

Naiadites 62, 85, 289
Neeroeteren-Rotem coalfield 222, 226
Netherlands, energy imports *196, 198*
Neuropteris 68
nickel in coal *151*
normal faults 237
North Staffordshire coalfield *18*
North Wales coalfield *18*
Northumberland coalfield *18*
Nottinghamshire coalfield *18*
Nova Scotia 222

opencasting
   dust monitoring
      methods 181–3
      results 183–7
   effect of sedimentary controls 117–19
   exploration methods 116–17
      channel identification 129
      coal quality prediction 131
      orientation measurements 131
      prediction of strata 128–9
      uniaxial compressive strength measures 130
   exploration stage 159
   stability case study
      site geology 241–2
      tectonic setting 238–41
      thrust structural analysis
         methods 242–4
         results 244–7

      results discussed 247–8
   stability factors 233–5
      bedding 235–6
      faults 236–8
      joints 236
   use of facies analysis
      groundwater 126
      highwall stability 122–5
      interseam intervals 126–7
Orbiculoidea 61
ostracods 62
overbank facies 129
   description 69
   effect on mine planning 34
   geometry *52*
   geotechnical properties *26*
   interpretation 69–70
   sedimentary discontinuities *124*, 125

palaeocurrent analysis 37–9, 128
palaeogeography
   Campine Basin 227–30
   Rough Rock Group *82*
   Westphalian *18*, *46*, 47–8
palaeosol (seat earth) facies
   description 58–9
   effect on mine planning 30–1
   geometry *52*
   geotechnical properties *25, 26*
   interpretation 61–2
palaeotopography study
   Kladno Basin
      methods 268–9
      results 269–71
peat accumulation, Kladno Basin 273
Pelecypodichnus 38, 65, 68, 70
Pennine Basin
   coal measure geotechnical properties *25*
   coalfield map *116*
   depositional environment 49–51, 117–19
   exploration history 17
   facies analysis 18, 19–21, 71–2, 117–19
      use in face cutting 33
      use in forecasting 35–9
      use in mine planning 24–34
   facies effects on opencasting
      groundwater 126
      highwall stability 122–5
      interseam interval 126–7
   geological setting 80–3
   palaeogeography *46*, 47–8
   regional studies
      Central *10*
      East *see* East Pennine coalfield
   Rough Rock Group
      facies analysis 83–91
      palaeogeography *82*
      sequence stratigraphy 91–5
      stratigraphy *81*, 83
   seam characters 120–2
   seam nomenclature *19*
   sedimentary controls 117–19
   stratigraphy *50*

tectonic setting 48–9
Permian coal 222
petrography, effect on friability of 105–6
Planolites 62, 65, 68, 70
playa environment 321
Point of Ayr coalfield 258
Portugal, energy imports *196, 198*
Posidonia 61
Pot Clay Coal 83
potholes 34
Powder River Basin 222
pressure in seams 256–8
progressive easy slip thrusting (PEST) 101–2
pyrite 29–30, 53, 122, *151, 154*
    Çayirhan lignite *311*, 315
        distribution 319–20
        origin 320–1

quality of coal 29–30, 122, 131
    Çayirhan lignite 315
    East Pennine coalfield 213

Radnice Member
    facies analysis
        basement–U Radnice seam 273
        Green Tuff–M Lubná seam 277–8
        L Lubná seam–Green Tuff 276–7
        M Lubná seam–U Lubná seam 278–9
        U Lubná seam–top of member 279–81
        U Radnice seam–L Lubná seam 273–6
    pre-sedimentary relief study
        methods 268–9
        results 269–71
    stratigraphy 267–8
    topography/facies relations 281–3
rank map *252*
reflectance 315
resinite *312*
Rhabdoderma 62
Rhadinichthys 62
Rhizocorallium 85, 86
Rhizodopsis 62
rib markings 103
roadway construction 22
    effect of facies on 30, 31, 32, 39–40
rock rolls 33
Rocky Mountains 222
Rough Rock Group *10*, 11, 79
    depositional environment 80–1
    facies analysis 83–91
    palaeogeography *82*
    sequence stratigraphy 91–5
    stratigraphy *81*, 83

San Juan Basin methane production 259
Sand Rock Mine Coal 83
sapropelic coal 55
sclerotinite *312*
Scotland 18
    see also Kilsyth Trough/Kincardine Basin
sea level relation to systems tracts 2
seams 55

splitting 120
    effect of channels on 33
    effect on mining of 28–9
    thickness 122
        East Pennine coalfield 213
        effect on mining of 25–8
seat earth facies *see* palaeosol facies
seismic reflection
    Campine Basin studies 222, 226–7
    opencast planning
        limitation of surface seismic 159
        methods 160, 163
        results 160–3, 163–4
        results discussed 164–7
Selby coalfield *18*
semifusinite *312*
sequence boundaries
    highstand systems tracts 9
    incised valleys 5–7
    interfluves 7–8
    transgressive systems tracts 9
sequence stratigraphy
    application 4–5
        UK coal measures 9–13
    development 1
    Limestone Coal Formation 293–6
    Rough Rock Group 91–5
shear strength 234–5
siderite (iron carbonate) 59, 62
Simon Unconformity 49
Six Inch Mine Coal 83
Skolithos 68
slabbing 30
slickenside fractures 105
South Derbyshire and Leicestershire coalfield *18*
South Staffordshire coalfield *18*
South Wales coalfield *18*
    fracture effects 104–5
    fracture patterns and styles 102–3
    fracture processes 101–2
    fracture study 106–8
        methods 108–11
        results 111–13
        significance of results 113–14
    fracture/petrography relations 105–6
    geological setting 99–101
    pressure in seams 258
Spain, energy imports *196, 198*
sphaerosiderite 59
Sphenophyllum 69
sporinite *312*, 314
squash outs 33
stability, solving problems of 122–5
staple shafts 24
Stigmaria ficoides 58
storage facilities, construction and design of 24
strata bunkers 24
stress, *in situ* crustal 259
strontium in coal *151, 154*
structural cross section 99
Subcrenatum marine band 61, 79
subsidence rate effects 119
sulphides *151*
sulphur 29–30, 31, 55, 121

Çayirhan lignite *311*, 315
   distribution 319–20
   origin 320–1
   Greek coals 177
swilleys 58, 121
Sydney Basin 5
systems tracts 2
   highstand 9, 93, 294, 295
   lowstand 11
   transgressive 9, 93, 294

Teke Volcanics 309
tenuidurain *54*, 56, 58
Tertiary coal 222
   Greece 171, 173, *174*
   Turkey 307
Teruel coal *151*
textinite 178, *312*
texto-ulminite *312*
thickness *see* seam thickness
Thin Coal 83
thorium in coal *151*, *154*
thrust faults 237
   deformation effects 99
      EST 101–2
      PEST 101–2
   structural analysis
      methods 242–3
      results 244–7
      results discussed 247–8
tissue preservation index (TPI) 314, 315
titanium in coal *154*
tonsteins 143–4
   bedforms 138
   defined 137
   field relations 138–9
   geochemistry 141–3
   history of research 137
   mineralogy 139–41
   origins 137–8
   texture 139
trace elements 177
   affinity measurement
      methods 147–9
      quantification techniques 151–3
      results 149–51

trace fossils 62, 65, 68, 70
transcurrent faults 237–8
transgressive systems tracts 9, 294
   Rough Rock Group 93
tree trunks 34
trona 310, 321
tungsten in coal *151*, *154*
Turkey *see* Beypazari Lignite Basin

uranium in coal *151*, *154*, 300–4
USA 5, 6, 8, 9
   coal environment analysis 222
   coalbed methane 259

vanadium in coal *151*, 300–4
Vanderbeckei (Clay Cross) marine band 49, 62, 238
vitrain 55–6, 58
vitrinite *151*
   friability 106
volcanism, Kladno Basin 273

Wales 18
   see also Ffos Las; Ffyndaff coalfield; South Wales coalfield
Warwickshire coalfield *18*, 117
wash-outs 33, 120
water table role in peat control 3
West Cumbrian coalfield 119
Western Interior Basin 5
Westphalian A/B *8*, 13, *19*
   depositional environment 49–51, 220, 253–4
   palaeogeography *46*, 47–8
   provenance *252*
   seams 19
Westphalian C *8*, 220–6
Widmerpool Gulf *80*

Yorkshire coalfield *18*

zeolites 316–17
zinc in coal *151*, *154*
Zwart-Opglabbeek coalfield 222–5